Rapid Guide to Hazardous Chemicals in the Workplace

FOURTH EDITION

Rapid Guide to Hazardous Chemicals *in the* Workplace

FOURTH EDITION

RICHARD J. LEWIS SR.

A John Wiley & Sons, Inc., Publication

New York • Chichester • Weinheim • Brisbane • Singapore • Toronto

This book is printed on acid-free paper.♾

Published simultaneously in Canada.

For ordering and customer service, call 1-800-CALL-WILEY.

Library of Congress Cataloging-in-Publication Data:

Rapid guide to hazardous chemicals in the workplace / edited by Richard J. Lewis, Sr.
p. cm.
ISBN 0-471-35542-9
1. Hazardous substances—Handbooks, manuals, etc. 2. Industrial safety—Handbooks, manuals, etc. I. Lewis, Richard J., Sr.

T55.3.H3 R37 2000
604.7'03—dc21 93-086808

10 9 8 7 6 5 4 3 2 1

Thanks to Gracie for help with everything.

My gratitude to Jerilynn Caliendo of John Wiley & Sons, Inc. for her professional assistance with this book.

Contents

Introduction

This book fulfills the need for a rapid reference to the most frequently encountered hazardous materials. This edition contains almost 760 entries. Almost 100 entries that were in the Third Edition were replaced with substances of greater importance to the millennium workplace. Each entry was selected because its dangerous properties prompted regulation by government agencies or consideration by advisory groups.

A court order has vacated the OSHA Air Standards set in 1989 and contained in 29CFR 1910.1000. OSHA has decided to enforce only pre-1989 air standards. We have elected to include the Final Rule limits that went into effect September 1, 1989 and those individually revised by OSHA since that date. These represent the current best judgment as to appropriate workplace air levels. While they may not be enforceable by OSHA, they are better guides than the OSHA Air Standards adopted in 1969. OSHA has stated that it "...continues to believe that many of the old limits which it will now be enforcing are out of date (they predate 1968) and are not sufficiently protective of employee health based on current scientific information and expert recommendations. In addition, many of the substances for which OSHA has no PELs present serious health hazards to employees."

Each entry has a recommended safe workplace air concentration or other workplace control recommendation from either the U.S. Occupational Safety and Health Administration (OSHA), the American Conference of Governmental Industrial Hygienists (ACGIH), or the German Research Society (DFG). These standards and recommendations constitute the most comprehensive set of workplace air level guidelines available and are presented in one convenient format in this book.

For assessment of transport hazards, the U.S. Department of Transportation (DOT) hazard class number and description are included. This information serves as an index to the transportation regulations of the United States and those regulating most international shipments. The hazard class number is in most cases an internationally agreed upon United Nations number. These values are useful guides to the control of workplace atmospheres, but they are not the complete solution.

The Safety Profile is a clear and concise summation of each entry's hazards. A chemical may present problems in occupational handling and use which is not reflected in the recommended air levels. Skin contact can cause irritation, skin corrosion and burns, allergenic reactions, or skin penetration leading to toxic effects in the body. Other problems arise from fire and explosion potential. Many substances present storage problems because they are incompatible with other commonly encountered chemicals. The Safety Profiles disclose the various types and degrees of dangerous or harmful effects reported in the literature, condensed in a compact, understandable paragraph. The Safety Profiles are designed to quickly define and clarify the hazard presented by a given substance.

The final section of each entry contains a physical description of the material and gives useful physical and flammability properties. This information can aid in the identification of unknown materials and in the design and selection of proper storage and handling facilities.

This guide is designed to afford easy access to information on the adverse properties of commonly encountered industrial materials. A book of this size cannot hope to present all the data necessary to completely assess the proper use of these substances. The information provided should allow a quick assessment of the relative hazards of the materials and the types and nature of the hazards likely to be encountered. Each entry is identified by a DPIM code, an alphanumeric code in the form of three letters and three numbers, for example AAA123. These codes provide an easy pointer to the complete entry in *Dangerous Properties of Industrial Materials, Tenth Edition.* In these entries are found the detailed data and literature citations that form the basis for the Safety Profiles.

Many publications on hazardous materials attempt to provide information on all aspects of hazardous material control. It is our belief that such subjects as fire control, first aid, and the selection and use of personal protective equipment and respirators should not be treated as briefly as would be required by this format. Decisions on such crucial matters must be made with careful consideration of specific workplace conditions. The information in this book should stimulate actions to provide safer work environments.

The standards and recommended air concentrations given in this publication are set by various mechanisms, which vary in frequency of change. While the current values at time of publication are listed, the reader is cautioned to *verify* the data with the appropriate agency before undertaking major control efforts based on the data given here. Many substances are under test for carcinogenic activity. When a positive finding is reported, the recommended or mandatory control values can change rapidly. Transportation regulations change in detail as transport experience dictates.

I have strived for accuracy and completeness in this presentation, but recognize that perfection is rarely achieved. Please bring any errors or suggestions to my attention.

Richard J. Lewis Sr.

How to Use This Book

Each entry consists of four sections:

1. identifying information
2. standards and recommendations
3. Safety Profiles
4. physical properties

1. Identifying Information

The first line of each entry contains the DPIM Code, the Chemical Abstracts Service (CAS:) number, the molecular formula, and, to the far right, the Hazard Rating (HR:). This rating varies from 3, indicating the highest hazard potential, to 1, indicating the lowest hazard potential.

The second line of each entry contains the index name used to alphabetize the entry. This is usually the common name used by OSHA, ACGIH, the DFG, DOT, or NIOSH.

The third line contains the U.S. Department of Transportation (DOT:) hazard code, when available. This code is recognized internationally and is in agreement with the United Nations coding system. The code is used on transport documents, labels, and placards. It is also used to determine the shipping regulations applicable to the material. Appendix III contains a cross-index of DOT hazard codes in numerical order.

Since chemicals are often known by several widely recognized names, a few useful synonyms are included to aid identification. Synonyms are listed within entries and listed alphabetically in Appendix I. If a name is not located in the entries, it may be a synonym for an entry and listed in Appendix I, with reference to the DPIM Code.

2. Standards and Recommendations

The five possible entries in this section are:

OSHA PEL, which is followed by the Permissible Exposure Limit (PEL) as defined by the U.S. Occupational Safety and Health Administration (OSHA), Department of Labor. These standards are either time-weighted average (TWA) concentrations for an 8-hour workday or ceiling levels (CL) indicating a value that must not be exceeded. Some entries have short-term exposure limit (STEL) values that represent a 15-minute concentration that should not be exceeded in an 8-hour workday. The notation "skin" indicates that the material penetrates intact skin, and skin contact should be avoided even if the PEL is not exceeded. These limits are found in 29 CFR (Code of Federal Regulations) 1910.1000. The CFR regulations also contain detailed requirements for control of some substances and special regulations for carcinogenic substances. Additional information is available from OSHA, Technical Data Center, U.S. Department of Labor, Washington, DC 20210, USA.

ACGIH TLV, which is followed by the Threshold Limit Value (TLV) of the American Conference of Governmental Industrial Hygienists (ACGIH). These standards are either time-weighted average (TWA) concentrations for an 8-hour workday or ceiling levels (CL) indicating a value that must not be exceeded. Some entries have short-term exposure limit (STEL) values that represent a 15-minute concentration that should not be exceeded in an 8-hour workday. The notation "skin" indicates that the material penetrates intact skin, and skin contact should be avoided even if the TLV concentration is not exceeded. Biological Exposure Indices (*BEI:*) are, according to the ACGIH, set to provide a warning level "...of biological response to the chemical, or warning levels of that chemical or its metabolic product(s) in tissues, fluids, or exhaled air of exposed workers..." The latest annual TLV list is contained in the publication *Threshold Limit Values and Biological Exposure Indices.* This publication should be consulted for future trends in recommendations. The ACGIH TLVs are adopted in whole or in part by many countries and local administrative agencies throughout the world. As a result, these recommendations have a major impact on the control of workplace contaminant concentrations. The ACGIH may be contacted for additional information at Kemper Woods Center, 1330 Kemper Meadow Drive, Cincinnati, OH 45240, USA.

DFG MAK, which is followed by the German Research Society's Maximum Allowable Concentration values. Those materials which are classified as to workplace hazard potential by the German Research Society are noted on this line. The MAK values are also revised annually and discussions of materials under consideration for MAK assignment are included in the annual publication together with the current values. *BAT* indicates Biological Tolerance Value for a Working Material which is defined as, "...the maximum permissible quantity of a chemical compound, its metabolites, or any deviation from the norm of biological parameters induced by these substances in exposed humans." *TRK* values are Technical Guiding Concentrations for workplace control of carcinogens. For additional information, write to Deutsche Forschungsgemeinschaft (German Research Society), Kennedyallée 40, D-53175 Bonn, Germany. The publication *Maximum Concentrations at the Workplace and Biological Tolerance Values for Working Materials* can be obtained from WILEY-VCH Verlag GmbH, P. O. Box 10 11 61, D-69451 Weinheim, Germany.

NIOSH REL, which is followed by the recommended level contained in a NIOSH criteria document. These documents contain extensive data, analysis, and references. The more recent publications can be obtained from the National Institute for Occupational Safety and Health, U.S. Department of Health and Human Services, 4676 Columbia Parkway, Cincinnati, OH 45226, USA.

DOT Class, which is followed by the hazard classification according to the U.S. Department of Transportation (DOT) or the International Maritime Organization (IMO). This classification gives an indication of the hazards expected in transportation and serves as a guide to the development of proper labels, placards, and shipping instructions. Many materials are regulated under general headings such as "pesticides" or "combustible liquids" as defined in the regulations. These are not noted here as their specific concentration or properties must be known for proper classification. Special regulations may govern shipment by air. This information should serve *only as a guide* since the regulation of transported materials is carefully controlled in most countries by federal and local agencies. U.S. transportation regulations are found in 40 CFR, Parts 100 to 189. Contact the U.S. Department of Transportation, Materials Transportation Bureau, Washington, DC 20590, USA.

3. Safety Profiles

This section contains a summary of the hazardous properties of the material. Acute immediate effects such as irritation, corrosion, or lethal action are reported in concise language. Chronic or delayed health effects are noted, including cancer and reproductive or allergenic effects. Reported human effects are specifically noted. The term "experimental" indicates that the effect was reported in experimental animals. The term "suspected carcinogen" indicates that a reviewing body such as the International Agency for Research on Cancer or the NTP Carcinogenesis Testing Program has indicated that there is some evidence of carcinogenic activity, either in animals or humans. An assessment is given of flammable and explosive properties. Incompatible materials and instabilities are listed to guide in the safe storage and use of materials.

4. Physical Properties

This part provides a physical description of the material in terms of form, color, and odor to aid in positive identification. Here are listed the physical properties that are used for determination of hazard potential and assessment of correct storage and handling practices. When available, the boiling point, melting point, density, vapor pressure, vapor density, and refractive index are given. The flash point, autoignition temperature, and lower and upper explosive limits are included to aid in fire protection and control. An indication is given of the solubility or miscibility of the material in water and common solvents.

Key to Abbreviations

abs – absolute
ACGIH – American Conference of Governmental Industrial Hygienists
alc – alcohol
alk – alkaline
amorph – amorphous
anhy – anhydrous
approx – approximately
aq – aqueous
atm – atmosphere
autoign – autoignition
aw – atomic weight
af – atomic formula
bp – boiling point
CAS – Chemical Abstracts Service
cc – cubic centimeter
CC – closed cup
CL – ceiling concentration
COC – Cleveland open cup
conc – concentrated
compd(s) – compound(s)
conc – concentration, concentrated
cryst, crys – crystal(s), crystalline
d – density
D – day(s)
decomp – decomposition
deliq – deliquescent
dil – dilute
DOT – U.S. Department of Transportation
EPA – U.S. Environmental Protection Agency
eth – ether
(F) – Fahrenheit
flash p – flash point
flam – flammable
fp – freezing point
g, gm – gram
gran – granular, granules
hygr – hygroscopic
H, hr – hour(s)
HR: – hazard rating
htd – heated
htg – heating
I – intermittent
IARC – International Agency for Research on Cancer
incomp – incompatible
insol – insoluble
IU – International Unit
kg – kilogram (one thousand grams)
L, l – liter
lel – lower explosive level
liq – liquid
M – minute(s)
m^3 – cubic meter
mg – milligram
misc – miscible
μ, u – micron
mL, ml – milliliter
mm – millimeter
mod – moderately
mp – melting point
mppcf – million particles per cubic foot
mf – molecular formula
mw – molecular weight
NIOSH – National Institute for Occupational Safety and Health
ng – nanogram
nonflam – nonflammable
OC – open cup
org – organic
OSHA – Occupational Safety and Health Administration
PEL – permissible exposure level
petr – petroleum
pg – picogram (one trillionth of a gram)

Pk – peak concentration
pmole – picomole
powd – powder
ppb – parts per billion (v!/v)
pph – parts per hundred (v!/v)(percent)
ppm – parts per million (v!/v)
ppt – parts per trillion (v!/v)
prep – preparation
PROP – properties
refr – refractive
rhomb – rhombic
S, sec – second(s)
sl, slt, sltly – slightly
sol – soluble
soln – solution
solv(s) – solvent(s)
spont – spontaneous(ly)
subl – sublimes
TCC – Tag closed cup
tech – technical
temp – temperature
TLV – Threshold Limit Value
TOC – Tag open cup
TWA – time-weighted average
U, unk – unknown, unreported
μ, u – micron
uel – upper explosive limits
μg, ug – microgram
ULC, ulc – Underwriters Laboratory Classification
vac – vacuum
vap – vapor
vap d – vapor density
vap press – vapor pressure
vol – volume
visc – viscosity
W – week(s)
Y – year(s)
% – percent(age)
> – greater than
< – less than
≤ – equal to or less than
≥ – equal to or greater than
° – degrees of temperature in Celsius (Centigrade)
°(F), °F – temperature in Fahrenheit

Rapid Guide to Hazardous Chemicals *in the* Workplace

FOURTH EDITION

AAG250 CAS: 75-07-0 C_2H_4O HR: 3

ACETALDEHYDE

DOT: UN 1089

SYNS: ACETIC ALDEHYDE ♦ ETHANAL ♦ ETHYL ALDEHYDE

OSHA PEL: TWA 100 ppm; STEL 150 ppm

ACGIH TLV: TWA 100 ppm; STEL 150 ppm (Proposed: CL 25, Animal Carcinogen)

DFG MAK: 50 ppm (90 mg/m^3), Suspected Carcinogen

DOT Classification: 3; Label: Flammable Liquid

SAFETY PROFILE: Confirmed carcinogen. A human systemic irritant by inhalation. An experimental teratogen. A skin and severe eye irritant. A narcotic. Highly flammable liquid. It can react violently with many substances.

PROP: Colorless, fuming liquid; pungent, fruity odor. Mp: −123.5°, bp: 20.8°, lel: 4.0%, uel: 57%, flash p: −36°F (CC), d: 0.804 @ 0°/20°, autoign temp: 347°F, vap d: 1.52. Misc in water, alc, and eth.

AAI000 CAS: 60-35-5 C_2H_5NO HR: 3

ACETAMIDE

SYNS: ACETIC ACID AMIDE ♦ ETHANAMIDE ♦ METHANECARBOXAMIDE

DFG MAK: Suspected Carcinogen

SAFETY PROFILE: Suspected carcinogen. An experimental teratogen.

PROP: Colorless crystals; mousy odor. Mp: 81°, bp: 221.2°, d: 1.159 @ 20°/4°, vap press: 1 mm @ 65°. Decomp in hot water.

AAT250 CAS: 64-19-7 $C_2H_4O_2$ HR: 3

ACETIC ACID

DOT: UN 2789/UN 2790

SYNS: ACETIC ACID (aqueous solution) (DOT) ♦ ETHANOIC ACID ♦ GLACIAL ACETIC ACID ♦ VINEGAR ACID

OSHA PEL: TWA 10 ppm

ACGIH TLV: TWA 10 ppm; STEL 15 ppm

DFG MAK: 10 ppm (25 mg/m^3)

DOT Classification: 8; Label: Corrosive

SAFETY PROFILE: A human poison. A severe eye and skin irritant. A flammable liquid. It can react violently with many substances.

PROP: Clear, colorless liquid; pungent odor. Mp: 16.7°, bp: 118.1°, flash p: 109°F (CC), lel: 5.4%, uel: 16.0% @ 212°F, d: 1.049 @ 20°/4°, autoign temp: 869°F, vap press: 11.4 mm @ 20°, vap d: 2.07. Misc in water, alc, and eth.

AAX500 CAS: 108-24-7 $C_4H_6O_3$ HR: 3

ACETIC ANHYDRIDE

DOT: UN 1715

SYNS: ACETANHYDRIDE ♦ ACETIC ACID, ANHYDRIDE (9CI) ♦ ACETYL ANHYDRIDE ♦ ACETYL OXIDE

OSHA PEL: CL 5 ppm

ACGIH TLV: CL 5 ppm (Proposed: TWA 5 ppm)

DFG MAK: 5 ppm (20 mg/m^3)
NIOSH REL: Acetic Anhydride: CL 5 ppm
DOT Classification: 8; Label: Corrosive
SAFETY PROFILE: Moderately toxic by inhalation, ingestion, and skin contact. A skin and severe eye irritant. A flammable liquid. It can react violently with many substances.
PROP: Colorless, very mobile, strongly refractive liquid; very strong, irritating, acetic odor. Mp: –73.1°, bp: 139.55°, flash p: 129°F (CC), d: 1.082 @ 20°/4°, lel: 2.9%, uel: 10.3%, autoign temp: 734°F, vap press: 10 mm @ 36.0°, vap d: 3.52. Sltly sol in water; sol in org solvs. Decomp in hot water and hot alc; misc in alc and eth.

ABC750 CAS: 67-64-1 C_3H_6O HR: 3
ACETONE
DOT: UN 1090/UN 1091
SYNS: ACETONE OILS (DOT) ♦ DIMETHYLFORMALDEHYDE ♦ DIMETHYL KETONE ♦ KETONE PROPANE ♦ METHYL KETONE ♦ 2-PROPANONE
OSHA PEL: TWA 750 ppm; STEL 1000 ppm
ACGIH TLV: TWA 500 ppm; STEL 750 ppm; Not Classifiable as a Human Carcinogen; BEI: 100 100 mg/L acetone in urine at end of shift (Proposed: BEI: 50 mg/L acetone in urine at end of shift)
DFG MAK: 500 ppm (1200 mg/m^3)
NIOSH REL: (Ketones) 10H TWA 590 mg/m^3
DOT Classification: 3; Label: Flammable Liquid
SAFETY PROFILE: Moderately toxic. A skin and severe eye irritant. Highly flammable liquid. It can react violently with many substances.
PROP: Volatile, colorless liquid; fragrant mintlike odor. Mp: –94.6°, bp: 56.2° @ 20 mm, refr index: 1.356, flash p: 0°F (CC), lel: 2.6%, uel: 12.8%, d: 0.7972 @ 15°, autoign temp: (color) 869°F, vap press: 240 hPa @ 20°, vap d: 2.00. Misc in water, alc, org solvs, and ether.

ABE500 CAS: 75-05-8 C_2H_3N HR: 3
ACETONITRILE
DOT: UN 1648
SYNS: CYANOMETHANE ♦ ETHANENITRILE ♦ ETHYL NITRILE ♦ METHANECARBONITRILE ♦ METHYL CYANIDE
OSHA PEL: TWA 40 ppm; STEL 60 ppm
ACGIH TLV: TWA 40 ppm; STEL 60 ppm (skin); Not Classifiable as a Human Carcinogen
DFG MAK: 40 ppm (70 mg/m^3)
NIOSH REL: (Nitriles) TWA 34 mg/m^3
DOT Classification: 3; Label: Flammable Liquid, Poison
SAFETY PROFILE: Poison by ingestion. An experimental teratogen. A skin and severe eye irritant. Dangerous fire hazard when exposed to heat, flame, or oxidizers. It can react violently with many substances.

PROP: Colorless liquid; almond-ethereal, aromatic odor. Mp: –45°, bp: 81.1°, flash p: 42°F (COC), d: 0.7868 @ 20°/20°, vap d: 1.42, vap press: 100 mm @ 27°, lel: 4.4%, uel: 16%, autoign temp: 975°F. Misc in water, alc, and org solvs. Immisc in pet eth.

ACI750 CAS: 74-86-2 C_2H_2 HR: 3

ACETYLENE

DOT: UN 1001

SYNS: ACETYLEN ♦ ETHINE ♦ ETHYNE

OSHA PEL: CL 2500 ppm

ACGIH TLV: Simple asphyxiant

NIOSH REL: (Acetylene) 10H TWA no exposure >2500 ppm

DOT Classification: Forbidden; DOT Class 2.1; Label: Flammable Gas

SAFETY PROFILE: Mildly toxic by inhalation. It is a very dangerous fire hazard when exposed to heat, flame, or oxidizers. It can react violently with many substances.

PROP: Colorless gas; garlic-like odor. Flammable. Bp: –84.0° (subl), lel: 2.5%, uel: 82%, mp: –81.8°, flash p: 0°F (CC), d: 1.173 g/L @ 0°, autoign temp: 581°F, vap press: 40 atm @ 16.8°, vap d: 0.91, d: (liquid) 0.613 @ –80°, d: (solid) 0.730 @ –85°. Sltly sol in water; mod sol in ethanol and acetic acid; very sol in Me_2CO; almost misc in ether.

ACK250 CAS: 79-27-6 $C_2H_2Br_4$ HR: 3

ACETYLENE TETRABROMIDE

SYNS: TBE ♦ TETRABROMOACETYLENE ♦ 1,1,2,2-TETRABROMOETHANE

OSHA PEL: TWA 1 ppm

ACGIH TLV: TWA 1 ppm

DFG MAK: 1 ppm (14 mg/m^3)

SAFETY PROFILE: Poison by inhalation, ingestion, An eye and skin irritant and a narcotic.

PROP: Colorless to yellow liquid. Bp: 151° @ 54 mm, fp: –1°, d: 2.9638 @ 20°/4°, mp: 0.1°, autoign temp: 635°F.

ACM000 CAS: 557-99-3 C_2H_3FO HR: 3

ACETYL FLUORIDE

SYNS: METHYLCARBONYL FLUORIDE

OSHA PEL: TWA 2.5 $mg(F)/m^3$

ACGIH TLV: TWA 2.5 $mg(F)/m^3$; BEI: 3 mg/g creatinine of fluorides in urine prior to shift; 10 mg/g creatinine of fluorides in urine at end of shift.

NIOSH REL: (Fluorides, Inorganic) TWA 2.5 $mg(F)/m^3$

SAFETY PROFILE: Poison by inhalation.

PROP: Liquid or gas. D: 1.002 @ 15°/4°, mp: –60°, bp: 20.8°. Sltly sol in alc, ether, acetone, and benzene.

ADA725 CAS: 50-78-2 $C_9H_8O_4$ HR: 3

ACETYLSALICYLIC ACID

SYNS: 2-ACETOXYBENZOIC ACID ♦ A.S.A. ♦ o-CARBOXYPHENYL ACETATE

OSHA PEL: TWA 5 mg/m^3
ACGIH TLV: TWA 5 mg/m^3
SAFETY PROFILE: Poison by ingestion. A human teratogen. An allergen; skin contact, inhalation, or ingestion can cause asthma, sneezing, irritation of eyes and nose, hives, and eczema. Combustible.
PROP: Colorless needles, crystals. Mp: 135°, fp: 118°. Very sltly sol in alc, sol in benzene. Solubility in water = 1% @ 37°, in ether = 5% @ 20°.

ADR000 CAS: 107-02-8 C_3H_4O HR: 3
ACROLEIN
DOT: UN 1092
SYNS: ACRYLALDEHYDE ♦ ACRYLIC ALDEHYDE ♦ ETHYLENE ALDEHYDE ♦ 2-PROPENAL
OSHA PEL: TWA 0.1 ppm; STEL 0.3 ppm
ACGIH TLV: STEL CL 0.1 ppm (skin); Not Classifiable as a Human Carcinogen
DFG MAK: Confirmed Animal Carcinogen with Unknown Relevance to Humans
DOT Classification: 6.1; Label: Poison, Flammable Liquid
SAFETY PROFILE: Human poison by inhalation. Severe eye and skin irritant. Dangerous fire hazard when exposed to heat, flame, or oxidizers.
PROP: Colorless or yellowish liquid; lachrymatory, disagreeable, choking odor. Mp: –87.7°, bp: 52.5°, flash p: <0°F, d: 0.841 @ 20°/4°, autoign temp: unstable (455°F), lel: 2.8%, uel: 31%, vap d: 1.94. Sol in water, alc, and ether.

ADS250 CAS: 79-06-1 C_3H_5NO HR: 3
ACRYLAMIDE
DOT: UN 2074
SYNS: ACRYLIC AMIDE ♦ PROPENAMIDE ♦ 2-PROPENAMIDE ♦ VINYL AMIDE
OSHA PEL: TWA 0.03 mg/m^3 (skin)
ACGIH TLV: Animal Carcinogen, TWA 0.03 mg/m^3 (skin)
DFG MAK: Animal Carcinogen, Suspected Human Carcinogen
NIOSH REL: TWA 0.3 mg/m^3
DOT Classification: 6.1; Label: KEEP AWAY FROM FOOD
SAFETY PROFILE: Confirmed carcinogen. Poison by ingestion and skin contact. A skin and eye irritant. It is dangerous because it can be absorbed through the unbroken skin.
PROP: White, crystalline solid. Leaflets from (C_6H_6). Mp: 84.5° ± 0.3°, bp: 125° @ 25 mm, d: 1.122 @ 30°, vap press: 1.6 mm @ 84.5°, vap d: 2.45. Very sol in water, alc, and ether.

ADS750 CAS: 79-10-7 $C_3H_4O_2$ HR: 3
ACRYLIC ACID
DOT: UN 2218
SYNS: ACROLEIC ACID ♦ ETHYLENECARBOXYLIC ACID ♦ GLACIAL ACRYLIC ACID ♦ PROPENOIC ACID ♦ VINYLFORMIC ACID
OSHA PEL: TWA 10 ppm (skin)

ACGIH TLV: 2 ppm (skin); Not Classifiable as a Human Carcinogen
DOT Classification: 8; Label: Corrosive
SAFETY PROFILE: Poison by ingestion and skin contact. An experimental teratogen. A severe skin and eye irritant. Corrosive. Flammable liquid.
PROP: Liquid with acrid odor. Misc in water, benzene, alc, chloroform, ether, and acetone. Mp: 13°, bp: 141° (polymerizes), d: 1.062, vap press: 10 mm @ 39.9°, flash p: 130°F (OC), vap d: 2.45.

ADX500 CAS: 107-13-1 C_3H_3N HR: 3

ACRYLONITRILE

DOT: UN 1093
SYNS: CYANOETHYLENE ♦ 2-PROPENENITRILE ♦ VINYL CYANIDE
OSHA PEL: TWA 2 ppm; CL 10 ppm/15M; Cancer Hazard
ACGIH TLV: Suspected Human Carcinogen, TWA 2 ppm (skin) (Proposed: Confirmed Animal Carcinogen, TWA 2 ppm (skin))
DFG TRK: Animal Carcinogen, Suspected Human Carcinogen
NIOSH REL: TWA 1 ppm; CL 10 ppm/15M
DOT Classification: 3; Label: Flammable Liquid, Poison
SAFETY PROFILE: Confirmed human carcinogen. Poison by inhalation, ingestion, and skin contact. Dangerous fire hazard when exposed to heat, flame, or oxidizers.
PROP: Colorless, mobile liquid; mild odor. Mp: –82°, bp: 77.3°, fp: –83°, flash p: 30°F (TCC), lel: 3.1%, uel: 17%, d: 0.806 @ 20°/4°, autoign temp: 898°F, vap press: 100 mm @ 22.8°, vap d: 1.83, flash p: (of 5% aq soln) <50°F. Sol in water.

AER250 CAS: 111-69-3 $C_6H_8N_2$ HR: 3

ADIPONITRILE

DOT: UN 2205
SYNS: ADIPIC ACID DINITRILE ♦ ADIPIC ACID NITRILE ♦ 1,4-DICYANOBUTANE ♦ HEXANEDINITRILE ♦ TETRAMETHYLENE CYANIDE
ACGIH TLV: TWA 2 ppm (skin)
NIOSH REL: TWA 18 mg/m^3
DOT Classification: 6.1; Label: KEEP AWAY FROM FOOD
SAFETY PROFILE: Poison by inhalation and ingestion. Flammable when exposed to heat or flame.
PROP: Water-white liquid; practically odorless. Mp: 2.3°, bp: 295°, flash p: 199.4°F (OC), d: 0.965 @ 20°/4°, vap d: 3.73. Sol in EtOH, $CHCl_3$; insol in H_2O, Et_2O, CS_2.

AFV500 CAS: 107-18-6 C_3H_6O HR: 3

ALLYL ALCOHOL

DOT: UN 1098
SYNS: 3-HYDROXYPROPENE ♦ PROPENOL ♦ PROPENYL ALCOHOL ♦ VINYLCARBINOL
OSHA PEL: TWA 2 ppm; STEL 4 ppm (skin)
ACGIH TLV: 0.5 ppm (skin); Not Classifiable as a Human Carcinogen
DFG MAK: Confirmed Animal Carcinogen with Unknown Relevance to Humans
DOT Classification: 6.1; Label: Poison, Flammable Liquid

SAFETY PROFILE: Suspected carcinogen. Poison by inhalation, ingestion and skin contact. A skin, severe eye, and systemic irritant.
PROP: Limpid liquid; pungent odor. Mp: –129°, fp: –50°, bp: 96–97°, lel: 2.5%, uel: 18%, flash p: 70°F (CC), d: 0.854 @ 20°/4°, autoign temp: 713°F, vap press: 10 mm @ 10.5°, vap d: 2.00. Misc in water, alc, and ether.

AGB250 CAS: 107-05-1 C_3H_5Cl HR: 3
ALLYL CHLORIDE
DOT: UN 1100
SYNS: CHLOROALLYLENE ♦ 3-CHLOROPRENE ♦ 3-CHLOROPROPENE ♦ 3-CHLORO-1-PROPYLENE ♦ 2-PROPENYL CHLORIDE
OSHA PEL: TWA 1 ppm; STEL 2 ppm
ACGIH TLV: TWA 1 ppm; STEL 2 ppm; Animal Carcinogen
DFG MAK: Confirmed Animal Carcinogen with Unknown Relevance to Humans
NIOSH REL: TWA 1 ppm; CL 3 ppm/15M
DOT Classification: 3; Label: Flammable Liquid, Poison
SAFETY PROFILE: Suspected carcinogen. Moderately toxic by ingestion, inhalation, and skin contact. Experimental teratogenic and reproductive effects. A skin and eye irritant. It can react violently with many substances.
PROP: Colorless liquid with pungent odor. Mp: –136.4°, bp: 44.6°, d: 0.938 @ 20°/4°, fp: –134.5°, flash p: –25°F, lel: 2.9%, uel: 11.2%, autoign temp: 905°F, vap d: 2.64. Misc in org solvs. Sltly sol in water.

AGH150 CAS: 106-92-3 $C_6H_{10}O_2$ HR: 3
ALLYL GLYCIDYL ETHER
DOT: UN 2219
SYNS: AGE ♦ ALLYL-2,3-EPOXYPROPYL ETHER ♦ ((2-PROPENYLOXY)METHYL)OXIRANE
OSHA PEL: TWA 5 ppm; STEL 10 ppm
ACGIH TLV: 1 ppm; Not Classifiable as a Human Carcinogen
DFG MAK: Confirmed Animal Carcinogen, Suspected Human Carcinogen
NIOSH REL: (Glycidyl Ethers) CL 45 mg/m^3/15M
DOT Classification: 3; Label: Flammable Liquid, Poison
SAFETY PROFILE: Confirmed animal carcinogen. Poison by ingestion. Moderately toxic by inhalation and skin contact. A severe skin and eye irritant. A flammable liquid.
PROP: Bp: 153.9°, fp: –100° (forms glass), flash p: 135°F (OC), d: 0.9698 @ 20°/4°, vap press: 21.59 mm @ 60°, vap d: 3.94.

AGR500 CAS: 2179-59-1 $C_6H_{12}S_2$ HR: 1
ALLYL PROPYL DISULFIDE
OSHA PEL: TWA 2 ppm; STEL 3 ppm
ACGIH TLV: TWA 2 ppm; STEL 3 ppm
DFG MAK: 2 ppm (12 mg/m^3)
NIOSH REL: (Allyl Propyl Disulfide): TWA 2 ppm; STEL 3 ppm
SAFETY PROFILE: A powerful irritant. Moderately flammable.
PROP: Liquid with pungent odor. Bp: 66–69° @ 16 mm.

AGX000 CAS: 7429-90-5 Al HR: 3

ALUMINUM

DOT: UN 1309/UN 1396/NA 9260

SYNS: ALUMINUM POWDER ♦ ALUMINUM PYRO POWDERS (OSHA) ♦ ALUMINUM WELDING FUMES (OSHA)

OSHA PEL: Total Dust: TWA 15 mg/m^3; Respirable Fraction: TWA 5 mg/m^3; Pyro Powders and Welding Fumes: 5 mg/m^3; Soluble Salts and Alkyls: 2 mg/m^3

ACGIH TLV: Metal and Oxide: TWA 10 mg/m^3 (dust); Pyro Powders and Welding Fumes: TWA 5 mg/m^3; Soluble Salts and Alkyls: TWA 2 mg/m^3

DFG MAK: 1.5 mg/m^3; BAT: 200 μg/L in urine at end of shift

DOT Classification: 9; Label: CLASS 9 (NA 9260); DOT Class: 4.1; Label: Flammable Solid (UN 1309); DOT Class: 4.3; Label: Dangerous When Wet (UN 1396)

SAFETY PROFILE: Inhalation of finely divided powder has been reported to cause lung damage. Dust is moderately flammable and explosive. It can react violently with many substances, especially in powdered form.

PROP: Hard, strong, silvery-white ductile metal: in bulk form protected from oxidation in air by coherent Al_2O_3 coating. Mp: 660°, bp: 2494° @ 24 mm, d: 2.702, vap press: 1 mm @ 1284°. Sol in HCl, H_2SO_4, hot water, and alkalies.

AGX750 CAS: 7727-15-3 $AlBr_3$ HR: 2

ALUMINUM BROMIDE

DOT: UN 1725/UN 2580

SYNS: ALUMINUM TRIBROMIDE ♦ TRIBROMOALUMINUM

ACGIH TLV: TWA 2 mg(Al)/m^3

DOT Classification: 8; Label: Corrosive

SAFETY PROFILE: A toxic, corrosive material.

PROP: White to yellow-red lumps. Mp: 97.5°, bp: 263.3° @ 748 mm, d: 3.2, vap press: 1 mm @ 81.3°.

AGY750 CAS: 7446-70-0 $AlCl_3$ HR: 3

ALUMINUM CHLORIDE

DOT: UN 1726/UN 2581

SYNS: ALUMINUM TRICHLORIDE ♦ TRICHLOROALUMINUM

ACGIH TLV: TWA 2 mg(Al)/m^3

DOT Classification: 8; Label: Corrosive

SAFETY PROFILE: Moderately toxic by ingestion. Experimental teratogenic and reproductive effects. The dust is an irritant by ingestion, inhalation, and skin contact.

PROP: White or colorless hexagonal deliquescent crystals or moisture sensitive plates. D: 2.44, mp: 192° @ 2.5 atm, bp: subl @ 181°, vap press: 1 mm @ 100.0°. Violently sol in water; sol in alc and ether.

AHB000 CAS: 7784-18-1 AlF_3 HR: 3

ALUMINUM FLUORIDE
SYNS: ALUMINUM TRIFLUORIDE
OSHA PEL: TWA 2.5 mg(F)/m^3
ACGIH TLV: TWA 2.5 mg(F)/m^3; BEI: 3 mg/g creatinine of fluorides in urine prior to shift; 10 mg/g creatinine of fluorides in urine at end of shift; TWA 2 mg(Al)/m^3
NIOSH REL: (Fluorides, Inorganic) TWA 2.5 mg(F)/m^3
SAFETY PROFILE: A poison by ingestion. A severe eye irritant.
PROP: Solid, colorless crystals. Mp: 1291°, subl @ 1260°, d: 2.88, vap press: 1 mm @ 1238°, bp: 1537°. Sparingly sol in water; insol in org solvs.

AHD750 CAS: 13473-90-0 $N_3O_9 \cdot Al$ HR: 3
ALUMINUM(III) NITRATE (1:3)
DOT: UN 1438
SYNS: ALUMINUM NITRATE (DOT) ♦ ALUMINUM TRINITRATE ♦ NITRIC ACID, ALUMINUM SALT
ACGIH TLV: TWA 2 mg(Al)/m^3
DOT Classification: 5.1; Label: Oxidizer
SAFETY PROFILE: A poison. A severe eye and mild skin irritant. A powerful oxidizer.
PROP: White crystals or very hygroscopic solid. Bp: 50° @ 0.01 mm.

AHE250 CAS: 1344-28-1 Al_2O_3 HR: 3
ALUMINUM OXIDE (2:3)
SYNS: ALUMINA ♦ α-ALUMINA (OSHA)
OSHA PEL: Total Dust: TWA 10 mg/m^3; Respirable Fraction: TWA 5 mg/m^3
ACGIH TLV: TWA (nuisance particulate) 10 mg/m^3 of total dust (when toxic impurities are not present, e.g., quartz <1%); Not Classifiable as a Human Carcinogen
DFG MAK: Animal Carcinogen, Suspected Human Carcinogen; 1.5 mg/m^3
SAFETY PROFILE: Suspected carcinogen. Inhalation of finely divided particles may cause lung damage (Shaver's disease).
PROP: White powder or solid. Mp: 2050°, bp: 2977°, d: 3.5–4.0, vap press: 1 mm @ 2158°. Sol in hot NaOH.

ALL750 CAS: 5307-14-2 $C_6H_7N_3O_2$ HR: 3
4-AMINO-2-NITROANILINE
SYNS: 1,4-DIAMINO-2-NITROBENZENE ♦ o-NITRO-p-PHENYLENEDIAMINE (MAK)
DFG MAK: Confirmed Animal Carcinogen with Unknown Relevance to Humans
SAFETY PROFILE: Suspected carcinogen. Moderately toxic by ingestion. An experimental teratogen.
PROP: Black needles with strong green reflection from water. Mp: 137°.

AMI000 CAS: 504-29-0 $C_5H_6N_2$ HR: 3
2-AMINOPYRIDINE
DOT: UN 2671

SYNS: o-AMINOPYRIDINE ♦ AMINO-2-PYRIDINE ♦ α-PYRIDYLAMINE
OSHA PEL: TWA 0.5 ppm
ACGIH TLV: TWA 0.5 ppm
DFG MAK: 0.5 ppm (2 mg/m^3)
DOT Classification: 6.1; Label: Poison
SAFETY PROFILE: Poison by ingestion and inhalation. Human central nervous system effects by inhalation.
PROP: White powder or crystals from ligroin. Mp: 58.1, bp: 210.6°. Sol in water and ether; very sol in alc; sltly sol in ligroin.

AMY500 CAS: 7664-41-7 H_3N HR: 3
AMMONIA
DOT: UN 1005
SYNS: AMMONIA GAS ♦ ANHYDROUS AMMONIA
OSHA PEL: TWA 35 ppm
ACGIH TLV: TWA 25 ppm; STEL 35 ppm
DFG MAK: 20 ppm (14 mg/m^3)
NIOSH REL: CL 50 ppm
DOT Classification: 2.3; Label: Poison Gas; DOT Class: 2.2; Label: Nonflammable Gas
SAFETY PROFILE: A human poison by inhalation. An eye, mucous membrane, and systemic irritant by inhalation. It can react violently with many substances.
PROP: Colorless, alkaline, nonflammable gas with extremely pungent odor; liquefied by compression. Mp: –77.7°, bp: –33.35°, lel: 16%, uel: 25%, d: 0.771 g/liter @ 0°, 0.817 g/liter @ –79°, autoign temp: 1204°F, vap press: 10 atm @ 25.7°, vap d: 0.6. Very sol in water; moderately sol in alc.

ANE500 CAS: 12125-02-9 $H_4N{\cdot}Cl$ HR: 3
AMMONIUM CHLORIDE
SYNS: AMMONIUM MURIATE
OSHA PEL: (Fume) TWA 10 mg/m^3; STEL 20 mg/m^3
ACGIH TLV: TWA 10 mg/m^3; STEL 20 mg/m^3
SAFETY PROFILE: Moderately toxic. A severe eye irritant.
PROP: White, hygroscopic solid or crystals; salty taste. Bp: 520°, mp: 337.8°, d: 1.520, vap press: 1 mm @ 160.4° (sublimes). Sol in water, alc, and glycerin.

ANF250 CAS: 16919-58-7 $Cl_6Pt{\cdot}2H_4N$ HR: 3
AMMONIUM CHLOROPLATINATE
SYNS: AMMONIUM HEXACHLOROPLATINATE(IV) ♦ AMMONIUM PLATINIC CHLORIDE ♦ PLATINIC AMMONIUM CHLORIDE
OSHA PEL: TWA 0.002 mg(Pt)/m^3
ACGIH TLV: TWA 0.002 mg(Pt)/m^3
SAFETY PROFILE: Poison by inhalation and ingestion. An explosively unstable compound.
PROP: Cubic, yellow crystals or solid. D: 3.065, mp: decomp. Aq solns slowly photoreduce with substitution. Sol in water.

ANH000 CAS: 13826-83-0 NH_4BF_4 HR: 3

AMMONIUM FLUOBORATE

SYNS: AMMONIUM BOROFLUORIDE ♦ AMMONIUM FLUOROBORATE ♦ AMMONIUM TETRAFLUOROBORATE

OSHA PEL: TWA 2.5 mg(F)/m^3

ACGIH TLV: TWA 2.5 mg(F)/m^3; BEI: 3 mg/g creatinine of fluorides in urine prior to shift; 10 mg/g creatinine of fluorides in urine at end of shift.

NIOSH REL: (Fluorides, Inorganic) TWA 2.5 mg(F)/m^3

SAFETY PROFILE: A poison and strong irritant.

PROP: White, colorless, rhombic crystals. D: 1.871 @ 15°, mp: sublimes. Sol in NH_4OH and water.

ANJ000 CAS: 1341-49-7 F_2H_5N HR: 3

AMMONIUM HYDROGEN FLUORIDE

DOT: UN 1727/UN 2817

SYNS: AMMONIUM BIFLUORIDE ♦ AMMONIUM DIFLUORIDE ♦ AMMONIUM HYDROGEN DIFLUORIDE

OSHA PEL: TWA 2.5 mg(F)/m^3

ACGIH TLV: TWA 2.5 mg(F)/m^3; BEI: 3 mg/g creatinine of fluorides in urine prior to shift; 10 mg/g creatinine of fluorides in urine at end of shift.

NIOSH REL: (Fluorides, Inorganic) TWA 2.5 mg(F)/m^3

DOT Classification: 8; Label: Corrosive (UN 1727); DOT Class: 8; Label: Corrosive, Poison (UN 2817)

SAFETY PROFILE: Caustic poison and strong irritant by all routes.

PROP: White, colorless crystals. D: 1.51, mp: 126°, bp: 239°. Will etch glass. Very sol in water; sltly sol in alc.

ANK250 CAS: 1336-21-6 $H_4N \cdot HO$ HR: 3

AMMONIUM HYDROXIDE

DOT: NA 2672

SYNS: AMMONIA AQUEOUS ♦ AMMONIA SOLUTIONS, with >10% but not >35% ammonia (UN 2672) (DOT) ♦ AMMONIA SOLUTIONS, with >35% but not >50% ammonia (UN 2073) (DOT)

NIOSH REL: (Ammonia) CL 50 ppm

DOT Classification: 8; Label: Corrosive (UN 2672); DOT Class: 2.2; Label: Nonflammable Gas (UN 2073)

SAFETY PROFILE: A poison by ingestion and inhalation. A severe eye irritant. It can react violently with many substances.

PROP: Clear, colorless liquid solution of ammonia; very pungent odor. D: 0.90, mp: –77°. Sol in water. Soln contains not more than 44% ammonia.

ANM750 CAS: 13106-76-8 $MoO_4 \cdot 2H_4N$ HR: 3

AMMONIUM MOLYBDATE

SYNS: DIAMMONIUM MOLYBDATE ♦ MOLYBDIC ACID DIAMMONIUM SALT
OSHA PEL: TWA 5 mg(Mo)/m^3
ACGIH TLV: Soluble Compounds: TWA 5 mg(Mo)/m^3; (Proposed: TWA Soluble Compounds: TWA 0.5 mg(Mo)/m^3 Confirmed Animal Carcinogen with Unknown Relevance to Humans)
SAFETY PROFILE: Poison by ingestion. An irritant.
PROP: White solid. Sol in water.

ANP625 CAS: 3825-26-1 $C_8F_{15}O_2 \cdot H_4N$ HR: 3
AMMONIUM PERFLUOROOCTANOATE
SYNS: AMMONIUM PERFLUOROCAPRILATE ♦ PERFLUOROAMMONIUM OCTANOATE
ACGIH TLV: 0.01 mg/m^3; Animal Carcinogen
SAFETY PROFILE: Confirmed carcinogen. Poison by inhalation. Moderately toxic by ingestion. An eye and skin irritant.
PROP: Solid.

ANR000 CAS: 7727-54-0 $O_8S_2 \cdot 2H_4N$ HR: 3
AMMONIUM PERSULFATE
DOT: UN 1444
SYNS: AMMONIUM PEROXYDISULFATE
ACGIH TLV: TWA 0.1 mg/m^3
DOT Classification: 5.1; Label: Oxidizer
SAFETY PROFILE: Moderately toxic by ingestion. It can react violently with many substances.
PROP: Colorless, white, monoclinic crystals. Mp: decomp @ 120°, d: 1.982. Stable as dry solid; decomposes in H_2O forming O_2.

ANU650 CAS: 7773-06-0 $H_2NO_3S \cdot H_4N$ HR: 2
AMMONIUM SULFAMATE
SYNS: AMMONIUM AMIDOSULPHATE ♦ AMMONIUM SULPHAMATE ♦ MONOAMMONIUM SULFAMATE
OSHA PEL: TWA 10 mg/m^3; Respirable Fraction: 5 mg/m^3
ACGIH TLV: TWA 10 mg/m^3
DFG MAK: 15 mg/m^3
SAFETY PROFILE: Moderately toxic by ingestion. A powerful oxidizer.
PROP: Deliquescent, hygroscopic, crystalline material (white crystalline solid). Bp: 160° (decomp), mp: 131°. Sol in water, liq NH_3, formamide, and glycerol.

ANY250 CAS: 7803-55-6 $O_3V \cdot H_4N$ HR: 3
AMMONIUM VANADATE
DOT: UN 2859
SYNS: AMMONIUM METAVANADATE (DOT) ♦ VANADIC ACID, AMMONIUM SALT
ACGIH TLV: TWA 0.05 mg(V_2O_5)/m^3

NIOSH REL: (Vanadium Compounds) CL 0.05 mg(V)/m^3/15M
DOT Classification: 6.1; Label: Poison
SAFETY PROFILE: Poison by ingestion. Moderately toxic by skin contact. An experimental teratogen.
PROP: Colorless to yellow crystals or solid. Mp: 200° (decomp), d: 2.326.

AOD725 CAS: 628-63-7 $C_7H_{14}O_2$ HR: 3
n-AMYL ACETATE
DOT: UN 1104
SYNS: ACETIC ACID, AMYL ESTER ♦ AMYL ACETIC ESTER ♦ n-PENTYL ACETATE
OSHA PEL: TWA 100 ppm
ACGIH TLV: TWA 100 ppm; (Proposed: TWA 50 ppm; STEL 100 ppm)
DFG MAK: 50 ppm
DOT Classification: 3; Label: Flammable Liquid
SAFETY PROFILE: A human eye irritant. Dangerous fire hazard.
PROP: Colorless liquid; pear- or banana-like odor. Mp: –78.5°, bp: 148° @ 737 mm, ULC: 55–60, lel: 1.1%, uel: 7.5%, flash p: 77°F (CC), d: 0.879 @ 20°/20°, autoign temp: 714°F, vap d: 4.5. Very sltly sol in water; misc in alc and ether.

AOD735 CAS: 626-38-0 $C_7H_{14}O_2$ HR: 3
sec-AMYL ACETATE
DOT: UN 1104
SYNS: 2-ACETOXYPENTANE ♦ 1-METHYLBUTYL ACETATE ♦ 2-PENTYL ACETATE
OSHA PEL: TWA 125 ppm
ACGIH TLV: TWA 125 ppm; (Proposed: TWA 50 ppm; STEL 100 ppm)
DFG MAK: 50 ppm
DOT Classification: 3; Label: Flammable Liquid
SAFETY PROFILE: Mildly toxic by inhalation. Dangerous fire hazard.
PROP: Colorless liquid. Bp: 120°, flash p: 73.4°F (CC), d: 0.862–0.866 @ 20°/20°, vap d: 4.48, lel: 1.1%, uel: 7.5%. Sltly sol in water; misc in alc and ether.

AOQ000 CAS: 62-53-3 C_6H_7N HR: 3
ANILINE
DOT: UN 1547
SYNS: AMINOBENZENE ♦ BENZENAMINE ♦ PHENYLAMINE
OSHA PEL: TWA 2 ppm (skin)
ACGIH TLV: TWA 2 ppm (skin); Animal Carcinogen; BEI: 50 mg/g creatinine of total p-aminophenol in urine at end of shift or 1.5% of hemoglobin for methemoglobin in blood during or end of shift.
DFG MAK: 2 ppm (7.7 mg/m^3), Confirmed Animal Carcinogen with Unknown Relevance to Humans; BAT: 1 mg/L in urine at end of shift
DOT Classification: 6.1; Label: Poison
SAFETY PROFILE: Suspected carcinogen. A poison inhalation and ingestion. A skin and severe eye irritant, and a mild sensitizer. A combustible liquid. It can react violently with many substances.

PROP: Colorless, oily liquid which darkens on exposure to light; characteristic odor. Mp: –6°, bp: 184.4°, lel: 1.3%, ULC: 20–25, flash p: 158°F (CC), fp: –6.2°, d: 1.02 @ 20°/4°, autoign temp: 1139°F, vap press: 1 mm @ 34.8°, vap d: 3.22.

AOV900 CAS: 90-04-0 C_7H_9NO HR: 3

o-ANISIDINE

DOT: UN 2431

SYNS: o-AMINOANISOLE ♦ 1-AMINO-2-METHOXYBENZENE ♦ 2-ANISIDINE ♦ o-ANISYLAMINE ♦ o-METHOXYANILINE ♦ 2-METHOXYBENZENAMINE

OSHA PEL: TWA 0.5 mg/m^3

ACGIH TLV: TWA 0.5 mg/m^3 (skin); Animal Carcinogen

DFG MAK: Animal Carcinogen, Suspected Human Carcinogen

SAFETY PROFILE: Confirmed carcinogen. Moderately toxic by ingestion.

PROP: Yellowish liquid. Mp: 5°, bp: 225°. Sol in acids; insol in H_2O; misc in EtOH, Et_2O, C_6H_6.

AOW000 CAS: 104-94-9 C_7H_9NO HR: 1

p-ANISIDINE

SYNS: p-AMINOANISOLE ♦ p-ANISYLAMINE ♦ p-METHOXYANILINE ♦ p-METHOXYPHENYLAMINE

OSHA PEL: TWA 0.5 mg/m^3

ACGIH TLV: TWA 0.5 mg/m^3 (skin); Not Classifiable as a Human Carcinogen

DFG MAK: 0.1 ppm (0.51 mg/m^3)

SAFETY PROFILE: Moderately toxic. A mild sensitizer. May cause a contact dermatitis.

PROP: Crystals, plates from aq soln. D: 1.089 @ 55°/55°, mp: 57°, bp: 246°, vap d: 4.28. Sol in alc, ether, and hot water (insol in cold water).

APG500 CAS: 120-12-7 $C_{14}H_{10}$ HR: 2

ANTHRACENE

SYNS: ANTHRACIN ♦ PARANAPHTHALENE

OSHA PEL: TWA 0.2 mg/m^3

SAFETY PROFILE: A skin irritant and allergen. Combustible.

PROP: Colorless crystals, monoclinic plates from EtOH, violet fluorescence when pure. Mp: 217°, lel: 0.6%, flash p: 250°F (CC), d: 1.24 @ 27°/4°, autoign temp: 1004°F, vap press: 1 mm @ 145.0° (subl), vap d: 6.15, bp: 339.9°. Insol in water. Solubility in alc @ 1.9/100 @ 20°; in ether 12.2/100 @ 20°.

AQB750 CAS: 7440-36-0 Sb HR: 3

ANTIMONY

DOT: UN 2871

SYNS: ANTIMONY POWDER (DOT) ♦ STIBIUM

OSHA PEL: TWA 0.5 mg(Sb)/m^3

ACGIH TLV: TWA 0.5 mg(Sb)/m^3
DFG MAK: 0.5 mg(Sb)/m^3
NIOSH REL: TWA 0.5 mg(Sb)/m^3
DOT Classification: 6.1; Label: KEEP AWAY FROM FOOD
SAFETY PROFILE: Moderate fire and explosion hazard in the forms of dust and vapor.
PROP: Silvery or gray, lustrous metalloid. Mp: 630°, bp: 1635°, d: 6.684 @ 25°, vap press: 1 mm @ 886°. Insol in water; sol in hot concentrated H_2SO_4.

AQC500 CAS: 10025-91-9 Cl_3Sb HR: 3
ANTIMONY(III) CHLORIDE
DOT: UN 1733
SYNS: ANTIMONOUS CHLORIDE (DOT) ♦ ANTIMONY TRICHLORIDE ♦ TRICHLOROSTIBINE
OSHA PEL: TWA 500 μg(Sb)/m^3
ACGIH TLV: TWA 0.5 mg(Sb)/m^3
NIOSH REL: (Antimony) TWA 0.5 mg(Sb)/m^3
DOT Classification: 8; Label: Corrosive
SAFETY PROFILE: Moderately toxic by ingestion.
PROP: Colorless, rhombic, deliq, hygroscopic crystals which fume in the air. D: 3.06, mp: 73.4°, bp: 220°, vap press: 1 mm @ 49.2° (subl). Sol in cold EtOH, CS_2, Et_2O, CCl_4, and H_2O (small amounts); insol in quinoline, other org bases.

AQD000 CAS: 7647-18-9 Cl_5Sb HR: 3
ANTIMONY(V) CHLORIDE
DOT: UN 1730/UN 1731
SYNS: ANTIMONY PENTACHLORIDE (DOT) ♦ ANTIMONY PERCHLORIDE ♦ PENTACHLOROANTIMONY
OSHA PEL: TWA 500 μg(Sb)/m^3
ACGIH TLV: TWA 0.5 mg(Sb)/m^3
NIOSH REL: (Antimony) TWA 0.5 mg(Sb)/m^3
DOT Classification: 8; Label: Corrosive
SAFETY PROFILE: Poison by ingestion. Corrosive.
PROP: Colorless or red-yellow oil or liquid; offensive odor. Mp: 4°, bp: 140°, d: 2.336, vap press: 1 mm @ 22.7°. Decomp in water; sol in HCl, HBr, CS_2, CCl_4, and $CHCl_3$.

AQF000 CAS: 1309-64-4 O_3Sb_2 HR: 3
ANTIMONY OXIDE
SYNS: ANTIMONY PEROXIDE ♦ ANTIMONY SESQUIOXIDE ♦ ANTIMONY TRIOXIDE ♦ ANTIMONY WHITE
OSHA PEL: TWA 0.5 mg(Sb)/m^3
ACGIH TLV: TWA 0.5 mg(Sb)/m^3; Suspected Carcinogen
DFG MAK: Animal Carcinogen, Suspected Human Carcinogen

NIOSH REL: TWA 0.5 mg(Sb)/m^3
SAFETY PROFILE: Confirmed carcinogen. Moderately toxic. An experimental teratogen.
PROP: White cubes. D: 5.2, mp: 650°, bp: 1550° (subl). Very sltly sol in water; sol in KOH and HCl.

AQF250 CAS: 7783-70-2 F_5Sb HR: 3
ANTIMONY(V) PENTAFLUORIDE
DOT: UN 1732
SYNS: ANTIMONY FLUORIDE ♦ PENTAFLUOROANTIMONY
OSHA PEL: TWA 0.5 mg(Sb)/m^3
ACGIH TLV: TWA 0.5 mg(Sb)/m^3
NIOSH REL: (Antimony) TWA 0.5 mg(Sb)/m^3
DOT Classification: 8; Label: Corrosive, Poison
SAFETY PROFILE: A poison by inhalation. A very reactive, corrosive liquid to skin, eyes, mucous membranes.
PROP: Oily, colorless liquid. Very reactive. Mp: 7.0°, bp: 149.5°, d: (liq) 2.99 @ 23°. Sol in water and KF.

ARA750 CAS: 7440-38-2 As HR: 3
ARSENIC
DOT: UN 1558
SYNS: COLLOIDAL ARSENIC ♦ METALLIC ARSENIC
OSHA PEL: TWA 0.01 mg(As)/m^3; Cancer Hazard
ACGIH TLV: TWA 0.01 mg/m^3; Confirmed Human Carcinogen; BEI: 35 μ (As)/L inorganic arsenic and methylated metabolites in urine; (Proposed: BEI: 50 μ (As)/L creatinine) inorganic arsenic and methylated metabolites in urine)
DFG TRK: 0.2 mg/m^3 calculated as arsenic in that portion of dust that can possibly be inhaled
NIOSH REL: CL 2 μg(As)/m^3
DOT Classification: 6.1; Label: Poison
SAFETY PROFILE: Confirmed human carcinogen. An experimental teratogen. Flammable in the form of dust. It can react violently with many substances.
PROP: Silvery to black, brittle, crystalline, or amorphous metalloid. Mp: 814° @ 36 atm, bp: subl @ 612°, d: black crystals 5.724 @ 14°, black amorphous 4.7, vap press: 1 mm @ 372° (subl). Insol in water; sol in HNO_3.

ARF750 HR: 3
ARSENIC COMPOUNDS
SYN: ARSENICALS
OSHA PEL: Inorganic: TWA 0.01 mg(As)/m^3; Cancer Hazard; Organic: TWA 0.5 mg(As)/m^3

ACGIH TLV: Inorganic: TWA 0.01 mg/m^3; Confirmed Human Carcinogen; BEI: 35 μ (As)/L inorganic arsenic and methylated metabolites in urine; (Proposed: BEI: 50 μ (As)/L creatinine) inorganic arsenic and methylated metabolites in urine)

NIOSH REL: CL 2 $\mu g(As)/m^3$/15M

SAFETY PROFILE: Inorganic compounds are confirmed human carcinogens. Poisoning from arsenic compounds may be acute or chronic. Acute poisoning usually results from swallowing arsenic compounds; chronic poisoning from either swallowing or inhaling. Inorganic arsenicals are more toxic than organics.

ARI750 CAS: 1327-53-3 As_2O_3 HR: 3

ARSENIC TRIOXIDE

DOT: UN 1561

SYNS: ARSENIC(III) OXIDE ♦ ARSENIOUS ACID ♦ ARSENOUS ACID ♦ ARSENOUS ACID ANHYDRIDE

OSHA PEL: TWA 0.01 $mg(As)/m^3$; Cancer Hazard

ACGIH TLV: TWA 0.01 mg/m^3; Confirmed Human Carcinogen; BEI: 35 μ (As)/L inorganic arsenic and methylated metabolites in urine; (Proposed: BEI: 50 μ (As)/L creatinine) inorganic arsenic and methylated metabolites in urine)

DFG MAK: Human Carcinogen

NIOSH REL: CL 2 $\mu g(As)/m^3$/15M

DOT Classification: 6.1; Label: Poison

SAFETY PROFILE: Confirmed human carcinogen. Poison by ingestion. An experimental teratogen.

PROP: Colorless, rhombic crystals (dimer, claudetite), or white powder. D: 4.15, mp: 312°, bp: 460°. Solubility in water: 1.82/100 @ 20°; sol in alc. Cubes: Colorless. D: 3.865, mp: 309°. Solubility in water: 1.2/100 @ 20°.

ARK250 CAS: 7784-42-1 AsH_3 HR: 3

ARSINE

DOT: UN 2188

SYNS: ARSENIC HYDRIDE ♦ ARSENIC TRIHYDRIDE

OSHA PEL: TWA 0.05 ppm

ACGIH TLV: TWA 0.05 ppm; (Proposed: 0.002 ppm) BEI: 35 μ (As)/L inorganic arsenic and methylated metabolites in urine; (Proposed: BEI: 50 μ (As)/L creatinine) inorganic arsenic and methylated metabolites in urine)

DFG MAK: 0.05 ppm (0.16 mg/m^3)

NIOSH REL: (Arsine) CL 2 $\mu g(As)/m^3$/15M

DOT Classification: 2.3; Label: Poison Gas, Flammable Gas

SAFETY PROFILE: Confirmed human carcinogen. Poison by inhalation. Flammable when exposed to flame.

PROP: Thermally unstable, colorless; gas with mild garlic odor. D: 2.695 g/L, bp: –62.5°, vap d: 2.66, mp: –116°. Readily oxidized to As_2O_3. Very little tendency to protonate. Solubility in water: 28 mg/100 @ 20°. Sol in benzene and chloroform.

ARM250 CAS: 1332-21-4 HR: 3

ASBESTOS

DOT: NA 2212

SYNS: ACTINOLITE ♦ AMOSITE ♦ AMPHIBOLE ♦ ANTHOPHYLITE ♦ CHRYSOTILE ♦ TREMOLITE

OSHA PEL: TWA 2 million fibers/m^3; CL 10 million fibers/m^3; Cancer and Lung Disease Hazard

ACGIH TLV: TWA 0.1 fibers/cc; Confirmed Human Carcinogen

DFG TRK: (Fine dust particles that are able to reach the alveolar area of the lung) crocidolite: 0.05 X 10^6 fibers/m^3 (0.025 mg/m^3) (definition of fiber: length greater than 5 µm; diameter less than 3 µm; length/diameter greater than 3:1, equivalent to 1 fiber/cc); chrysotile, amosite, anthophyllite, tremolite, actinolite: 1 X 10^6 fibers/m^3 (0.05 mg/m^3), applicable when there is more than 2.5% asbestos in the dust; 2.0 mg/m^3, applicable when there is less than or equal to 2.5 weight percent asbestos in fine dust

NIOSH REL: (asbestos): 0.1 fb/cc in a 400 L air sample

DOT Classification: 9; Label: CLASS 9

SAFETY PROFILE: Confirmed human carcinogen. Human pulmonary system effects by inhalation. Usually at least 4 to 7 years of exposure are required before serious lung damage (fibrosis) results.

ARO500 CAS: 8052-42-4 HR: 3

ASPHALT

DOT: NA 1999

SYNS: ASPHALT FUMES (ACGIH) ♦ BITUMEN (MAK) ♦ PETROLEUM ASPHALT ♦ ROAD ASPHALT (DOT) ♦ ROAD TAR (DOT)

ACGIH TLV: TWA 5 mg/m^3; Not Classifiable as a Human Carcinogen; (Proposed: 0.5 mg/m^3; Not Classifiable as a Human Carcinogen)

DFG MAK: Confirmed Animal Carcinogen with Unknown Relevance to Humans

NIOSH REL: (Asphalt Fumes) CL 5 mg/m^3/15M

DOT Classification: 3; Label: Flammable Liquid

SAFETY PROFILE: Suspected carcinogen. A moderate irritant. Combustible.

PROP: Black or dark-brown mass. Bp: <470°, flash p: 400°F (CC), d: 0.95–1.1, autoign temp: 905°F.

BAB750 CAS: 1395-21-7 HR: 3

BACILLUS SUBTILIS BPN

SYNS: BACILLOMYCIN (8CI, 9CI) ♦ SUBTILISINS (ACGIH)

OSHA PEL: CL 0.00006 mg/m^3

ACGIH TLV: CL 0.00006 mg/m^3

SAFETY PROFILE: A severe eye irritant.

PROP: A commercial raw proteolytic enzyme used in laundry detergents (FCTXAV 7,581,69).

BAH250 CAS: 7440-39-3 Ba HR: 3

BARIUM

DOT: UN 1400

OSHA PEL: TWA 0.5 mg(Ba)/m^3

ACGIH TLV: TWA 0.5 mg(Ba)/m^3; Not Classifiable as a Human Carcinogen

DFG MAK: 0.5 mg(Ba)/m^3

DOT Classification: 4.3; Label: Dangerous When Wet

SAFETY PROFILE: Water and stomach acids solubilize barium salts and can cause poisoning. Dust is dangerous and explosive when exposed to heat, flame, or chemical reaction.

PROP: Silver-white, sltly lustrous, somewhat malleable metal. Mp: 727°, bp: 1640°, d: 3.5 @ 20°, vap press: 10 mm @ 1049°. Dissolves in H_2O forming $Ba(OH)_2$ solns. Solution in NH_3(l) blue-black soln.

BAI000 CAS: 18810-58-7 BaN_6 HR: 3

BARIUM AZIDE

DOT: UN 0224/UN 1571

SYNS: BARIUM AZIDE, dry or wetted with <50% water, by weight (UN 0224) (DOT) ♦ BARIUM AZIDE, wetted with not <50% water, by weight (UN 1571) (DOT)

OSHA PEL: TWA 0.5 mg(Ba)/m^3

ACGIH TLV: TWA 0.5 mg(Ba)/m^3; Not Classifiable as a Human Carcinogen

DFG MAK: 0.5 mg(Ba)/m^3

DOT Classification: EXPLOSIVE 1.1A; Label: EXPLOSIVE 1.1A, Poison (UN 0224); DOT Class: 4.1; Label: Flammable Solid, Poison (UN 1571)

SAFETY PROFILE: A poison. Moderate explosion hazard when shocked or heated to 275°. Spontaneously flammable in air. Very unstable.

PROP: Monoclinic prisms or crystals, decomp on heating with loss of N_2 at about 12°. Mp: evolves N_2 at about 120°, bp: explodes, d: 2.936. Very sol in H_2O; sltly sol in EtOH; insol in Et_2O.

BAK250 CAS: 10294-40-3 $Ba \cdot CrO_4$ HR: 3

BARIUM CHROMATE(VI)

SYNS: BARIUM CHROMATE OXIDE

OSHA PEL: TWA 0.1 mg (C_3O_3)m^3; 0.5 mg(Ba)/m^3

ACGIH TLV: TWA 0.5 mg(Ba)/m^3; Not Classifiable as a Human Carcinogen; 0.05 mg(Cr)/m^3; Confirmed Human Carcinogen

DFG MAK: 0.5 mg(Ba)/m^3

NIOSH REL: TWA 0.001 mg(Cr(VI))/m^3

SAFETY PROFILE: Confirmed human carcinogen. A poison. Reacts vigorously with reducing materials. Used in pyrotechnics and as an explosive initiator.

PROP: Heavy, pale-yellow, crystalline powder; darkens on heating. D: 4.498 @ 15°. Sol in strong acids; insol in org solvents.

BAK500 HR: 3
BARIUM COMPOUNDS (soluble)
OSHA PEL: Soluble Compounds: TWA 0.5 mg(Ba)/m^3
ACGIH TLV: Soluble Compounds:TWA 0.5 mg/m^3
DFG MAK: Soluble Compounds: 0.5 mg/m^3
DOT Classification: 6.1; Label: Poison
SAFETY PROFILE: The chromate is a human carcinogen. The soluble barium salts, such as the chloride and sulfide, are poisonous when ingested. The insoluble sulfate used in radiography is not acutely toxic.

BAM000 CAS: 7787-32-8 BaF_2 HR: 3
BARIUM FLUORIDE
OSHA PEL: TWA 0.5 mg(Ba)/m^3; 2.5 mg(F)/m^3
ACGIH TLV: TWA 0.5 mg(Ba)/m^3; Not Classifiable as a Human Carcinogen; TWA 2.5 mg(F)/m^3; BEI: 3 mg/g creatinine of fluorides in urine prior to shift; 10 mg/g creatinine of fluorides in urine at end of shift.
DFG MAK: 0.5 mg(Ba)/m^3
NIOSH REL: (Fluorides, Inorganic) TWA 2.5 mg(F)/m^3
SAFETY PROFILE: A poison by ingestion. An experimental teratogen.
PROP: White, colorless powder or cubic crystals. Mp: 1368°, bp: 2137°, d: 4.89. Sltly sol in H_2O.

BAN250 CAS: 10022-31-8 $N_2O_6 \cdot Ba$ HR: 3
BARIUM(II) NITRATE (1:2)
DOT: UN 1446
SYNS: BARIUM DINITRATE ♦ BARIUM NITRATE (DOT) ♦ NITRIC ACID, BARIUM SALT
OSHA PEL: TWA 0.5 mg(Ba)/m^3
ACGIH TLV: TWA 0.5 mg(Ba)/m^3; Not Classifiable as a Human Carcinogen
DFG MAK: 0.5 mg(Ba)/m^3
DOT Classification: 5.1; Label: Oxidizer, Poison
SAFETY PROFILE: A poison by ingestion. An irritant to skin and eyes. An oxidizer.
PROP: Lustrous, colorless, cubic crystals. Mp: 592°, bp: decomp, d: 3.24 @ 23°. Decomp on heating with evolution of NO_2 and O_2 and formation of BaO. Insol in EtOH.

BAP000 CAS: 7727-43-7 $O_4S \cdot Ba$ HR: 2
BARIUM SULFATE
SYNS: BAYRITES ♦ SULFURIC ACID, BARIUM SALT (1:1)
OSHA PEL: Total Dust: TWA 10 mg/m^3; Respirable Fraction: 5 mg/m^3
ACGIH TLV: TWA (nuisance particulate) 10 mg/m^3 of total dust (when toxic impurities are not present, e.g., quartz <1%)

SAFETY PROFILE: A relatively insoluble salt used as an opaque medium in radiography. Soluble impurities can lead to toxic reactions.
PROP: White, heavy, orthorhombic, odorless powder or crystals. Undergoes orthorhombic to monoclinic phase transition at 11°. D: 4.50 @ 15°, mp: 1580°. Sltly sol in H_2O. Insol in water or dilute acids.

BAY300 CAS: 98-87-3 $C_7H_6Cl_2$ HR: 3
BENZAL CHLORIDE
DOT: UN 1886
SYNS: BENZYLENE CHLORIDE ♦ BENZYLIDENE CHLORIDE (DOT) ♦ CHLOROBENZAL
DFG MAK: Confirmed Human Carcinogen
DOT Classification: 6.1; Label: Poison
SAFETY PROFILE: Confirmed carcinogen. Poison by inhalation. Moderately toxic by ingestion. A strong irritant and lachrymator.
PROP: Very refractive liquid. Mp: –16°, bp: 214°, d: 1.29.

BBL250 CAS: 71-43-2 C_6H_6 HR: 3
BENZENE
DOT: UN 1114
SYNS: BENZOL (DOT) ♦ MINERAL NAPHTHA ♦ PHENYL HYDRIDE
OSHA PEL: TWA 1 ppm; STEL 5 ppm; Pk 5 ppm/15M/8H; Cancer Hazard
ACGIH TLV: TWA 0.5 ppm; STEL 2.5 ppm (skin); Confirmed Human Carcinogen; BEI: 25 μ/g creatinine of Sphenylmercapturic acid in urine at end of shift
DFG TRK: Human Carcinogen
NIOSH REL: TWA 0.32 mg/m^3; CL 3.2 mg/m^3/15M
DOT Classification: 3; Label: Flammable Liquid
SAFETY PROFILE: Confirmed human carcinogen. A poison by inhalation and skin contact. Moderately toxic by ingestion. A severe eye and moderate skin irritant. Experimental teratogenic effects. A dangerous fire hazard when exposed to heat or flame. It can react violently with many substances.
PROP: Clear, colorless liquid. Mp: 5.51°, bp: 80.093–80.094°, flash p: 12°F (CC), d: 0.8794 @ 20°, autoign temp: 1044°F, lel: 1.4%, uel: 8.0%, vap press: 100 mm @ 26.1°, vap d: 2.77, ULC: 95–100. Very sltly sol in H_2O; misc in most org solvs.

BBP000 CAS: 123-61-5 $C_8H_4N_2O_2$ HR: 3
BENZENE-1,3-DIISOCYANATE
SYNS: 1,3-DIISOCYANATOBENZENE ♦ m-PHENYLENE ISOCYANATE
NIOSH REL: TWA (Diisocyanates) 0.005 ppm; CL 0.02 ppm/10M
SAFETY PROFILE: A sensitizer at very low concentrations.
PROP: Crystals. Mp: 51–55°, bp: 102–104° @ 8 mm.

BBQ000 CAS: 319-84-6 $C_6H_6Cl_6$ HR: 3
BENZENE HEXACHLORIDE-α-isomer

SYNS: α-BENZENEHEXACHLORIDE ♦ α-BHC ♦ α-HEXACHLOROCYCLOHEXANE ♦ α-LINDANE
DFG MAK: 0.5 mg/m^3
SAFETY PROFILE: Confirmed carcinogen. Poison by ingestion.
PROP: Solid. Mp: 158°.

BBQ500 CAS: 58-89-9 $C_6H_6Cl_6$ HR: 3
BENZENE HEXACHLORIDE-γ-isomer
SYNS: γ-BENZENE HEXACHLORIDE ♦ BHC ♦ γ-HEXACHLORAN ♦ LINDANE (ACGIH, DOT, USDA)
OSHA PEL: TWA 0.5 mg/m^3 (skin)
ACGIH TLV: TWA 0.5 mg/m^3 (skin)
SAFETY PROFILE: Confirmed carcinogen. A poison by ingestion and skin contact. Experimental teratogenic effects.
PROP: Solid. Mp: 112.5°.

BBR000 CAS: 319-85-7 $C_6H_6Cl_6$ HR: 3
trans-α-BENZENEHEXACHLORIDE
SYNS: β-BHC ♦ β-HEXACHLOROBENZENE ♦ β-LINDANE
DFG MAK: 0.5 mg/m^3
SAFETY PROFILE: Confirmed carcinogen. Mildly toxic by ingestion.
PROP: Solid. Mp: 297°.

BBX000 CAS: 92-87-5 $C_{12}H_{12}N_2$ HR: 3
BENZIDINE
DOT: UN 1885
SYNS: 4,4′-BIANILINE ♦ 4,4′-BIPHENYLDIAMINE ♦ 4,4′-BIPHENYLENEDIAMINE ♦ 4,4′-DIAMINOBIPHENYL
OSHA: Cancer Suspect Agent
ACGIH TLV: Confirmed Human Carcinogen
DFG MAK: Human Carcinogen
DOT Classification: 6.1; Label: Poison
SAFETY PROFILE: Confirmed human carcinogen. Poison by ingestion.
PROP: Grayish-yellow, crystalline powder; white or sltly reddish crystals, powder, or leaf from water or alc. Mp: 127.5–128.7° @ 740 mm, bp: 401.7°, d: 1.250 @ 20°/4°.

BCS750 CAS: 50-32-8 $C_{20}H_{12}$ HR: 3
BENZO(a)PYRENE
SYNS: BENZO(d,e,f)CHRYSENE ♦ 3,4-BENZOPYRENE ♦ 6,7-BENZOPYRENE ♦ B(a)P
OSHA PEL: TWA 0.2 mg/m^3
SAFETY PROFILE: Confirmed carcinogen. Experimental teratogenic effects. A skin irritant.
PROP: Pale-yellow crystals. Mp: 177°, bp: 312° @ 10 mm. Insol in water; sol in benzene, toluene, and xylene.

BDS000 CAS: 94-36-0 $C_{14}H_{10}O_4$ HR: 3

BENZOYL PEROXIDE

SYNS: BENZOIC ACID, PEROXIDE ♦ DIBENZOYL PEROXIDE (MAK)

OSHA PEL: TWA 5 mg/m^3

ACGIH TLV: TWA 5 mg/m^3; Not Classifiable as a Human Carcinogen

DFG MAK: 5 mg/m^3; Weak allergin and skin irritant

NIOSH REL: (Benzoyl Peroxide) TWA 5 mg/m^3

SAFETY PROFILE: Can cause dermatitis, asthmatic effects, testicular atrophy, and vasodilation. An allergen and eye irritant. Moderate fire hazard by spontaneous chemical reaction in contact with reducing agents. It can react violently with many substances.

PROP: White, granular, tasteless, odorless powder or prisms. Mp: 106–108.6° (decomp), bp: decomposes explosively, autoign temp: 176°F. Sol in benzene, acetone, chloroform; sltly sol in alc; insol in water.

BDX000 CAS: 140-11-4 $C_9H_{10}O_2$ HR: 3

BENZYL ACETATE

SYNS: ACETIC ACID BENZYL ESTER ♦ α-ACETOXYTOLUENE

ACGIH TLV: (Proposed: TWA 10 ppm, Animal Carcinogen)

SAFETY PROFILE: A poison by inhalation. Moderately toxic by ingestion. Combustible liquid.

PROP: Colorless liquid; sweet, floral fruity odor. Mp: –51.5°, bp: 134° @ 102 mm, flash p: 216°F (CC), d: 1.06, autoign temp: 862°F, vap press: 1 mm @ 45°, vap d: 5.1, refr index: 1.501. Sol in alc, most fixed oils, propylene glycol; insol in glycerin and water @ 214°.

BEE375 CAS: 100-44-7 C_7H_7Cl HR: 3

BENZYL CHLORIDE

DOT: UN 1738

SYNS: CHLOROMETHYLBENZENE ♦ CHLOROPHENYLMETHANE ♦ α-CHLOROTOLUENE ♦ TOLYL CHLORIDE

OSHA PEL: TWA 1 ppm

ACGIH TLV: TWA 1 ppm; Animal Carcinogen

DFG MAK: Confirmed Human Carcinogen

NIOSH REL: (Benzyl Chloride) CL 5 mg/m^3/15M

DOT Classification: 6.1; Label: Poison, Corrosive

SAFETY PROFILE: Confirmed carcinogen. Poison by inhalation. Moderately toxic by ingestion. A corrosive irritant to skin, eyes, and mucous membranes. Flammable and moderately explosive.

PROP: Colorless liquid, very refractive; irritating, unpleasant odor. Mp: –48°, bp: 99° @ 62 mm, lel: 1.1%, flash p: 153°F, d: 1.11 @ 4°/4°, autoign temp: 1085°F, vap d: 4.36.

BFL250 CAS: 98-07-7 $C_7H_5Cl_3$ HR: 3

BENZYL TRICHLORIDE

DOT: UN 2226

SYNS: BENZENYL CHLORIDE ♦ BENZOIC TRICHLORIDE ♦ BENZOTRICHLORIDE (DOT, MAK) ♦ BENZYLIDYNE CHLORIDE ♦ PHENYLTRICHLOROMETHANE ♦ TRICHLOROMETHYLBENZENE
ACGIH TLV: (Proposed: CL 0.1 (skin); Suspected Human Carcinogen)
DFG MAK: Confirmed Human Carcinogen
DOT Classification: 8; Label: Corrosive
SAFETY PROFILE: Confirmed carcinogen. Experimental poison by inhalation. Corrosive to the skin, eyes, and mucous membranes.
PROP: Clear, colorless to yellowish liquid; penetrating odor. Mp: –5°, bp: 221°, d: 1.38 @ 15.5°/15.5°, vap d: 6.77.

BFO750 CAS: 7440-41-7 Be HR: 3
BERYLLIUM
DOT: UN 1966/UN 1567
SYNS: BERYLLIUM, powder (UN 1567) (DOT)
OSHA PEL: TWA 0.002 mg(Be)/m^3; STEL 0.005 mg(Be)/m^3/30M; CL 0.025 mg(Be)/m^3
ACGIH TLV: TWA 0.002 (Be)mg/m^3; Confirmed Human Carcinogen; (Proposed: TWA 0.0002 (Be)mg/m^3 (skin, sensitizer); Confirmed Human Carcinogen)
DFG TRK: Animal Carcinogen, Suspected Human Carcinogen. Grinding of beryllium metal and alloys: 0.005 mg/m^3 calculated as beryllium in that portion of dust that can possibly be inhaled; other beryllium compounds: 0.002 mg/m^3 calculated as beryllium in that portion of dust that can possibly be inhaled
NIOSH REL: CL not to exceed 0.0005 mg(Be)/m^3
DOT Classification: 6.1; Label: Poison (UN 1566); DOT Class: 6.1; Label: Poison, Flammable Solid (UN 1567)
SAFETY PROFILE: Confirmed carcinogen. A moderate fire hazard in the form of dust or powder.
PROP: A silvery-white, relatively soft, lustrous metal, ductile at red heat. Unreactive to H_2O and air; dissolves vigorously in dil acids. Be reacts with aq alkalies or H_2. Mp: 1287–1292°, bp: 2970°, d: 1.85.

BFQ500 HR: 3
BERYLLIUM COMPOUNDS
OSHA PEL: TWA 0.002 mg(Be)/m^3; STEL 0.005 mg(Be)/m^3/30M; CL 0.025 mg(Be)/m^3
ACGIH TLV: TWA 0.002 (Be)mg/m^3; Confirmed Human Carcinogen; (Proposed: TWA 0.0002 (Be)mg/m^3 (sensitizer); Confirmed Human Carcinogen)
DFG TRK: Animal Carcinogen, Suspected Human Carcinogen. Grinding of beryllium metal and alloys: 0.005 mg/m^3 calculated as beryllium in that portion of dust that can possibly be inhaled; other beryllium compounds: 0.002 mg/m^3 calculated as beryllium in that portion of dust that can possibly be inhaled

SAFETY PROFILE: Confirmed carcinogens. Beryllium compounds can enter the body through inhalation of dusts and fumes and may act locally on the skin. Even alloys of low beryllium content have been shown to be dangerous.

BFW750 CAS: 128-37-0 $C_{15}H_{24}O$ HR: 2

BHT (food grade)

SYNS: BUTYLATED HYDROXYTOLUENE ♦ DBPC (technical grade) ♦ 2,6-DI-tert-BUTYL-p-CRESOL (OSHA, ACGIH) ♦ 3,5-DI-tert-BUTYL-4-HYDROXYTOLUENE ♦ 2,6-DI-tert-BUTYL-4-METHYLPHENOL

OSHA PEL: TLV 10 mg/m^3

ACGIH TLV: TLV 10 mg/m^3; Not Classifiable as a Human Carcinogen

SAFETY PROFILE: Moderately toxic by ingestion. An experimental teratogen. A human skin irritant. A skin and eye irritant. Combustible.

PROP: White, crystalline solid; faint characteristic odor. Bp: 265°, fp: 68°, flash p: 260°F (TOC), d: 1.048 @ 20°/4°, vap d: 7.6, mp: 71°. Sol in alc; insol in water and propylene glycol.

BGE000 CAS: 92-52-4 $C_{12}H_{10}$ HR: 3

BIPHENYL

SYNS: DIPHENYL (OSHA) ♦ PHENYLBENZENE

OSHA PEL: TWA 0.2 ppm

ACGIH TLV: TWA 0.2 ppm

DFG MAK: 0.16 ppm (1 mg/m^3)

SAFETY PROFILE: Moderately toxic by ingestion. A powerful irritant by inhalation. Combustible.

PROP: Monoclinic, white scales, with a pleasant odor. Mp: 71°, bp: 255°, flash p: 235°F (CC), d: 0.991 @ 75°/4°, autoign temp: 1004°F, vap d: 5.31, lel: 0.6% @ 232°, uel: 5.8% @ 331°F.

BIK000 CAS: 542-88-1 $C_2H_4Cl_2O$ HR: 3

BIS(CHLOROMETHYL) ETHER

DOT: UN 2249

SYNS: BCME ♦ BIS-CME ♦ CHLORO(CHLOROMETHOXY)METHANE ♦ sym-DICHLORODIMETHYL ETHER (DOT)

OSHA: Cancer Suspect Agent

ACGIH TLV: TWA 0.001 ppm; Confirmed Human Carcinogen

DFG MAK: Human Carcinogen

DOT Classification: 6.1; Label: Poison

SAFETY PROFILE: Confirmed human carcinogen. Poison by inhalation, ingestion, and skin contact. A dangerous fire hazard.

PROP: Volatile liquid. Bp: 105°, d: 1.315 @ 20°, vap d: 4.0, flash p: <19°, fp: –41.5°.

BJH750 CAS: 3033-62-3 $C_8H_{20}N_2O$ HR: 3

BIS(2-DIMETHYLAMINOETHYL) ETHER

SYN: DMAEE

ACGIH TLV: (Proposed: TWA 0.05 ppm; STEL 0.33 ppm (skin))
SAFETY PROFILE: Poison by skin contact. Moderately toxic by ingestion. A severe skin and eye irritant.
PROP: Bp: 180–182°.

BJQ250 CAS: 2781-10-4 $C_{24}H_{48}O_4Sn$ HR: 3
BIS(2-ETHYLHEXANOYLOXY)DIBUTYL STANNANE
SYNS: DIBUTYLTIN BIS(α-ETHYLHEXANOATE) ♦ DIBUTYLTIN DI(2-ETHYLHEXANOATE) ♦ DI-n-BUTYLTIN DI-2-ETHYLHEXANOATE ♦ DIBUTYLTIN DI(2-ETHYLHEXOATE)
OSHA PEL: TWA 0.1 mg(Sn)/m^3 (skin)
ACGIH TLV: TWA 0.1 mg(Sn)/m^3 (skin) (Proposed: TWA 0.1 mg(Sn)/m^3; STEL 0.2 mg(Sn)/m^3 (skin))
NIOSH REL: (Organotin Compounds) TWA 0.1 mg(Sn)/m^3
SAFETY PROFILE: Poison by ingestion.

BKY000 CAS: 1304-82-1 Bi_2Te_3 HR: 3
BISMUTH TELLURIDE
SYNS: BISMUTH SESQUITELLURIDE
OSHA PEL: Total Dust: TWA 0.1 mg(Te)/m^3; Respirable Fraction: TWA 5 mg/m^3; Se doped: 5 mg/m^3
ACGIH TLV: TWA 10 mg/m^3; Not Classifiable as a Human Carcinogen; Se doped: 5 mg/m^3; Not Classifiable as a Human Carcinogen
SAFETY PROFILE: Moderate fire hazard by spontaneous chemical reaction with powerful oxidizers. Reacts with moisture to evolve a toxic gas. Slight explosion hazard by chemical reaction with powerful oxidizers; reacts with moisture.
PROP: Gray crystals or solid. D: 7.7.

BMG000 CAS: 1303-86-2 B_2O_3 HR: 2
BORON OXIDE
SYNS: BORIC ANHYDRIDE ♦ BORON SESQUIOXIDE ♦ BORON TRIOXIDE
OSHA PEL: Total Dust: TWA 10 mg/m^3; Respirable Fraction: TWA 5 mg/m^3
ACGIH TLV: TWA 10 mg/m^3
DFG MAK: 15 mg/m^3
SAFETY PROFILE: Moderately toxic by ingestion. An eye and skin irritant.
PROP: Vitreous or colorless. Two crystalline forms. Bp: 2250°, mp: 450° (approx), d: 2.46.

BMG400 CAS: 10294-33-4 BBr_3 HR: 3
BORON TRIBROMIDE
DOT: UN 2692
SYNS: BORON BROMIDE ♦ TRONA
OSHA PEL: CL 1 ppm

BUTANE SULTONE
SYNS: BUTANESULFONE ♦ 1,4-BUTANESULTONE (MAK) ♦ 1,4-BUTYLENE SULFONE ♦ Δ-VALEROSULTONE
DFG MAK: Animal Carcinogen, Suspected Human Carcinogen
SAFETY PROFILE: Suspected carcinogen. Moderately toxic by ingestion.
PROP: Liquid. D: 1.33 @ 20°/4°, mp: 12.5–14.5°, bp: 134–136° @ 4 mm.

BOY500 CAS: 78-94-4 C_4H_6O HR: 3
3-BUTEN-2-ONE
DOT: UN 1251
SYNS: ACETYL ETHYLENE ♦ 3-BUTENE-2-ONE ♦ METHYLENE ACETONE ♦ METHYL VINYL KETONE ♦ VINYL METHYL KETONE
ACGIH TLV: STEL CL 0.2 ppm (skin, sensitizer)
DOT Classification: 3; Label: Flammable Liquid
SAFETY PROFILE: Poison by ingestion and inhalation. A severe irritant to skin, eyes, and mucous membranes. A lachrymator. Dangerous fire hazard.
PROP: Colorless liquid; powerfully irritating odor. Bp: 81.4°, flash p: 20°F (CC), d: 0.8393 @ 25°/4°, vap d: 2.41.

BPJ850 CAS: 111-76-2 $C_6H_{14}O_2$ HR: 3
2-BUTOXYETHANOL
DOT: UN 2369
SYNS: n-BUTOXYETHANOL ♦ BUTYL CELLOSOLVE ♦ ETHYLENE GLYCOL MONOBUTYL ETHER (MAK, DOT)
OSHA PEL: TWA 25 ppm (skin)
ACGIH TLV: 20 ppm (skin)
DFG MAK: 20 ppm (98 mg/m^3)
DOT Classification: 6.1; Label: KEEP AWAY FROM FOOD
SAFETY PROFILE: Poison by ingestion and skin contact. Moderately toxic via inhalation. Experimental teratogenic effects. A skin irritant. Combustible.
PROP: Clear, mobile liquid; pleasant odor. Fp: –74.8°, bp: 171–172°, flash p: 160°F (COC), d: 0.9012 @ 20°/20°, vap press: 300 mm @ 140°.

BPM000 CAS: 112-07-2 $C_8H_{16}O_3$ HR: 3
2-BUTOXYETHYL ACETATE
SYNS: 2-BUTOXYETHYL ESTER ACETIC ACID ♦ BUTYL CELLOSOLVE ACETATE ♦ ETHYLENE GLYCOL MONOBUTYL ETHER ACETATE (MAK)
DFG MAK: 20 ppm (130 mg/m^3)
SAFETY PROFILE: Moderately toxic by ingestion and skin contact. Mild skin irritant. Flammable.
PROP: Colorless liquid; fruity odor. Bp: 192.3°, d: 0.9424 @ 20°/20°, fp: –63.5°, flash p: 190°F. Sol in hydrocarbons and org solvs; insol in water.

BPU750 CAS: 123-86-4 $C_6H_{12}O_2$ HR: 3
n-BUTYL ACETATE
DOT: UN 1123

SYNS: ACETIC ACID n-BUTYL ESTER ♦ BUTYL ACETATE ♦ 1-BUTYL ACETATE
OSHA PEL: TWA 150 ppm; STEL 200 ppm
ACGIH TLV: Proposed: 150 ppm; STEL 200 ppm
DFG MAK: 100 ppm (480 mg/m^3)
DOT Classification: 3; Label: Flammable Liquid
SAFETY PROFILE: Mildly toxic by inhalation and ingestion. An experimental teratogen. A skin and severe eye irritant. A mild allergen. Flammable liquid.
PROP: Colorless liquid; strong fruity odor. Fp: −77°, bp: 126°, ULC: 50–60, lel: 1.4%, uel: 7.5%, flash p: 72°F, d: 0.88 @ 20°/20°, refr index: 1.393–1.396, autoign temp: 797°F, vap press: 15 mm @ 25°. Misc with alc, ether, and propylene glycol. Sol in EtOH, Et_2CO, and Me_2CO; insol in H_2O.

BPV000 CAS: 105-46-4 $C_6H_{12}O_2$ HR: 3

sec-BUTYL ACETATE

DOT: UN 1123
SYNS: ACETIC ACID-1-METHYLPROPYL ESTER (9CI) ♦ 2-BUTANOL ACETATE ♦ 2-BUTYL ACETATE ♦ sec-BUTYL ALCOHOL ACETATE
OSHA PEL: TWA 200 ppm
ACGIH TLV: TWA 200 ppm
DFG MAK: 100 ppm (480 mg/m^3)
DOT Classification: 3; Label: Flammable Liquid
SAFETY PROFILE: An irritant and allergen. Flammable liquid.
PROP: Colorless liquid; mild odor. Bp: 112°, flash p: 18°, d: 0.862–0.866 @ 20°/20°, vap d: 4.00, lel: 1.3%, uel: 7.5%.

BPV100 CAS: 540-88-5 $C_6H_{12}O_2$ HR: 3

tert-BUTYL ACETATE

DOT: UN 1123
SYNS: ACETIC ACID-tert-BUTYL ESTER ♦ ACETIC ACID-1,1-DIMETHYLETHYL ESTER ♦ TLA
OSHA PEL: TWA 200 ppm
ACGIH TLV: TWA 200 ppm
DFG MAK: 100 ppm (480 mg/m^3)
DOT Classification: 3; Label: Flammable Liquid
SAFETY PROFILE: Poison by inhalation and ingestion. Flammable.
PROP: Liquid. Bp: 97–98°.

BPW100 CAS: 141-32-2 $C_7H_{12}O_2$ HR: 3

n-BUTYL ACRYLATE

DOT: UN 2348
SYNS: ACRYLIC ACID BUTYL ESTER ♦ ACRYLIC ACID n-BUTYL ESTER (MAK) ♦ BUTYLACRYLATE, INHIBITED (DOT) ♦ BUTYL-2-PROPENOATE
OSHA PEL: TWA 10 ppm
ACGIH TLV: TWA 2 ppm (sensitizer); Not Classifiable as a Carcinogen
DFG MAK: 2 ppm (11 mg/m^3)

DOT Classification: 3; Label: Flammable Liquid
SAFETY PROFILE: Moderately toxic by ingestion, inhalation, and skin contact. A skin and eye irritant. A flammable liquid.
PROP: Water-white, extremely reactive monomer. Bp: 69° @ 50 mm, fp: –64.6°, flash p: 120°F (OC), d: 0.89 @ 25°/25°, vap press: 10 mm @ 35.5°, vap d: 4.42.

BPW500 CAS: 71-36-3 $C_4H_{10}O$ HR: 3
n-BUTYL ALCOHOL
SYNS: n-BUTANOL ♦ BUTANOL (DOT) ♦ BUTYL ALCOHOL (DOT) ♦ 1-HYDROXYBUTANE ♦ PROPYLMETHANOL
OSHA PEL: CL 50 ppm (skin)
ACGIH TLV: CL 50 ppm (skin) (Proposed: CL 25 ppm)
DFG MAK: 100 ppm (310 mg/m^3)
SAFETY PROFILE: Moderately toxic by skin contact and ingestion. A severe skin and eye irritant. Flammable liquid.
PROP: Colorless liquid; vinous odor. Bp: 117.4°, ULC: 40, lel: 1.4%, uel: 11.2%, fp: –90°, flash p: 95–100°F, d: 0.80978 @ 20°/4°, autoign temp: 689°F, vap press: 5.5 mm @ 20°, vap d: 2.55. Misc in alc, ether, and org solvs. Mod sol in water.

BPW750 CAS: 78-92-2 $C_4H_{10}O$ HR: 3
sec-BUTYL ALCOHOL
SYNS: sec-BUTANOL (DOT) ♦ 2-BUTANOL ♦ 2-HYDROXYBUTANE ♦ METHYLETHYLCARBINOL
OSHA PEL: TWA 100 ppm
ACGIH TLV: TWA 100 ppm
DFG MAK: 100 ppm (310 mg/m^3)
SAFETY PROFILE: Mildly toxic by ingestion. A skin and eye irritant. Dangerous fire hazard.
PROP: Colorless liquid. Mp: –89°, bp: 99.5°, flash p: 14°, d: 0.808 @ 20°/4°, autoign temp: 763°F, vap press: 10 mm @ 20°, vap d: 2.55, lel: 1.7% @ 212°F, uel: 9.8% @ 212°F.

BPX000 CAS: 75-65-0 $C_4H_{10}O$ HR: 3
tert-BUTYL ALCOHOL
SYNS: tert-BUTANOL ♦ 1,1-DIMETHYLETHANOL ♦ TRIMETHYLCARBINOL
OSHA PEL: TWA 100 ppm; STEL 150 ppm
ACGIH TLV: TWA 100 ppm; STEL 150 ppm (Proposed: TWA 100 ppm)
DFG MAK: 100 ppm (310 mg/m^3)
SAFETY PROFILE: Moderately toxic by ingestion. An experimental teratogen. Dangerous fire hazard.
PROP: Colorless liquid or rhombic prisms or plates with camphoraceous odor. Mp: 25.5°, bp: 82.8°, flash p: 50°F (CC), d: 0.781 @ 25°/4°, autoign temp: 896°F, vap press: 40 mm @ 24.5°, vap d: 2.55, lel: 2.4%, uel: 8.0%. Misc in H_2O.

BPX750 CAS: 109-73-9 $C_4H_{11}N$ HR: 3

n-BUTYLAMINE
DOT: UN 1125
SYNS: 1-AMINOBUTANE ♦ MONOBUTYLAMINE
OSHA PEL: CL 5 ppm (skin)
ACGIH TLV: CL 5 ppm
DFG MAK: 5 ppm (15 mg/m^3)
DOT Classification: 3; Label: Flammable Liquid; DOT Class: 3; Label: Flammable Liquid, Corrosive
SAFETY PROFILE: Poison by ingestion and skin contact. Moderately toxic by inhalation. A corrosive and severe skin irritant. A flammable liquid.
PROP: Liquid; ammonia-like odor. Mp: –50°, bp: 78°, flash p: 10°F (OC), 10°F (CC), d: 0.74–0.76 @ 20°/20°, autoign temp: 594°F, vap d: 2.52, lel: 1.7%, uel: 9.8%.

BPY000 CAS: 13952-84-6 $C_4H_{11}N$ HR: 3
sec-BUTYLAMINE
DOT: UN 2733/UN 2734
SYNS: 2-AMINOBUTANE ♦ 1-METHYLPROPYLAMINE
DFG MAK: 5 ppm (15 mg/m^3)
DOT Classification: 8; Label: Corrosive, Flammable Liquid (UN 2734); DOT Class: 3; Label: Flammable Liquid, Corrosive (UN 2733)
SAFETY PROFILE: A poison by ingestion. A powerful irritant. Moderately toxic by skin contact. Dangerous fire hazard.
PROP: Liquid. Mp: –104°, bp: 63°, flash p: 15°F, d: 0.724 @ 20°.

BPY250 CAS: 75-64-9 $C_4H_{11}N$ HR: 3
tert-BUTYLAMINE
DOT: UN 2733/UN 2734
SYNS: 2-AMINOISOBUTANE ♦ 1,1-DIMETHYLETHYLAMINE ♦ TRIMETHYLAMINOMETHANE
DFG MAK: 5 ppm (15 mg/m^3)
DOT Classification: 8; Label: Corrosive, Flammable Liquid (UN 2734); DOT Class: 3; Label: Flammable Liquid, Corrosive (UN 2733)
SAFETY PROFILE: Poison by ingestion. Moderately toxic by inhalation. A corrosive liquid. Very dangerous fire hazard.
PROP: Colorless liquid. Mp: –67.5°, bp: 46.4°, fp: –72.65°, d: 0.700 @ 15°, lel: 1.7% @ 212°F, uel: 8.9% @ 212°F, vap d: 2.5, autoign temp: 716°F.

BRK750 CAS: 2426-08-6 $C_7H_{14}O_2$ HR: 3
n-BUTYL GLYCIDYL ETHER
SYNS: BGE ♦ 2,3-EPOXYPROPYL BUTYL ETHER ♦ GLYCIDYL BUTYL ETHER
OSHA PEL: TWA 25 ppm
ACGIH TLV: TWA 25 ppm
DFG MAK: Confirmed Animal Carcinogen with Unknown Relevance to Humans
NIOSH REL: (Glycidyl Ethers) CL 30 mg/m^3/15M

SAFETY PROFILE: Moderately toxic by ingestion and skin contact. Mildly toxic by inhalation. An experimental teratogen. A skin and severe eye irritant.

BRM250 CAS: 75-91-2 $C_4H_{10}O_2$ HR: 2

tert-BUTYLHYDROPEROXIDE

SYNS: 1,1-DIMETHYLETHYL HYDROPEROXIDE ♦ 2-HYDROPEROXY-2-METHYLPROPANE

DFG MAK: Moderate skin effects

DOT Classification: Forbidden

SAFETY PROFILE: Moderately toxic by ingestion and inhalation. A severe skin and eye irritant. Very dangerous fire hazard.

PROP: Water-white liquid. Flash p: 80°F or above, fp: –35°, d: 0.860, mp: –8°, bp: 40° @ 23 mm, vap d: 2.07. Sltly sol in water; very sol in esters and alc.

BRR600 CAS: 138-22-7 $C_7H_{14}O_3$ HR: 3

n-BUTYL LACTATE

SYNS: BUTYL α-HYDROXYPROPIONATE ♦ LACTIC ACID, BUTYL ESTER

OSHA PEL: TWA 5 ppm

ACGIH TLV: TWA 5 ppm

SAFETY PROFILE: A skin irritant. Flammable.

PROP: Liquid. Sltly sol in water; misc in alc and ether. Mp: –43°, bp: 188°, flash p: 160°F (OC), d: 0.968, autoign temp: 720°F, vap d: 5.04, vap press: 0.4 mm @ 20°.

BRR900 CAS: 109-79-5 $C_4H_{10}S$ HR: 3

n-BUTYL MERCAPTAN

DOT: UN 2347

SYNS: BUTANETHIOL (OSHA) ♦ BUTYL MERCAPTAN

OSHA PEL: TWA 0.5 ppm

ACGIH TLV: TWA 0.5 ppm

DFG MAK: 0.5 ppm (1.9 mg/m^3)

NIOSH REL: (n-Alkane Mono Thiols) CL 0.5 ppm/15M

DOT Classification: 3; Label: Flammable Liquid

SAFETY PROFILE: Moderately toxic by ingestion. An eye irritant. Dangerous fire hazard.

PROP: Colorless liquid; skunk-like odor. Mp: –116°, bp: 98°, d: 0.8365 @ 25°/4°, flash p: 35°F, vap d: 3.1.

BRY500 CAS: 924-16-3 $C_8H_{18}N_2O$ HR: 3

n-BUTYL-N-NITROSO-1-BUTAMINE

SYNS: DBN ♦ DBNA ♦ DIBUTYLNITROSOAMINE ♦ N,N-DIBUTYLNITROSOAMINE ♦ NDBA ♦ N-NITROSODI-n-BUTYLAMINE (MAK)

DFG MAK: Animal Carcinogen, Suspected Human Carcinogen

SAFETY PROFILE: Confirmed carcinogen. Moderately toxic by ingestion. Experimental teratogenic effects.

PROP: Pale-yellow liquid. Bp: 235°.

BSC250 CAS: 107-71-1 $C_6H_{12}O_3$ HR: 3

tert-BUTYL PERACETATE

SYNS: t-BUTYL PEROXYACETATE

DFG MAK: Moderate skin irritant

DOT Classification: Forbidden

SAFETY PROFILE: Moderately toxic by ingestion. Mildly toxic by inhalation. Moderate skin and eye irritant. A shock- and heat-sensitive explosive. Dangerous fire.

PROP: Clear, colorless, benzene solution; insol in water; sol in org solvs. D: 0.923, vap press: 50 mm @ 26°, flash p: <80°F (COC).

BSC750 CAS: 110-05-4 $C_8H_{18}O_2$ HR: 3

tert-BUTYL PEROXIDE

SYNS: DI-tert-BUTYL PEROXIDE (MAK) ♦ DTBP ♦ (TRIBUTYL)PEROXIDE

DFG MAK: Mild skin irritant

SAFETY PROFILE: A powerful irritant by ingestion and inhalation. A mild skin and eye irritant. Flammable liquid.

PROP: Clear, water-white liquid. Mp: –40°, bp: 80° @ 284 mm, flash p: 65°F (OC), d: 0.79, vap press: 19.51 mm @ 20°, vap d: 5.03. Very sltly sol in H_2O.

BSE000 CAS: 89-72-5 $C_{10}H_{14}O$ HR: 3

o-sec-BUTYLPHENOL

OSHA PEL: TWA 5 ppm (skin)

ACGIH TLV: TWA 5 ppm (skin)

SAFETY PROFILE: Moderately toxic by ingestion and skin contact. A severe skin and eye irritant. Combustible.

PROP: Colorless liquid. Bp: 226–228° @ 25 mm, fp: 12°, flash p: 225°F, d: 0.981 @ 25°/25°.

BSE500 CAS: 98-54-4 $C_{10}H_{14}O$ HR: 3

4-t-BUTYLPHENOL

SYNS: p-tert-BUTYLPHENOL (MAK) ♦ 1-HYDROXY-4-tert-BUTYLBENZENE

DFG MAK: 0.08 ppm (0.5 mg/m^3)

SAFETY PROFILE: Moderately toxic by skin contact and ingestion. A skin and severe eye irritant. Combustible.

PROP: Crystals, needles, or practically white flakes. Mp: 99°, bp: 236–238°, d: 0.9081 @ 114°/4°, vap press: 1 mm @ 70.0°, vap d: 5.1.

BSP500 CAS: 98-51-1 $C_{11}H_{16}$ HR: 2

p-tert-BUTYLTOLUENE

SYNS: p-METHYL-tert-BUTYLBENZENE ♦ 1-METHYL-4-tert-BUTYLBENZENE ♦ TBT

OSHA PEL: TWA 10 ppm; STEL 20 ppm

ACGIH TLV: TWA 10 ppm; STEL 20 ppm (Proposed: TWA 1 ppm)

DFG MAK: 10 ppm (60 mg/m^3)

SAFETY PROFILE: Moderately toxic by inhalation and ingestion. A skin and human eye irritant. Flammable.

PROP: Colorless liquid. D: 0.861 @ 20°/4°, mp: –54°, bp: 189–192°.

BSX250 CAS: 109-74-0 C_4H_7N HR: 3
BUTYRONITRILE
DOT: UN 2411
SYNS: n-BUTANENITRILE ♦ BUTYRIC ACID NITRILE ♦ BUTYRONITRILE (DOT) ♦ 1-CYANOPROPANE ♦ PROPYL CYANIDE
NIOSH REL: (Nitriles) TWA 22 mg/m^3
DOT Classification: 3; Label: Flammable Liquid, Poison
SAFETY PROFILE: A poison by ingestion and skin contact. Moderately toxic by inhalation. A skin irritant. Dangerous fire hazard.
PROP: Colorless liquid. D: 0.796 @ 15°, mp: –112.6°, bp: 117°, flash p: 79°F (OC). Sltly sol in water; sol in alc and ether.

CAD000 CAS: 7440-43-9 Cd HR: 3
CADMIUM
OSHA PEL: TWA 5 $\mu g(Cd)/m^3$
ACGIH TLV: Dust and Salts: TWA 0.05 $mg(Cd)/m^3$ (Proposed: TWA 0.01 $mg(Cd)/m^3$ (dust), Suspected Human Carcinogen; 0.002 $mg(Cd)/m^3$ (respirable dust), Suspected Human Carcinogen); BEI: 10 μg/g creatinine in urine; 10 μg/L in blood (Proposed: 5 μg/g creatinine in urine; 5 μg/L in blood)
DFG BAT: Blood 1.5 μg/dL; Urine 15 μg/dL. MAK: Animal Carcinogen, Suspected Human Carcinogen
NIOSH REL: (Cadmium) Reduce to lowest feasible level
SAFETY PROFILE: Confirmed human carcinogen. Poison by ingestion and inhalation. Experimental teratogenic effects. The dust ignites spontaneously in air and is flammable and explosive. It can react violently with many substances.
PROP: Hexagonal, ductile crystals or soft, silver-white, lustrous, malleable metal. Tarnishes in air, particularly moist air. Mp: 321°, bp: 767°, d: 8.642, vap press: 1 mm @ 394°. Sol in dil acids (H_2 evolved).

CAE250 CAS: 10108-64-2 $CdCl_2$ HR: 3
CADMIUM CHLORIDE
SYNS: CADMIUM DICHLORIDE
OSHA PEL: TWA 5 $\mu g(Cd)/m^3$
ACGIH TLV: TWA 0.05 $mg(Cd)/m^3$ (Proposed: TWA 0.01 $mg(Cd)/m^3$ (dust), Suspected Human Carcinogen; 0.002 $mg(Cd)/m^3$ (respirable dust), Suspected Human Carcinogen); BEI: 5 μg/g creatinine in urine; 5 μg/L in blood
DFG MAK: Animal Carcinogen, Suspected Human Carcinogen
NIOSH REL: (Cadmium) Reduce to lowest feasible level
SAFETY PROFILE: Confirmed human carcinogen. Poison by ingestion, inhalation and skin contact. Experimental teratogenic effects.
PROP: Hexagonal, colorless crystals. Mp: 568°, bp: 969.6°, d: 4.047 @ 25°, vap press: 10 mm @ 656°. Sol in H_2O: sltly sol in EtOH.

CAE750 HR: 3
CADMIUM COMPOUNDS
OSHA PEL: TWA 5 µg(Cd)/m^3
ACGIH TLV: Dust and Salts: TWA 0.05 mg(Cd)/m^3 (Proposed: TWA 0.01 mg(Cd)/m^3 (dust), Suspected Human Carcinogen; 0.002 mg(Cd)/m^3 (respirable dust), Suspected Human Carcinogen); BEI: 5 µg/g creatinine in urine; 5 µg/L in blood
DFG BAT: Blood 1.5 µg/dL; Urine 15 µg/dL. MAK: Suspected Carcinogen
NIOSH REL: (Cadmium, dust and fume) Reduce to lowest feasible level
SAFETY PROFILE: Confirmed human carcinogens. Poison by ingestion. Inhalation of fumes or dusts affects the respiratory tract and the kidneys. Cadmium oxide fumes can cause metal fume fever resembling that caused by zinc oxide fumes.

CAO000 CAS: 1317-65-3 CO_3•Ca HR: 1
CALCIUM CARBONATE
SYNS: AGRICULTURAL LIMESTONE ♦ CALCITE ♦ CARBONIC ACID, CALCIUM SALT (1:1) ♦ CHALK ♦ DOLOMITE ♦ LIMESTONE (FCC)
OSHA PEL: Total Dust: 15 mg/m^3; Respirable Fraction: 5 mg/m^3
ACGIH TLV: TWA (nuisance particulate) 10 mg/m^3 of total dust (when toxic impurities are not present, e.g., quartz <1%)
SAFETY PROFILE: A nuisance dust. An eye and skin irritant.
PROP: White microcrystalline powder. Mp: 825° (α), 1339° (β) @ 102.5 atm, d: 2.7–2.95. Found in nature as the minerals limestone, marble, aragonite, calcite, and vaterite. Odorless, tasteless powder or crystals. Two crystalline forms are of commercial importance: aragonite, orthorhombic, mp: 825° (decomp), d: 2.83, formed at temperatures above 30°; calcite, hexagonal-rhombohedral, mp: 1339° (102.5 atm), d: 2.711, formed at temperatures below 30°. At about 825° it decomposes into CaO and CO_2. Practically insol in water, alc; sol in dilute acids.

CAQ250 CAS: 156-62-7 CN_2•Ca HR: 3
CALCIUM CYANAMIDE
DOT: UN 1403
SYNS: CALCIUM CARBIMIDE ♦ CALCIUM CYANAMID ♦ LIME-NITROGEN (DOT) ♦ NITROGEN LIME
OSHA PEL: TWA 0.5 mg/m^3
ACGIH TLV: TWA 0.5 mg/m^3; Not Classifiable as a Human Carcinogen
DFG MAK: 1 mg/m^3
DOT Classification: 4.3; Label: Dangerous When Wet
SAFETY PROFILE: Poison by ingestion, inhalation and skin contact. Moderately toxic to humans by ingestion. Flammable.
PROP: Hexagonal, rhombohedral, colorless, moisture-sensitive crystals. Mp: 1300°, subl @ >1500°. Decomposes in water. Compound not hydrated; compound contains more than 0.1% calcium (FEREAC 41,15972,76).

CAQ500 CAS: 592-01-8 C_2CaN_2 HR: 3

CALCIUM CYANIDE

DOT: UN 1575

SYNS: CALCYANIDE ♦ CYANOGAS

OSHA PEL: TWA 5 mg(CN)/m^3

ACGIH TLV: CL 5 mg(CN)/m^3 (skin)

DFG MAK: 5 mg/m^3

NIOSH REL: (Cyanide) CL 5 mg(CN)/m^3/10M

DOT Classification: 6.1; Label: Poison

SAFETY PROFILE: A deadly poison by ingestion and probably other routes.

PROP: Rhombohedral crystals or white powder. Mp: decomp >350°.

CAT225 CAS: 1305-62-0 CaH_2O_2 HR: 2

CALCIUM HYDROXIDE

SYNS: CALCIUM HYDRATE ♦ CALCIUM HYDROXIDE (ACGIH, OSHA) ♦ HYDRATED LIME ♦ MILK OF LIME ♦ SLAKED LIME

OSHA PEL: TWA 5 mg/m^3

ACGIH TLV: TWA 5 mg/m^3

SAFETY PROFILE: Mildly toxic by ingestion. A severe eye irritant. A skin, mucous membrane, and respiratory system irritant. Causes dermatitis.

PROP: Rhombic, trigonal, colorless crystals or white power; sltly bitter taste. Mp: loses H_2O @ 580°, bp: decomp, d: 2.343. Sltly sol in water and glycerin; insol in alc.

CAU500 CAS: 1305-78-8 CaO HR: 3

CALCIUM OXIDE

DOT: UN 1910

SYNS: BURNT LIME ♦ LIME ♦ LIME, BURNED ♦ LIME, UNSLAKED (DOT) ♦ QUICKLIME (DOT)

OSHA PEL: TWA 5 mg/m^3

ACGIH TLV: TWA 2 mg/m^3

DFG MAK: 5 mg/m^3

DOT Classification: 8; Label: Corrosive

SAFETY PROFILE: A caustic and irritating material. It can react violently with many substances.

PROP: Cubic, colorless, white crystals. Mp: 2580°, d: 3.37, bp: 2850°. Sol in water and glycerin; insol in alc.

CAW850 CAS: 1344-95-2 HR: 1

CALCIUM SILICATE

SYNS: CALCIUM HYDROSILICATE ♦ CALCIUM MONOSILICATE ♦ CALCIUM SILICATE, synthetic nonfibrous (ACGIH)

OSHA PEL: Total Dust: 15 mg/m^3; Respirable Fraction: 5 mg/m^3

ACGIH TLV: TWA (nuisance particulate) 10 mg/m^3 of total dust (when toxic impurities are not present, e.g., quartz <1%); Not Classifiable as a Human Carcinogen
SAFETY PROFILE: A nuisance dust.
PROP: Varying proportions of CaO and SiO_2. White powder. Insol in water.

CAX500 CAS: 7778-18-9 $CaSO_4$ HR: 1
CALCIUM SULFATE
SYNS: ANHYDROUS CALCIUM SULFATE ♦ DRIERITE ♦ PLASTER of PARIS
OSHA PEL: Total Dust: 15 mg/m^3; Respirable Fraction: 5 mg/m^3
ACGIH TLV: TWA (nuisance particulate) 10 mg/m^3 of total dust (when toxic impurities are not present, e.g., quartz <1%)
DFG MAK: 6 mg/m^3
SAFETY PROFILE: A nuisance dust.
PROP: Pure anhydrous, colorless or white powder or odorless crystals. D: 2.964, mp: 1570°. Dissolves in acids. Sltly sol in H_2O.

CAX750 CAS: 10101-41-4 $O_4S{\cdot}Ca{\cdot}2H_2O$ HR: 1
CALCIUM(II) SULFATE DIHYDRATE (1:1:2)
SYNS: ANNALINE ♦ GYPSUM ♦ MAGNESIA WHITE ♦ SULFURIC ACID, CALCIUM(2+) SALT, DIHYDRATE
OSHA PEL: Total Dust: 15 mg/m^3; Respirable Fraction: 5 mg/m^3
ACGIH TLV: TWA (nuisance particulate) 10 mg/m^3 of total dust (when toxic impurities are not present, e.g., quartz <1%)
SAFETY PROFILE: A nuisance dust (depending on silica content).
PROP: Colorless, monoclinic, hygroscopic crystals. D: 2.32, mp: 128°, bp: 163°. Sltly sol in water.

CBA750 CAS: 76-22-2 $C_{10}H_{16}O$ HR: 3
CAMPHOR
DOT: UN 2717
SYNS: 2-BORNANONE ♦ 2-CAMPHANONE ♦ CAMPHOR, synthetic (ACGIH, DOT) ♦ LAUREL CAMPHOR ♦ 2-OXOBORNANE ♦ 1,7,7-TRIMETHYLBICYCLO(2.2.1)-2-HEPTANONE
OSHA PEL: TWA 2 mg/m^3
ACGIH TLV: TWA 2 ppm; STEL 3 ppm; Not Classifiable as a Human Carcinogen
DFG MAK: 2 ppm (13 mg/m^3)
DOT Classification: 4.2; Label: Flammable Solid
SAFETY PROFILE: A poison by inhalation. Flammable liquid.
PROP: White, transparent, crystalline masses; penetrating odor; pungent, aromatic taste. Mp: 180°, bp: 204°, lel: 0.6%, uel: 3.5%, flash p: 150°F (CC), d: 0.992 @ 25°/4°, autoign temp: 871°F, vap d: 5.24.

CBF700 CAS: 105-60-2 $C_6H_{11}NO$ HR: 2

CAPROLACTAM

SYNS: 6-AMINOHEXANOIC ACID CYCLIC LACTAM ♦ 6-CAPROLACTAM ♦ ω-CAPROLACTAM (MAK) ♦ HEXAHYDRO-2H-AZEPIN-2-ONE ♦ 1,6-HEXOLACTAM ♦ 2-OXOHEXAMETHYLENIMINE ♦ 2-PERHYDROAZEPINONE

OSHA PEL: Dust: 1 mg/m^3; STEL 3 mg/m^3; Vapor: 5 ppm; STEL 10 ppm

ACGIH TLV: TWA Dust: 1 mg/m^3; Not Classifiable as a Human Carcinogen; Vapor: 5 ppm; STEL 10 ppm; Not Classifiable as a Human Carcinogen (Proposed: TWA (aerosol and vapor) 5 mg/m^3; Not Suspected as a Human Carcinogen)

DFG MAK: 5 mg/m^3

NIOSH REL: (Caprolactam, dust) TWA 1 mg/m^3; STEL 3 mg/m^3; (Caprolactam, vapor) TWA 0.22 ppm; STEL 0.66 ppm

SAFETY PROFILE: Moderately toxic by ingestion and skin contact. A skin and eye irritant.

PROP: White crystals or leaflets from ligroin. Mp: 69°, bp: 139° @ 12 mm, vap press: 6 mm @ 120°.

CBT500 CAS: 7440-44-0 C HR: 1

CARBON

DOT: UN 1361/UN 1362

SYNS: ACTIVATED CARBON ♦ CARBON, activated (DOT) ♦ GRAPHITE SYNTHETIC (ACGIH,OSHA)

OSHA PEL: (Natural graphite) TWA 2.5 mg/m^3; (Synthetic graphite) TWA Total Dust: 10 mg/m^3; Respirable Fraction: 5 mg/m^3

ACGIH TLV: TWA 2 mg/m^3 (respirable dust)

DFG MAK: 1.5 mg/m^3

DOT Classification: 4.2; Label: Spontaneously Combustible

SAFETY PROFILE: It can cause a dust irritation, particularly to the eyes and mucous membranes. Combustible. It can react violently with many substances.

PROP: Black crystals, powder or diamond form. Mp: 3652–3697° (subl), bp: approx 4200°, d (amorph): 1.8–2.1, d (graphite): 2.25, d (diamond): 3.51, vap press: 1 mm @ 3586°.

CBT750 CAS: 1333-86-4 HR: 1

CARBON BLACK

SYNS: ACETYLENE BLACK ♦ CARBON BLACK, CHANNEL ♦ CARBON BLACK, FURNACE ♦ CARBON BLACK, LAMP ♦ CARBON BLACK, THERMAL ♦ CHANNEL BLACK ♦ THERMAL ACETYLENE BLACK

OSHA PEL: TWA 3.5 mg/m^3

ACGIH TLV: TWA 3.5 mg/m^3; Not Classifiable as a Human Carcinogen

NIOSH REL: (Carbon Black) TWA 3.5 mg/m^3

SAFETY PROFILE: Mildly toxic by ingestion, inhalation, and skin contact.

PROP: A generic term applied to a family of high-purity colloidal carbons commercially produced by carefully controlled pyrolysis of gaseous or liquid hydrocarbons. Carbon blacks, including commercial colloidal carbons such as furnace blacks, lampblacks and acetylene blacks, usually contain less than several tenths percent of extractable organic matter and less than one percent ash.

CBU250 CAS: 124-38-9 CO_2 HR: 1

CARBON DIOXIDE

DOT: UN 1013/UN 1845/UN 2187

SYNS: CARBON DIOXIDE, refrigerated liquid (UN 2187) (DOT) ♦ CARBON DIOXIDE, solid (UN 1845) (DOT) ♦ CARBONIC ACID ANHYDRIDE ♦ CARBONIC ACID GAS ♦ DRY ICE (UN 1845) (DOT)

OSHA PEL: TWA 10,000 ppm; STEL 30,000 ppm

ACGIH TLV: TWA 5000 ppm; STEL 30,000 ppm

DFG MAK: 5000 ppm (9100 mg/m^3)

NIOSH REL: (Carbon Dioxide) TWA 10,000 ppm; CL 30,000 ppm/10M

DOT Classification: 2.2; Label: Nonflammable Gas; DOT Class: 9; Label: None (UN 1845)

SAFETY PROFILE: An asphyxiant. Contact of solid carbon dioxide snow with the skin can cause burns.

PROP: Colorless, odorless gas. Mp: 57° (sublimes @ –78.5°), vap d: 1.53 @ 78.2°. Sltly sol in water, forming H_2CO_3.

CBV500 CAS: 75-15-0 CS_2 HR: 3

CARBON DISULFIDE

DOT: UN 1131

SYNS: CARBON BISULFIDE (DOT) ♦ CARBON SULPHIDE (DOT) ♦ DITHIOCARBONIC ANHYDRIDE ♦ SULPHOCARBONIC ANHYDRIDE

OSHA PEL: TWA 4 ppm; STEL 12 ppm (skin)

ACGIH TLV: TWA 10 ppm (skin); BEI: 5 mg (2-thiothiazolidine-4-carboxylic acid (TTCA))/g creatinine in urine

DFG MAK: 5 ppm (16 mg/m^3); BAT: 8 mg/L of 4-thio-4-thiazolidine carboxylic acid (TTCA) at end of shift

NIOSH REL: (Carbon Disulfide) TWA 1 ppm; CL 10 ppm/15M

DOT Classification: 3; Label: Flammable Liquid, Poison

SAFETY PROFILE: Mildly toxic to humans by inhalation. Effects on HUMAN spermatogenesis by inhalation. Experimental teratogenic effects. Flammable liquid. It can react violently with many substances.

PROP: Highly refracting, clear, colorless liquid; nearly odorless when pure. Mp: –111.6°, d: 1.293 @ 0°/4°, bp: 46.5°, lel: 1.3%, uel: 50%, flash p: –22°F (CC), autoign temp: 257°F, vap press: 400 mm @ 28°, vap d: 2.64. Misc in EtOH, Et_2O, and C_6H_6; sltly sol in H_2O.

CBW750 CAS: 630-08-0 CO HR: 3

CARBON MONOXIDE

DOT: UN 1016/NA 9202

SYNS: CARBON MONOXIDE, refrigerated liquid (cryogenic liquid) (NA 9202) (DOT) ♦ CARBON OXIDE (CO) ♦ FLUE GAS

OSHA PEL: TWA 35 ppm; CL 200 ppm
ACGIH TLV: 25 ppm; BEI: 3% of hemoglobin indicating carboxyhemoglobin in blood at end of shift; 20 ppm CO in end-exhaled air at end of shift.
DFG MAK: 30 ppm (35 mg/m^3); BAT: 5% carboxyhemoglobin in blood at end of shift
NIOSH REL: (Carbon Monoxide) TWA 35 ppm; CL 200 ppm
DOT Classification: 2.3; Label: Poison Gas, Flammable Gas
SAFETY PROFILE: Mildly toxic by inhalation in humans but has caused many fatalities. Experimental teratogenic effects. A dangerous fire hazard when exposed to flame. It can react violently with many substances.
PROP: Colorless, odorless, tasteless gas. Mp: –213°, bp: –190°, lel: 12.5%, uel: 74.2%, d: (gas) 1.250 g/L @ 0°, (liquid) 0.793, autoign temp: 1128°F. Very sltly sol in H_2O; sol in AcOH, MeOH, and EtOH.

CBX750 CAS: 558-13-4 CBr_4 HR: 3
CARBON TETRABROMIDE
DOT: UN 2516
SYNS: METHANE, TETRABROMIDE ♦ METHANE, TETRABROMO- ♦ TETRABROMIDE METHANE ♦ TETRABROMOMETHANE
OSHA PEL: TWA 0.1 ppm; STEL 0.3 ppm
ACGIH TLV: TWA 0.1 ppm; STEL 0.3 ppm
DOT Classification: 6.1; Label: KEEP AWAY FROM FOOD
SAFETY PROFILE: Narcotic in high concentration.
PROP: Colorless, monoclinic tablets. Mp: (α) 48.4°, (β) 90.1°, bp: 102° @ 50 mm, d: 2.961 @ 99.5°/4°, vap press: 40 mm @ 96.3°. Sol in EtOH, Et_2O, and $CHCl_3$; insol in H_2O.

CBY000 CAS: 56-23-5 CCl_4 HR: 3
CARBON TETRACHLORIDE
DOT: UN 1846
SYNS: CARBON TET ♦ METHANE TETRACHLORIDE ♦ PERCHLOROMETHANE ♦ TETRACHLOROMETHANE
OSHA PEL: TWA 2 ppm
ACGIH TLV: TWA 5 ppm; STEL 10 (skin); Suspected Human Carcinogen
DFG MAK: 10 ppm (64 mg/m^3); BEI: 1.6 mL/m^3 in alveolar air 1 hour after exposure; Suspected Carcinogen
NIOSH REL: (Carbon Tetrachloride) CL 2 ppm/60M
DOT Classification: 6.1; Label: Poison
SAFETY PROFILE: Confirmed carcinogen. A human poison by ingestion. Mildly toxic by inhalation. Experimental teratogenic effects. An eye and skin irritant. Contact dermatitis can result from skin contact. It can react violently with many substances.
PROP: Colorless liquid; heavy, ethereal odor. Mp: –22.6°, bp: 76.8°, flash p: none, d: 1.632 @ 0°/4°, vap press: 100 mm @ 23.0°. Sol in EtOH and Et_2O; practically insol in H_2O.

CCA500 CAS: 353-50-4 CF_2O HR: 3

CARBONYL FLUORIDE

DOT: UN 2417

SYNS: CARBON DIFLUORIDE OXIDE ♦ CARBON FLUORIDE OXIDE ♦ CARBON OXYFLUORIDE ♦ DIFLUOROFORMALDEHYDE ♦ FLUOROFORMYL FLUORIDE ♦ FLUOROPHOSGENE

OSHA PEL: TWA 2 ppm; STEL 5 ppm

ACGIH TLV: TWA 2 ppm; STEL 5 ppm

DOT Classification: 2.3; Label: Poison Gas

SAFETY PROFILE: A poison. Moderately toxic by inhalation. A powerful irritant. Hydrolyzes instantly to form HF on contact with moisture.

PROP: Colorless gas; pungent; hygroscopic. Readily hydrolyzes to CO_2 and HF. Mp: –114°, bp: –83°, d: 1.139 @ –114°.

CCP850 CAS: 120-80-9 $C_6H_6O_2$ HR: 3

CATECHOL

SYNS: o-BENZENEDIOL ♦ 1,2-BENZENEDIOL ♦ 1,2-DIHYDROXYBENZENE ♦ o-DIOXYBENZENE ♦ o-HYDROXYPHENOL ♦ o-PHENYLENEDIOL ♦ PYROCATECHOL

OSHA PEL: TWA 5 ppm (skin)

ACGIH TLV: TWA 5 ppm (skin); Animal Carcinogen

SAFETY PROFILE: Poison by ingestion. Moderately toxic by skin contact. Can cause dermatitis. An allergen. Combustible

PROP: Colorless crystals or needles from water. Mp: 105°, bp: 240°, flash p: 261°F (CC), d: 1.341 @ 15°, vap press: 10 mm @ 118.3°, vap d: 3.79. Sol in water, chloroform, and benzene; very sol in alc and ether.

CDD750 CAS: 21351-79-1 CsHO HR: 3

CESIUM HYDROXIDE

DOT: UN 2681/UN 2682

SYNS: CESIUM HYDRATE

OSHA PEL: TWA 2 mg/m^3

ACGIH TLV: TWA 2 mg/m^3

DOT Classification: 8; Label: Corrosive

SAFETY PROFILE: Moderately toxic by ingestion. A powerful caustic. A corrosive skin and eye irritant.

PROP: Colorless to yellowish, very deliquescent crystals. Undergoes transition from orthorhombic to cubic at 2°. Mp: 315°, d: 3.675. Very sol in H_2O and EtOH.

CDN200 CAS: 78-95-5 C_3H_5ClO HR: 3

CHLORACETONE

DOT: UN 1695

SYNS: ACETONYL CHLORIDE ♦ CHLOROACETONE, stabilized (DOT) ♦ CHLOROPROPANONE ♦ 1-CHLORO-2-PROPANONE ♦ MONOCHLOROACETONE

ACGIH TLV: CL 1 ppm (skin)

DOT Classification: 6.1; Label: Poison (UN 1695); DOT Class: Forbidden

SAFETY PROFILE: Poison by inhalation, ingestion, and skin contact. A lachrymator poison gas. Flammable.
PROP: Colorless, lachrymatory liquid with pungent odor. Mp: –44.5°, bp: 119°, d: 1.162.

CDV100 CAS: 8001-35-2 $C_{10}H_{10}Cl_8$ HR: 3
CHLORINATED CAMPHENE
SYNS: CAMPHECHLOR ♦ CAMPHOCHLOR ♦ OCTACHLOROCAMPHENE ♦ POLYCHLORCAMPHENE ♦ TOXAPHENE
OSHA PEL: TWA 0.5 mg/m^3; STEL 1 mg/m^3 (skin)
ACGIH TLV: TWA 0.5 mg/m^3; STEL 1 mg/m^3 (skin); Animal Carcinogen
DFG MAK: Animal Carcinogen, Suspected Human Carcinogen
SAFETY PROFILE: Confirmed carcinogen. Human poison by ingestion. Moderately toxic by inhalation and skin contact. A skin irritant; absorbed through the skin. Experimental teratogenic effects.
PROP: Yellow, waxy solid; pleasant piney odor. Mp: 65–90°. Almost insol in water; very sol in aromatic hydrocarbons.

CDV175 CAS: 31242-93-0 $C_{12}H_4Cl_6O$ HR: 2
CHLORINATED DIPHENYL OXIDE
SYNS: HEXACHLORODIPHENYL ETHER ♦ PHENYL ETHER, HEXACHLORO derivative (8CI) ♦ TRICHLORO DIPHENYL OXIDE
OSHA PEL: TWA 0.5 mg/m^3
ACGIH TLV: TWA 0.5 mg/m^3
DFG MAK: 0.5 mg/m^3
SAFETY PROFILE: Moderately toxic by ingestion and probably by inhalation. Combustible.
PROP: Light-yellow, very viscous liquid. Bp: 230–260° @ 8 mm, d: 1.60 @ 20°/60°, autoign temp: 1148°F, vap d: 13.0.

CDV750 CAS: 7782-50-5 Cl_2 HR: 3
CHLORINE
DOT: UN 1017
OSHA PEL: TWA 0.5 ppm; STEL 1 ppm
ACGIH TLV: TWA 0.5 ppm; STEL 1 ppm; Not Classifiable as a Human Carcinogen
DFG MAK: 0.5 ppm (1.5 mg/m^3)
NIOSH REL: (Chlorine) CL 0.5 ppm/15M
DOT Classification: 2.3; Label: Poison Gas
SAFETY PROFILE: Moderately toxic and very irritating to humans by inhalation. A strong irritant to eyes and mucous membranes. It can react violently with many substances.
PROP: Greenish-yellow gas, liquid, or rhombic crystals. Mp: –101°, bp: –34.9°, d: (liquid) 1.47 @ 0° (3.65 atm), vap press: 4800 mm @ 20°, vap d: 2.49. Sol in water.

CDW450 CAS: 10049-04-4 ClO_2 HR: 3

CHLORINE DIOXIDE

SYNS: CHLORINE DIOXIDE, not hydrated (DOT) ♦ CHLORINE OXIDE ♦ CHLORINE(IV) OXIDE

OSHA PEL: TWA 0.1 ppm; STEL 0.3 ppm

ACGIH TLV: TWA 0.1 ppm; STEL 0.3 ppm

DFG MAK: 0.1 ppm (0.28 mg/m^3)

DOT Classification: Forbidden

SAFETY PROFILE: Moderately toxic by inhalation. An eye irritant. A powerful explosive sensitive to spark, impact, sunlight, or heating rapidly to 100°C. It can react violently with many substances.

PROP: Red-yellow or orange-green gas or orange-red crystals. Unstable in light, stable in the dark. Mp: –59°, bp: 11°, d: 3.09 g/L @ 11°. Insol in water.

CDX750 CAS: 7790-91-2 ClF_3 HR: 3

CHLORINE TRIFLUORIDE

DOT: UN 1749

SYNS: CHLORINE FLUORIDE

OSHA PEL: CL 0.1 ppm

ACGIH TLV: CL 0.1 ppm

DFG MAK: 0.1 ppm (0.38 mg/m^3)

DOT Classification: 2.3; Label: Poison Gas, Oxidizer, Corrosive

SAFETY PROFILE: Human poison by inhalation. An eye irritant. Spontaneously flammable. It can react violently with many substances.

PROP: Colorless gas to yellow liquid; sweet odor. One of the most reactive chemical compds known. Mp: –83°, bp: 11.8°, d: 1.77 @ 13°.

CDY500 CAS: 107-20-0 C_2H_3ClO HR: 3

CHLOROACETALDEHYDE

DOT: UN 2232

SYNS: 2-CHLOROACETALDEHYDE ♦ 2-CHLOROETHANAL ♦ 2-CHLORO-1-ETHANAL ♦ MONOCHLOROACETALDEHYDE

OSHA PEL: CL 1 ppm

ACGIH TLV: CL 1 ppm

DFG MAK: Confirmed Animal Carcinogen with Unknown Relevance to Humans

DOT Classification: 6.1; Label: Poison

SAFETY PROFILE: Suspected carcinogen. Poison by ingestion and skin contact. Combustible.

PROP: Clear, colorless liquid; pungent odor. Bp: 90.0–100.1° (40% soln), fp: –16.3° (40% soln), flash p: 190°F, d: 1.19 @ 25°/25° (40% soln), vap press: 100 mm @ 45° (40% soln).

CEA750 CAS: 532-27-4 C_8H_7ClO HR: 3

α-CHLOROACETOPHENONE

DOT: UN 1697

SYNS: 1-CHLOROACETOPHENONE ♦ CHLOROMETHYL PHENYL KETONE ♦ 2-CHLORO-1-PHENYLETHANONE ♦ MACE (lachrymator) ♦ PHENACYL CHLORIDE
OSHA PEL: TWA 0.05 ppm
ACGIH TLV: TWA 0.05 ppm; Not Classifiable as a Human Carcinogen
DOT Classification: 6.1; Label: Poison
SAFETY PROFILE: A poison by inhalation and ingestion. A severe eye and moderate skin irritant.
PROP: Leaflets from pet ether. Mp: 54°, bp: 139–141° @ 14 mm.

CEC250 CAS: 79-04-9 $C_2H_2Cl_2O$ HR: 3
CHLOROACETYL CHLORIDE
DOT: UN 1752
SYNS: CHLORACETYL CHLORIDE ♦ CHLOROACETIC CHLORIDE ♦ MONOCHLOROACETYL CHLORIDE
OSHA PEL: TWA 0.05 ppm
ACGIH TLV: TWA 0.05 ppm; STEL 0.15 ppm
DOT Classification: 8; Label: Corrosive, Poison
SAFETY PROFILE: Poison by ingestion. Mildly toxic by inhalation. Corrosive. A lachrymator.
PROP: Water-white or sltly yellow liquid. Bp: 108–110°, fp: –22.5°, flash p: none, d: 1.495 @ 0°.

CEJ125 CAS: 108-90-7 C_6H_5Cl HR: 3
CHLOROBENZENE
DOT: UN 1134
SYNS: BENZENE CHLORIDE ♦ CHLOROBENZOL (DOT) ♦ MCB ♦ MONOCHLOROBENZENE ♦ PHENYL CHLORIDE
OSHA PEL: TWA 75 ppm
ACGIH TLV: TWA 10 ppm; Animal Carcinogen; BEI: 150 mg/g creatinine of total 4-chlorocatechol in urine at end of shift; 25 mg/g creatinine of total p-chloropohenol in urine at end of shift
DFG MAK: 10 ppm (47 mg/m^3)
DOT Classification: 3; Label: Flammable Liquid
SAFETY PROFILE: Suspected carcinogen. Moderately toxic by ingestion. Experimental teratogenic effects. Dangerous fire hazard.
PROP: Clear, colorless liquid with faint odor. Bp: 131.7°, lel: 1.3%, uel: 7.1% @ 150°, mp: –45°, flash p: 85°F (CC), d: 1.11 @ 20°/4°, autoign temp: 1180°F, vap press: 10 mm @ 22.2°, vap d: 3.88.

CEQ600 CAS: 2698-41-1 $C_{10}H_5ClN_2$ HR: 3
o-CHLOROBENZYLIDENE MALONONITRILE
SYNS: o-CHLOROBENZAL MALONONITRILE ♦ 2-CHLOROBENZYLIDENE MALONONITRILE ♦ PROPANEDINITRILE((2-CHLOROPHENYL)METHYLENE)
OSHA PEL: CL 0.05 ppm (skin)
ACGIH TLV: CL 0.05 ppm (skin); Not Classifiable as a Human Carcinogen

SAFETY PROFILE: Poison by ingestion. Moderately toxic by inhalation. A human skin and eye irritant.
PROP: White crystals. Mp: 95°, bp: 313°.

CES650 CAS: 74-97-5 CH_2BrCl HR: 2
CHLOROBROMOMETHANE
DOT: UN 1887
SYNS: BROMOCHLOROMETHANE ♦ HALON 1011 ♦ MONO-CHLORO-MONO-BROMO-METHANE
OSHA PEL: TWA 200 ppm
ACGIH TLV: TWA 200 ppm
DFG MAK: 200 ppm (1100 mg/m^3)
DOT Classification: 6.1; Label: KEEP AWAY FROM FOOD
SAFETY PROFILE: Mildly toxic by ingestion and inhalation.
PROP: Clear, colorless liquid; sweet odor. Fp: –88°, bp: 67.8°, flash p: none, d: 1.930 @ 25°/25°, vap d: 4.46. Sol in organic solvs; insol in water.

CFX500 CAS: 75-45-6 $CHClF_2$ HR: 1
CHLORODIFLUOROMETHANE
DOT: UN 1018
SYNS: DIFLUOROCHLOROMETHANE ♦ FLUOROCARBON-22 ♦ FREON 22 ♦ MONOCHLORODIFLUOROMETHANE ♦ R22 (DOT)
OSHA PEL: TWA 1000 ppm
ACGIH TLV: TWA 1000 ppm; Not Classifiable as a Human Carcinogen
DFG MAK: 500 ppm (1800 mg/m^3)
DOT Classification: 2.2; Label: Nonflammable Gas
SAFETY PROFILE: Mildly toxic by inhalation.
PROP: Gas. D: 1.49 @ 69°/4°, mp: –146°, bp: –40.8°, fp: –160°, autoign temp: 1170°F. Sltly sol in water.

CHI900 CAS: 593-70-4 CH_2ClF HR: 3
CHLOROFLUOROMETHANE
SYNS: CFC 31 ♦ FREON 31 ♦ MONOCHLOROMONOFLUOROMETHANE
DFG MAK: Animal Carcinogen, Suspected Human Carcinogen
SAFETY PROFILE: Confirmed carcinogen. Moderately toxic by inhalation.
PROP: Gas. Bp: –9°.

CHJ500 CAS: 67-66-3 $CHCl_3$ HR: 3
CHLOROFORM
DOT: UN 1888
SYNS: FORMYL TRICHLORIDE ♦ METHANE TRICHLORIDE ♦ R 20 (refrigerant) ♦ TRICHLOROMETHANE
OSHA PEL: TWA 2 ppm
ACGIH TLV: TWA 10 ppm; Suspected Human Carcinogen; Animal Carcinogen

DFG MAK: 10 ppm (50 mg/m^3); Confirmed Animal Carcinogen with Unknown Relevance to Humans)
NIOSH REL: (Waste Anesthetic Gases and Vapors) CL 2 ppm/1H; (Chloroform) CL 2 ppm/60M
DOT Classification: 6.1; Label: Poison
SAFETY PROFILE: Confirmed carcinogen. A human poison by ingestion and inhalation. Experimental teratogenic effects. It can react violently with many substances.
PROP: Colorless liquid; heavy, ethereal odor. Mp: –63.2°, bp: 61.3°, flash p: none, d: 1.481 @ 25°/4°, vap press: 100 mm @ 10.4°, vap d: 4.12. Sltly sol in H_2O.

CIO250 CAS: 107-30-2 C_2H_5ClO HR: 3

CHLOROMETHYL METHYL ETHER

DOT: UN 1239
SYNS: CMME ♦ METHYL CHLOROMETHYL ETHER, anhydrous (DOT) ♦ MONOCHLORODIMETHYL ETHER (MAK)
OSHA: Cancer Suspect Agent
ACGIH TLV: Suspected Human Carcinogen
DFG MAK: Human Carcinogen
NIOSH REL: (Methyl Chloromethyl Ether) TWA use 29 CFR 1910.1006
DOT Classification: 6.1; Label: Poison, Flammable Liquid
SAFETY PROFILE: Confirmed human carcinogen. Poison by inhalation. Moderately toxic by ingestion. A very dangerous fire.
PROP: Liquid. Flash p: <73.4°F, d: 1.070 @ 25 mm, bp: 59.5°.

CJE000 CAS: 600-25-9 $C_3H_6ClNO_2$ HR: 3

1-CHLORO-1-NITROPROPANE

SYN: CHLORONITROPROPANE
OSHA PEL: TWA 2 ppm
ACGIH TLV: TWA 2 ppm
DFG MAK: 20 ppm (100 mg/m^3)
SAFETY PROFILE: Poison by ingestion. Moderately toxic by inhalation. Flammable liquid.
PROP: Liquid. Bp: 139.5°, flash p: 144°F (OC), d: 1.209 @ 20°/20°, vap d: 4.26.

CJI500 CAS: 76-15-3 C_2ClF_5 HR: 1

CHLOROPENTAFLUOROETHANE

DOT: UN 1020
SYNS: FREON 115 ♦ HALOCARBON 115 ♦ MONOCHLOROPENTAFLUOROETHANE (DOT)
OSHA PEL: TWA 1000 ppm
ACGIH TLV: TWA 1000 ppm
DOT Classification: 2.2; Label: Nonflammable Gas
SAFETY PROFILE: Mildly toxic by inhalation. A nonflammable gas.
PROP: Colorless gas. Bp: –37.7°, mp: –106°, d: 1.5678 @ –42°. Insol in water; sol in alc and ether.

CKN500 CAS: 76-06-2 CCl_3NO_2 HR: 3

CHLOROPICRIN

DOT: UN 1580/UN 1583

SYNS: NITROTRICHLOROMETHANE ♦ TRICHLORONITROMETHANE

OSHA PEL: TWA 0.1 ppm

ACGIH TLV: TWA 0.1 ppm; Not Classifiable as a Human Carcinogen

DFG MAK: 0.1 ppm (0.68 mg/m^3)

DOT Classification: 6.1; Label: Poison (UN 1580); DOT Class: 6.1; Label: Poison, KEEP AWAY FROM FOOD (UN 1583)

SAFETY PROFILE: Poison by ingestion. Moderately toxic by inhalation. A powerful irritant that affects all body surfaces.

PROP: Sltly oily, colorless liquid. D: 1.692 @ 0.4°, fp: –69°, mp: –64°, bp: 112.3° @ 766 mm, vap press: 40 mm @ 33.80, vap d: 6.69. Sol in water, alc, and ether.

CKS750 CAS: 598-78-7 $C_3H_5ClO_2$ HR: 3

α-CHLOROPROPIONIC ACID

DOT: UN 2511

ACGIH TLV: TWA 0.1 ppm (skin)

DOT Classification: 8; Label: Corrosive

SAFETY PROFILE: Poison by skin contact. A corrosive. Combustible.

PROP: Sol in water. D: 1.260–1.268 @ 20°, bp: 183–187°, flash p: 225°F.

CLE750 CAS: 2039-87-4 C_8H_7Cl HR: 1

o-CHLOROSTYRENE

OSHA PEL: TWA 50 ppm; STEL 75 ppm

ACGIH TLV: TWA 50 ppm; STEL 75 ppm

SAFETY PROFILE: A skin and eye irritant.

PROP: A solid. Mp: –63.15°, bp: 188.6°, d: 1.100 @ 20°/4°.

CLK100 CAS: 95-49-8 C_7H_7Cl HR: 3

o-CHLOROTOLUENE

DOT: UN 2238

SYNS: 2-CHLORO-1-METHYLBENZENE (9CI) ♦ 1-METHYL-2-CHLOROBENZENE ♦ o-TOLYL CHLORIDE

OSHA PEL: TWA 50 ppm

ACGIH TLV: TWA 50 ppm

DOT Classification: 3; Label: Flammable Liquid

SAFETY PROFILE: Moderately toxic. Flammable when exposed to heat or flame.

PROP: Liquid. Mp: –34°, bp: 159°, d: 1.08 @ 20°/4°. Volatile with steam. Sltly sol in water; freely sol in alc, benzene, chloroform, ether.

CLK220 CAS: 95-69-2 C_7H_8ClN HR: 3

4-CHLORO-o-TOLUIDINE

SYNS: 2-AMINO-5-CHLOROTOLUENE ♦ 4-CHLORO-2-METHYLANILINE

DFG MAK: Human Carcinogen

DOT Classification: 6.1; Label: KEEP AWAY FROM FOOD
SAFETY PROFILE: Confirmed carcinogen. Poison by ingestion.
PROP: Leaflets from EtOH. Mp: 29–30°, bp: 236–238° @ 730 mm.

CLK225 CAS: 95-79-4 C_7H_8ClN HR: 3
5-CHLORO-o-TOLUIDINE
SYNS: 2-AMINO-4-CHLOROTOLUENE
DFG MAK: Confirmed Animal Carcinogen with Unknown Relevance to Humans
DOT Classification: 6.1; Label: KEEP AWAY FROM FOOD
SAFETY PROFILE: Suspected carcinogen. Moderately toxic by ingestion.
PROP: Solid. Bp: 237° @ 722 mm, mp: 21–22°.

CLP750 CAS: 1929-82-4 $C_6H_3Cl_4N$ HR: 3
2-CHLORO-6-(TRICHLOROMETHYL)PYRIDINE
SYNS: NITRAPYRIN (ACGIH)
OSHA PEL: Total Dust: 15 mg/m^3; Respirable Fraction: 5 mg/m^3
ACGIH TLV: TWA 10 mg/m^3; STEL 20 mg/m^3; Not Classifiable as a Human Carcinogen
SAFETY PROFILE: Poison by ingestion. Moderately toxic by skin contact.
PROP: Crystals. Mp: 62–63°. Very sltly sol in H_2O.

CMH250 CAS: 7738-94-5 CrH_2O_4 HR: 3
CHROMIC(VI) ACID
OSHA PEL: CL 0.1 $mg(CrO_3)/m^3$
ACGIH TLV: TWA 0.05 $mg(Cr)/m^3$, Confirmed Human Carcinogen
DFG MAK: Animal Carcinogen, Suspected Human Carcinogen
NIOSH REL: (Chromium(VI)) TWA 0.025 $mg(Cr(VI))/m^3$; CL 0.05/15M
SAFETY PROFILE: Confirmed human carcinogen. A powerful irritant of skin, eyes, and mucous membranes. Dangerously reactive with many substances.
PROP: Found in solution.

CMI750 CAS: 7440-47-3 Cr HR: 3
CHROMIUM
SYNS: CHROME ♦ CHROMIUM METAL (OSHA)
OSHA PEL: TWA 1 mg/m^3
ACGIH TLV: TWA 0.5 $(Cr)mg/m^3$; Not Classifiable as a Carcinogen
SAFETY PROFILE: Confirmed human carcinogen. Powder will explode spontaneously in air.
PROP: Hard, ductile, blue-white metal. Resists oxidation in air. Bp: 26° @ 2690 mm. More reactive to acids than Mo or W and can be rendered passive. Rapidly attacked by fused NaOH + KNO_3 or $KClO_4$.

CMY800 CAS: 8007-45-2 HR: 3

COAL TAR
OSHA PEL: TWA 0.2 mg/m^3; Carcinogen
DFG MAK: Human Carcinogen
NIOSH REL: (Coal Tar Products) TWA 0.1 mg/m^3
DOT Classification: 3; Label: Flammable Liquid
SAFETY PROFILE: Confirmed human carcinogen. A human and experimental skin irritant. A flammable liquid.

CMY825 CAS: 8001-58-9 HR: 3
COAL TAR CREOSOTESYNS: COAL TAR OIL ♦ COAL TAR OIL (DOT) ♦ CREOSOTE ♦ CRESYLIC CREOSOTE ♦ NAPHTHALENE OIL ♦ TAR OIL
NIOSH REL: (Coal Tar Products) TWA 0.1 mg/m^3 CHE fraction
DOT Classification: 3; Label: Flammable Liquid
SAFETY PROFILE: Confirmed carcinogen. Poison by ingestion. A flammable liquid.

CMZ100 CAS: 65996-93-2 HR: 3
COAL TAR PITCH VOLATILES
OSHA PEL: TWA 0.2 mg/m^3; Carcinogen
ACGIH TLV: TWA 0.2 mg/m^3 (volatile), Confirmed Human Carcinogen
NIOSH REL: (Coal Tar Products) TWA 0.1 mg/m^3 CHE fraction
DOT Classification: 3; Label: Flammable Liquid
SAFETY PROFILE: Confirmed carcinogen by skin contact.

CNA250 CAS: 7440-48-4 Co HR: 3
COBALT
OSHA PEL: TWA 0.05 mg/m^3
ACGIH TLV: (metal, dust, and fume) TWA 0.02 $mg(Co)/m^3$; Animal Carcinogen
DFG TRK: Animal Carcinogen, Suspected Human Carcinogen
NIOSH REL: (Cobalt) Insufficient evidence for recommending limit
SAFETY PROFILE: Confirmed carcinogen. Moderately toxic by ingestion. Inhalation of the dust may cause pulmonary damage. The powder may cause dermatitis. Powdered cobalt ignites spontaneously in air. Flammable.
PROP: Gray, hard, magnetic, lustrous, ductile, somewhat malleable, silvery-blue metal. Mp: 1495°, bp: 28° @ 3100 mm, d: 8.92, Brinell hardness: 125, latent heat of fusion: 62 cal/g, latent heat of vaporization: 1500 cal/g, specific heat (15–100°): 0.1056 cal/g/°C. Exists in two allotropic forms. At room temperature, the hexagonal form is more stable than the cubic form; both forms can exist at room temperature. Stable in air or toward water at ordinary temperatures. Readily sol in dil HNO_3; very slowly attacked by HCl or cold H_2SO_4. The hydrated salts of cobalt are red, and the sol salts form red solns that become blue on adding conc HCl.

CNB500 CAS: 10210-68-1 $C_8Co_2O_8$ HR: 3
COBALT CARBONYL

SYNS: COBALT OCTACARBONYL ♦ COBALT TETRACARBONYL ♦ DICOBALT CARBONYL ♦ DICOBALT OCTACARBONYL ♦ OCTACARBONYLDICOBALT
OSHA PEL: TWA 0.1 mg(Co)/m^3
ACGIH TLV: TWA 0.1 mg(Co)/m^3
SAFETY PROFILE: Poison by inhalation. Decomposes in air to form a product that ignites spontaneously in air.
PROP: Air-sensitive orange-red platelets or crystals. D: 1.87, mp: 51°, decomp above 52°. Decomp on exposure to air. Insol in water; sol in org solvs.

CNC230 CAS: 16842-03-8 C_4HCoO_4 HR: 3
COBALT HYDROCARBONYL
OSHA PEL: TWA 0.1 mg(Co)/m^3
ACGIH TLV: TWA 0.1 mg(Co)/m^3
SAFETY PROFILE: Poison by inhalation.
PROP: Light-yellow liquid or gas. Unstable, but distillable in current of CO. Mp: –33°, bp: 10°. Sol org solvs; spar sol in H_2O.

CNI000 CAS: 7440-50-8 Cu HR: 2
COPPER
OSHA PEL: TWA (dust, mist) 1 mg(Cu)/m^3; (fume and respirable particles) 0.1 mg/m^3
ACGIH TLV: TWA (fume) 0.2 mg/m^3; (dust, mist) 1 mg(Cu)/m^3
DFG MAK: (dust) 1 mg/m^3; (fume) 0.1 mg/m^3
SAFETY PROFILE: Toxic by inhalation. Experimental teratogenic effects.
PROP: Reddish, malleable and ductile metal. Slowly weathers to green patina. Mp: 1083°, bp: 25° @ 2595 mm, d: 8.92, vap press: 1 mm @ 1628°.

CNK750 HR: 3
COPPER COMPOUNDS
ACGIH TLV: Inorganic: TWA (fume) 0.2 mg/m^3; (dust, mist) 1 mg(Cu)/m^3
SAFETY PROFILE: As the sublimed oxide, copper may be responsible for one form of metal fume fever. As regards local effect, copper chloride and sulfate have been reported as causing irritation of the skin and conjunctiva, possibly on an allergic basis. Cuprous oxide is irritating to the eyes and upper respiratory tract. Discoloration of the skin is often seen in persons handling copper, but this does not indicate any actual injury. Many copper salts form highly unstable acetylides.

CNL000 CAS: 544-92-3 CCuN HR: 3
COPPER CYANIDE
DOT: UN 1587
SYNS: COPPER(I) CYANIDE ♦ CUPROUS CYANIDE
ACGIH TLV: TWA 1 mg(Cu)/m^3
DOT Classification: 6.1; Label: Poison
SAFETY PROFILE: A poison. Reacts violently with magnesium.

PROP: Monoclinic, white prisms; white-cream powder; dark-green orthorhombic crystals; dark red monoclinic crystals. Mp: 474° in N_2, bp: decomp, d: 2.92. Sol in NH_3 (aq); insol in H_2O and alcohols.

CNT750 HR: 2
COTTON DUST
OSHA PEL: TWA 1 mg/m^3 (raw dust); 0.2 mg/m^3 (yarn manufacturing); 0.75 mg/m^3 (slashing and weaving); 0.5 mg/m^3 (other operations)
ACGIH TLV: TWA 0.2 mg/m^3 (raw dust)
DFG MAK: 1.5 mg/m^3 (raw cotton)
NIOSH REL: (Cotton Dust) CL 0.200 mg/m^3 lint-free
SAFETY PROFILE: Human pulmonary effects. Causes a mild febrile condition of the lungs resembling metal fume fever. It can cause some illness, due to the allergens or fungi in the cotton or on the dust. Moderate fire and explosion hazard when exposed to heat or flame.

CNW500 CAS: 1319-77-3 C_7H_8O HR: 3
CRESOL
DOT: UN 2022
SYNS: CRESYLIC ACID ♦ TRICRESOL
OSHA PEL: TWA 5 ppm (skin)
ACGIH TLV: TWA 5 ppm
DFG MAK: (all isomers) 5 ppm (22 mg/m^3)
NIOSH REL: (Cresol) TWA 10 mg/m^3
DOT Classification: 6.1; Label: Poison
SAFETY PROFILE: A poison by ingestion. Moderately toxic by skin contact. Corrosive to skin and mucous membranes. Flammable.
PROP: Mixture of isomeric cresols obtained from coal tar, colorless or yellowish to brown-yellow or pinkish liquid; phenolic odor. Mp: 10.9–35.5°, bp: 191–203°, flash p: 178°F, d: 1.030–1.038 @ 25°/25°, vap press: 1 mm @ 38–53°, vap d: 3.72.

CNX000 CAS: 95-48-7 C_7H_8O HR: 3
o-CRESOL
DOT: UN 2076
SYNS: o-CRESYLIC ACID ♦ 1-HYDROXY-2-METHYLBENZENE ♦ o-HYDROXYTOLUENE ♦ 2-METHYLPHENOL ♦ o-OXYTOLUENE ♦ o-TOLUOL
OSHA PEL: TWA 5 ppm (skin)
ACGIH TLV: TWA 5 ppm
NIOSH REL: (Cresol) TWA 10 mg/m^3
DOT Classification: 6.1; Label: Poison
SAFETY PROFILE: Poison by ingestion and inhalation. Moderately toxic by skin contact. A severe eye and skin irritant. Flammable.

PROP: Crystals or liquid darkening with exposure to air and light. Mp: 30°, bp: 191°, flash p: 178°F, d: 1.05 @ 20°/4°, autoign temp: 1110°F, vap press: 1 mm @ 38.2°, vap d: 3.72, lel: 1.4% @ 300°F.

COB250 CAS: 4170-30-3 C_4H_6O HR: 3

CROTONALDEHYDE

DOT: UN 1143

SYNS: 2-BUTENAL ♦ CROTONIC ALDEHYDE ♦ β-METHYLACROLEIN

OSHA PEL: TWA 2 ppm

ACGIH TLV: STEL CL 0.3 ppm (skin); Animal Carcinogen

DFG MAK: Suspected Carcinogen

DOT Classification: 3; Label: Flammable Liquid, Poison

SAFETY PROFILE: Suspected carcinogen. Poison by ingestion and inhalation. An eye, skin, and mucous membrane irritant. Dangerous fire hazard.

PROP: Water-white, mobile liquid; pungent suffocating odor. Bp: 104°, fp: –76.0°, lel: 2.1%, uel: 15.5%, flash p: 55°F, d: 0.853 @ 20°/20°, vap d: 2.41, autoign temp: 405°F.

COE750 CAS: 98-82-8 C_9H_{12} HR: 3

CUMENE

DOT: UN 1918

SYNS: ISOPROPYL BENZENE ♦ 2-PHENYLPROPANE

OSHA PEL: TWA 50 ppm (skin)

ACGIH TLV: TWA 50 ppm

DFG MAK: 50 ppm (250 mg/m^3)

DOT Classification: 3; Label: Flammable Liquid

SAFETY PROFILE: Moderately toxic by ingestion. Mildly toxic by inhalation and skin contact. An eye and skin irritant. Flammable liquid.

PROP: Colorless liquid. Mp: –96.0°, bp: 152°, flash p: 111°F, d: 0.864 @ 20°/4°, vap press: 10 mm @ 38.3°, autoign temp: 795°F, lel: 0.9%, uel: 6.5%, vap d: 4.1.

COH500 CAS: 420-04-2 CH_2N_2 HR: 3

CYANAMIDE

SYNS: CARBAMONITRILE ♦ CYANOGENAMIDE ♦ CYANOGEN NITRIDE ♦ HYDROGEN CYANAMIDE

OSHA PEL: TWA 2 mg/m^3

ACGIH TLV: TWA 2 mg/m^3

SAFETY PROFILE: Poison by ingestion and inhalation. Moderately toxic by skin contact. Combustible.

PROP: Deliquescent crystals. Mp: 45°, bp: 260°, flash p: 285°F, d: 1.282, vap d: 1.45.

COI500 CAS: 57-12-5 CN^- HR: 3

CYANIDE

DOT: UN 1935

SYNS: CYANIDE ION ♦ CYANIDE SOLUTIONS (DOT)

OSHA PEL: TWA 5 $mg(CN)/m^3$

ACGIH TLV: CL 5 mg/m^3 (skin)
DFG MAK: 5 mg/m^3
NIOSH REL: (Cyanide) TWA CL 5 mg/m^3/10M
DOT Classification: 6.1; Label: Poison, KEEP AWAY FROM FOOD
SAFETY PROFILE: Very poisonous. Flammable by chemical reaction with heat, moisture, acid. Many cyanides rather easily evolve HCN, a flammable gas that is highly toxic.

COO000 CAS: 460-19-5 C_2N_2 HR: 3
CYANOGEN
DOT: UN 1026
SYNS: CARBON NITRIDE ♦ CYANOGEN GAS (DOT) ♦ NITRILOACETONITRILE ♦ OXALIC ACID DINITRILE ♦ OXALONITRILE ♦ PRUSSITE
OSHA PEL: TWA 10 ppm
ACGIH TLV: TWA 10 ppm
DFG MAK: 10 ppm (22 mg/m^3)
DOT Classification: 2.3; Label: Poison Gas, Flammable Gas
SAFETY PROFILE: Moderately toxic by inhalation. A systemic irritant by inhalation. A human eye irritant. Very dangerous fire hazard.
PROP: Colorless gas; pungent odor. Mp: –34.4°, bp: –21.0°, d: 0.866 @ 17°/4°, lel: 6.6%, uel: 32%, vap d: 1.8.

COO750 CAS: 506-77-4 CClN HR: 3
CYANOGEN CHLORIDE
DOT: UN 1589
SYNS: CHLORINE CYANIDE ♦ CHLOROCYANIDE ♦ CHLOROCYANOGEN
OSHA PEL: CL 0.3 ppm
ACGIH TLV: CL 0.3 ppm
DOT Classification: 2.3; Label: Poison Gas, Flammable Gas
SAFETY PROFILE: Poison by ingestion. Toxic by inhalation. A primary irritant. A severe human eye irritant. Flammable.
PROP: Colorless liquid or gas; lachrymatory and irritating odor. Mp: –6.5°, bp: 13.1°, d: 1.218 @ 4°/4°, vap press: 1010 mm @ 20°, vap d: 1.98.

CPB000 CAS: 110-82-7 C_6H_{12} HR: 3
CYCLOHEXANE
DOT: UN 1145
SYNS: HEXAHYDROBENZENE ♦ HEXAMETHYLENE ♦ HEXANAPHTHENE
OSHA PEL: TWA 300 ppm
ACGIH TLV: TWA 300 ppm (Proposed: 200 ppm; STEL 400 ppm)
DFG MAK: 200 ppm (720 mg/m^3)
DOT Classification: 3; Label: Flammable Liquid
SAFETY PROFILE: Moderately toxic by ingestion. A systemic irritant by inhalation and ingestion. A skin irritant. Flammable liquid.

PROP: Colorless, mobile liquid; pungent odor. Mp: 6.5°, bp: 81°, fp: 4.6°, flash: p: 1.4°F, ULC: 90–95, lel: 1.3%, uel: 8.4%, d: 0.7791 @ 20°/4°, autoign temp: 473°F, vap press: 100 mm @ 60.8°, vap d: 2.90. Prac insol in H_2O; sol in MeOH; misc in most org solvs.

CPB625 CAS: 1569-69-3 $C_6H_{12}S$ HR: 3

CYCLOHEXANETHIOL

DOT: UN 3054

SYNS: CYCLOHEXYL MERCAPTAN (DOT) ♦ CYKLOHEXANTHIOL

NIOSH REL: (Cyclohexanethiol) CL 0.5 ppm/15M

DOT Classification: 3; Label: Flammable Liquid

SAFETY PROFILE: An eye and severe skin irritant.

PROP: Oil. D: 0.991, bp: 158–160°. Sol in EtOH, $CHCl_3$; insol in H_2O.

CPB750 CAS: 108-93-0 $C_6H_{12}O$ HR: 3

CYCLOHEXANOL

SYNS: CYCLOHEXYL ALCOHOL ♦ HEXAHYDROPHENOL ♦ HEXALIN ♦ HYDROPHENOL ♦ HYDROXYCYCLOHEXANE

OSHA PEL: TWA 50 ppm (skin)

ACGIH TLV: TWA 50 ppm (skin)

DFG MAK: 50 ppm (210 mg/m^3)

SAFETY PROFILE: Moderately toxic by ingestion. Mildly toxic by skin contact. A severe eye irritant. Flammable.

PROP: Colorless needles or viscous liquid; hygroscopic, camphor-like odor. Mp: 24°, bp: 161.5°, flash p: 154°F (CC), d: 0.9449 @ 25°/4°, vap press: 1 mm @ 21.0°, vap d: 3.45, autoign temp: 572°F. Sol in EtOH, Et_2O; mod sol in H_2O; misc in nonpolar solvents.

CPC000 CAS: 108-94-1 $C_6H_{10}O$ HR: 3

CYCLOHEXANONE

DOT: UN 1915

SYNS: KETOHEXAMETHYLENE ♦ PIMELIC KETONE

OSHA PEL: TWA 25 ppm (skin)

ACGIH TLV: TWA 25 ppm (skin); Not Classifiable as a Human Carcinogen

DFG MAK: Confirmed Animal Carcinogen with Unknown Relevance to Humans

NIOSH REL: (Ketone (Cyclohexanone)) TWA 100 mg/m^3

DOT Classification: 3; Label: Flammable Liquid

SAFETY PROFILE: Suspected carcinogen. Moderately toxic by ingestion and inhalation. A skin and severe eye irritant. Human irritant by inhalation. Flammable liquid.

PROP: Colorless oily liquid; acetone-like odor. Mp: –45.0°, bp: 155°, ULC: 35–40, lel: 1.1% @ 100°, flash p: 111°F, d: 0.9478 @ 20°/4°, autoign temp: 788°F, vap press: 10 mm @ 38.7°, vap d: 3.4. Mod sol in H_2O.

CPC579 CAS: 110-83-8 C_6H_{10} HR: 3

CYCLOHEXENE
DOT: UN 2256
SYNS: BENZENETETRAHYDRIDE ♦ 1,2,3,4-TETRAHYDROBENZENE
OSHA PEL: 300 ppm
ACGIH TLV: 300 ppm
DFG MAK: 300 ppm (1000 mg/m^3)
DOT Classification: 3; Label: Flammable Liquid
SAFETY PROFILE: Moderately toxic by inhalation and ingestion. A very dangerous fire hazard.
PROP: Colorless liquid. Bp: 83°, fp: –103.7°, flash p: <21.2°F, d: 0.8102 @ 20°/4°, vap press: 160 mm @ 38°, autoign temp: 590°F, vap d: 2.8, lel: 1.2%.

CPD750 CAS: 100-40-3 C_8H_{12} HR: 3
CYCLOHEXENYLETHYLENE
SYNS: BUTADIENE DIMER ♦ 4-ETHENYL-1-CYCLOHEXENE ♦ 1,2,3,4-TETRAHYDROSTYRENE ♦ 4-VINYLCYCLOHEXENE ♦ 4-VINYL-1-CYCLOHEXENE
ACGIH TLV: TWA 0.1 ppm (skin); Animal Carcinogen
DFG MAK: Animal Carcinogen, Suspected Human Carcinogen
SAFETY PROFILE: Confirmed carcinogen. Moderately toxic by ingestion and inhalation. Mildly toxic by skin contact. Dangerous fire hazard.
PROP: Liquid. Bp: 128°, fp: –109°, flash p: 60°F (TOC), d: 0.832 @ 20°/4°, autoign temp: 517°F, vap press: 25.8 mm @ 38°, vap d: 3.76.

CPF500 CAS: 108-91-8 $C_6H_{13}N$ HR: 3
CYCLOHEXYLAMINE
DOT: UN 2357
SYNS: AMINOCYCLOHEXANE ♦ AMINOHEXAHYDROBENZENE ♦ CHA ♦ CYCLOHEXANAMINE ♦ HEXAHYDROANILINE ♦ HEXAHYDROBENZENAMINE
OSHA PEL: TWA 10 ppm
ACGIH TLV: TWA 10 ppm; Not Classifiable as a Human Carcinogen
DFG MAK: 10 ppm (41 mg/m^3)
DOT Classification: 8; Label: Corrosive, Flammable Liquid
SAFETY PROFILE: A poison by ingestion, skin contact, and routes. Experimental teratogenic effects. A severe human skin irritant. Can cause dermatitis and convulsions. Flammable liquid.
PROP: Liquid; strong, fishy odor. Mp: –17.7°, bp: 134.5°, flash p: 69.8°F, d: 0.865 @ 25°/25°, autoign temp: 560°F, vap d: 3.42. Misc in H_2O, org solvs.

CPR800 CAS: 121-82-4 $C_3H_6N_6O_6$ HR: 3
CYCLONITE
DOT: UN 0072/UN 0118/UN 0391/UN 0483
SYNS: CYCLONITE, desensitized (UN 0483) (DOT) ♦ CYCLONITE, wetted (UN 0072) (DOT) ♦ CYCLOTRIMETHYLENETRINITRAMINE, desensitized (UN 0483) (DOT) ♦ CYCLOTRIMETHYLENETRINITRAMINE, wetted (UN 0072) (DOT) ♦ HEXOGEN (Explosive) ♦ HEXOLITE ♦ RDX ♦ T4 ♦ 1,3,5-TRIAZINE, HEXAHYDRO-1,3,5-TRINITRO-(9CI) ♦ TRIMETHYLENETRINITRAMINE ♦ sym-TRIMETHYLENETRINITRAMINE ♦

TRINITROCYCLOTRIMETHYLENE TRIAMINE ♦ 1,3,5-TRINITRO-1,3,5-TRIAZACYCLOHEXANE

OSHA PEL: TWA 1.5 mg/m^3 (skin)

ACGIH TLV: TWA 0.5 mg/m^3 (skin); Not Classifiable as a Human Carcinogen

DOT Classification: EXPLOSIVE 1.1D; Label: EXPLOSIVE 1.1D

SAFETY PROFILE: Poison by ingestion. An experimental teratogen. A corrosive irritant to skin, eyes, and mucous membranes. It is one of the most powerful high explosives in use today.

PROP: White, crystalline powder. Mp: 202°.

CPU500 CAS: 542-92-7 C_5H_6 HR: 3

1,3-CYCLOPENTADIENE

SYNS: CYCLOPENTADIENE ♦ PENTOLE ♦ PYROPENTYLENE

OSHA PEL: TWA 75 ppm

ACGIH TLV: TWA 75 ppm

DFG MAK: 75 ppm (210 mg/m^3)

SAFETY PROFILE: Low toxicity by ingestion. A dangerous fire hazard when exposed to heat or flame It can react violently with many substances.

PROP: Colorless liquid. Mp: –85°, bp: 41–42°, d: 0.80475 @ 19°/4°, flash p: 77°F. Misc in EtOH and C_6H_6.

CPV000 CAS: 12079-65-1 $C_8H_5MnO_3$ HR: 3

CYCLOPENTADIENYLMANGANESE TRICARBONYL

SYNS: MANGANESE CYCLOPENTADIENYL TRICARBONYL ♦ MCT

OSHA PEL: TWA 0.1 $mg(Mn)/m^3$ (skin)

ACGIH TLV: TWA 0.1 $mg(Mn)/m^3$

SAFETY PROFILE: A poison by ingestion and inhalation.

PROP: Pale-yellow crystals with camphoraceous odor. Mp: 76.8–77.1°.

CPV750 CAS: 287-92-3 C_5H_{10} HR: 3

CYCLOPENTANE

DOT: UN 1146

SYN: PENTAMETHYLENE

OSHA PEL: TWA 600 ppm

ACGIH TLV: TWA 600 ppm

DOT Classification: 3; Label: Flammable Liquid

SAFETY PROFILE: Mildly toxic by ingestion and inhalation. A very dangerous fire hazard.

PROP: Colorless liquid. Bp: 49.3°, fp: –93.7°, flash p: 19.4°F, autoign temp: 716°F, d: 0.745 @ 20°/4°, vap press: 400 mm @ 31.0°, vap d: 2.42.

DAE400 CAS: 17702-41-9 $B_{10}H_{14}$ HR: 3

DECABORANE

DOT: UN 1868

SYN: DECABORANE(14)
OSHA PEL: TWA 0.05 ppm; STEL 0.15 ppm (skin)
ACGIH TLV: TWA 0.05 ppm; STEL 0.15 ppm (skin)
DFG MAK: 0.05 ppm (0.25 mg/m^3)
DOT Classification: 4.1; Label: Flammable Solid, Poison
SAFETY PROFILE: Poison by inhalation, ingestion, and skin contact.
PROP: Colorless needles or crystals. Solid by sublimation. Mp: 99.6°, d: 0.94 (solid), d: 0.78 (liquid @ 100°), bp: 213°, vap press: 19 mm @ 100°. Sol in CS_2.

DBF750 CAS: 123-42-2 $C_6H_{12}O_2$ HR: 3
DIACETONE ALCOHOL
DOT: UN 1148
SYNS: DIKETONE ALCOHOL ♦ 4-HYDROXY-2-KETO-4-METHYLPENTANE ♦ 2-METHYL-2-PENTANOL-4-ONE ♦ PYRANTON ♦ TYRANTON
OSHA PEL: TWA 50 ppm
ACGIH TLV: TWA 50 ppm
DFG MAK: 50 ppm (240 mg/m^3)
NIOSH REL: (Ketones) TWA 240 mg/m^3
DOT Classification: 3; Label: Flammable Liquid
SAFETY PROFILE: Moderately toxic by ingestion. Mildly toxic by skin contact. A skin, mucous membrane, and severe eye irritant. Flammable liquid.
PROP: Liquid; oily; faint pleasant odor. Mp: –47 to –54°, bp: 164°, flash p: 148°F, d: 0.9306 @ 25°/4°, autoign temp: 1118°F, vap d: 4.00, vap press: 1.1 mm @ 20°, lel: 1.8%, uel: 6.9%, flash p: (acetone free) 136°F. Sol in water.

DBO000 CAS: 615-05-4 $C_7H_{10}N_2O$ HR: 3
2,4-DIAMINOANISOLE
SYNS: 2,4-DAA ♦ 2,4-DIAMINO-1-METHOXYBENZENE ♦ FURRO L ♦ p-METHOXY-m-PHENYLENEDIAMINE ♦ 4-MMPD
DFG MAK: Animal Carcinogen, Suspected Human Carcinogen
NIOSH REL: (2,4-diaminoanisole) Reduce to lowest feasible level
SAFETY PROFILE: Confirmed carcinogen. Moderately toxic by ingestion. A skin irritant.
PROP: Needles. Mp: 68°.

DCJ200 CAS: 119-90-4 $C_{14}H_{16}N_2O_2$ HR: 3
o-DIANISIDINE
SYNS: C.I. 24110 ♦ 3,3′-DIANISIDINE ♦ 3,3′-DIMETHOXYBENZIDINE
DFG MAK: Animal Carcinogen, Suspected Human Carcinogen
NIOSH REL: (o-Dianisidine-Based Dyes) Reduce to lowest feasible concentration
SAFETY PROFILE: Confirmed carcinogen. Moderately toxic by ingestion. Combustible.
PROP: Colorless leaflets or crystals. Mp: 137–138°, flash p: 403°F, vap d: 8.5. Sol in C_6H_6 and AcOH; sltly sol in H_2O.

DCJ400 CAS: 91-93-0 $C_{16}H_{12}N_2O_4$ HR: 3

DIANISIDINE DIISOCYANATE

SYNS: 3,3′-DIMETHOXYBENZIDINE-4,4′-DIISOCYANATE ♦ 3,3′-DIMETHOXY-4,4′-BIPHENYLENE DIISOCYANATE

NIOSH REL: (Diisocyanates) TWA 0.005 ppm; CL 0.02 ppm/10M

SAFETY PROFILE: A strong sensitizer.

DCJ800 CAS: 61790-53-2 HR: 1

DIATOMACEOUS EARTH

SYNS: AMORPHOUS SILICA ♦ DIATOMACEOUS SILICA ♦ KIESELGUHR ♦ SILICA, AMORPHOUS-DIATOMACEOUS EARTH (UNCALCINED) (ACGIH)

OSHA PEL: TWA 6 mg/m^3

ACGIH TLV: TWA (nuisance particulate) 10 mg/m^3 of total dust (when toxic impurities are not present, e.g., quartz <1%)

DFG MAK: 4 mg/m^3 as fine dust

SAFETY PROFILE: A nuisance dust that may cause fibrosis of the lungs.

PROP: Composed of skeletons of small aquatic plants related to algae and contains as much as 88% amorphous silica (DTLVS* 4,120,80). White to buff-colored solid. Insol in water; sol in hydrofluoric acid.

DCP800 CAS: 334-88-3 CH_2N_2 HR: 3

DIAZOMETHANE

SYNS: AZIMETHYLENE ♦ DIAZIRINE

OSHA PEL: TWA 0.2 ppm

ACGIH TLV: TWA 0.2 ppm; Suspected Human Carcinogen

DFG MAK: Animal Carcinogen, Suspected Human Carcinogen

SAFETY PROFILE: Confirmed carcinogen A poisonous irritant by inhalation. A powerful allergen. Highly explosive when shocked, exposed to heat, or by chemical reaction. It can react violently with many substances.

PROP: Yellow gas at ordinary temp which forms yellow solns in ethereal solvs. Mp: –145°, bp: –23°, d: 1.45.

DDI450 CAS: 19287-45-7 B_2H_6 HR: 3

DIBORANE

DOT: UN 1911/NA 1911

SYNS: BORON HYDRIDE ♦ DIBORANE MIXTURES (NA 1911) ♦ DIBORON HEXAHYDRIDE

OSHA PEL: TWA 0.1 ppm

ACGIH TLV: TWA 0.1 ppm

DFG MAK: 0.1 ppm (0.1 mg/m^3)

DOT Classification: 2.1; Label: Flammable Gas (NA 1911); DOT Class: 2.3; Label: Poison Gas, Flammable Gas

SAFETY PROFILE: Poison by inhalation. An irritant to skin, eyes, and mucous membranes. Dangerously flammable. It can react violently with many substances.

PROP: Colorless air- and moisture-sensitive gas; sickly-sweet odor. Mp: –165.5°, bp: –92.5°, d: 0.447 (liquid @ –112°), 0.577 (solid @ –183°), vap press: 224 mm @ –112°, autoign temp: 38–52°, lel: 0.9%, uel: 98%, flash p: –90°F. Sol in THF as BH_3-THF complex.

DDL800 CAS: 96-12-8 $C_3H_5Br_2Cl$ HR: 3

1,2-DIBROMO-3-CHLOROPROPANE

DOT: UN 2872
SYNS: 1-CHLORO-2,3-DIBROMOPROPANE ♦ DBCP
OSHA PEL: TWA 0.001 ppm; Cancer Hazard
DFG MAK: Animal Carcinogen, Suspected Human Carcinogen
NIOSH REL: (Dibromochloropropane) CL 0.01 ppm/30M
DOT Classification: 6.1; Label: KEEP AWAY FROM FOOD
SAFETY PROFILE: Confirmed human carcinogen. Poison by ingestion and inhalation. Moderately toxic by skin contact. An eye and severe skin irritant. Combustible.
PROP: Bp: 196°, flash p: 170°F (TOC).

DDU600 CAS: 102-81-8 $C_{10}H_{23}NO$ HR: 3

2-N-DIBUTYLAMINOETHANOL

DOT: UN 2873
SYNS: 2-DIBUTYLAMINOETHANOL ♦ N,N-DI-n-BUTYLAMINOETHANOL (DOT) ♦ β-N-DIBUTYLAMINOETHYL ALCOHOL ♦ N,N-DIBUTYL-N-(2-HYDROXYETHYL)AMINE
OSHA PEL: TWA 2 ppm
ACGIH TLV: TWA 0.5 ppm (skin)
DOT Classification: 6.1; Label: KEEP AWAY FROM FOOD
SAFETY PROFILE: Moderately toxic by ingestion and skin contact. A severe eye and skin irritant. Combustible.
PROP: Liquid. Bp: 222°, flash p: 220°F (OC), d: 0.85, vap d: 6.0.

DEF400 CAS: 818-08-6 $C_8H_{18}OSn$ HR: 3

DIBUTYLOXOSTANNANE

SYNS: DBOT ♦ DIBUTYLOXIDE of TIN ♦ DIBUTYLOXOTIN ♦ DIBUTYLSTANNANE OXIDE
OSHA PEL: TWA 0.1 mg(Sn)/m^3 (skin)
ACGIH TLV: TWA 0.1 mg(Sn)/m^3 (skin) (Proposed: TWA 0.1 mg(Sn)/m^3; STEL 0.2 mg(Sn)/m^3 (skin))
NIOSH REL: (Organotin Compounds) TWA 0.1 mg(Sn)/m^3
SAFETY PROFILE: Poison by ingestion. A skin and eye irritant. Flammable.
PROP: White, amorphous powder or polymeric infusible solid. Mp: decomp without melting, bulk density: 0.5, vap d: 8.6.

DEG600 CAS: 2528-36-1 $C_{14}H_{23}O_4P$ HR: 2

DIBUTYL PHENYL PHOSPHATE

ACGIH TLV: TWA 0.3 ppm (skin)
SAFETY PROFILE: Moderately toxic by ingestion.

DEG700 CAS: 107-66-4 $C_8H_{19}PO_4$ HR: 2

DIBUTYL PHOSPHATE

SYNS: DIBUTYL ACID PHOSPHATE ♦ DIBUTYL HYDROGEN PHOSPHATE ♦ DI-n-BUTYL PHOSPHATE

OSHA PEL: TWA 1 ppm; STEL 2 ppm

ACGIH TLV: TWA 1 ppm; STEL 2 ppm

SAFETY PROFILE: Moderately toxic by ingestion.

PROP: Pale-amber liquid or oil. Bp: 135–138° @ 0.05 mm, decomp >100°. Sol in butanol and CCl_4.

DEH200 CAS: 84-74-2 $C_{16}H_{22}O_4$ HR: 3

DIBUTYL PHTHALATE

SYNS: DBP ♦ DIBUTYL-1,2-BENZENEDICARBOXYLATE

OSHA PEL: TWA 5 mg/m^3

ACGIH TLV: TWA 5 mg/m^3

SAFETY PROFILE: Mildly toxic by ingestion. Experimental teratogenic effects. Combustible.

PROP: Oily liquid; mild odor. Mp: –35°, bp: 340°, flash p: 315°F (CC), d: 1.047–1.049 @ 20°/20°, autoign temp: 757°F, vap d: 9.58.

DEN600 CAS: 7572-29-4 C_2Cl_2 HR: 3

DICHLOROACETYLENE

SYNS: DICHLOROETHYNE

OSHA PEL: CL 0.1 ppm

ACGIH TLV: CL 0.1 ppm

DFG MAK: Animal Carcinogen, Suspected Human Carcinogen

DOT Classification: Forbidden

SAFETY PROFILE: Confirmed carcinogen. Poison by inhalation. Strong explosive when shocked or exposed to heat or air.

PROP: Volatile liquid. Mp: –66 to –64°, bp: 33°.

DEP600 CAS: 95-50-1 $C_6H_4Cl_2$ HR: 3

o-DICHLOROBENZENE

DOT: UN 1591

SYNS: CHLOROBEN ♦ DCB ♦ 1,2-DICHLOROBENZENE ♦ ODB ♦ ODCB ♦ ORTHODICHLOROBENZENE ♦ ORTHODICHLOROBENZOL

OSHA PEL: CL 50 ppm

ACGIH TLV: TWA 25 ppm; STEL 50 ppm; Not Classifiable as a Human Carcinogen

DFG MAK: 50 ppm (300 mg/m^3)

DOT Classification: 6.1; Label: KEEP AWAY FROM FOOD

SAFETY PROFILE: Poison by ingestion. Moderately toxic by inhalation. An experimental teratogen. An eye, skin, and mucous membrane irritant. Flammable.

PROP: Clear liquid. Mp: –17.5°, bp: 180.5°, fp: –22°, flash p: 151°F, d: 1.307 @ 20°/20°, vap d: 5.05, autoign temp: 1198°F, lel: 2.2%, uel: 9.2%.

DEP800 CAS: 106-46-7 $C_6H_4Cl_2$ HR: 3

p-DICHLOROBENZENE

DOT: UN 1592

SYNS: 1,4-DICHLOROBENZENE (MAK) ♦ DICHLOROBENZENE, PARA, solid (DOT) ♦ p-DICHLOROBENZOL ♦ PARA CRYSTALS ♦ PARADICHLOROBENZENE ♦ PDB ♦ PDCB

OSHA PEL: TWA 75 ppm; STEL 110 ppm

ACGIH TLV: TWA 75 ppm; STEL 110 ppm (Proposed: 10 ppm; Animal Carcinogen)

DFG MAK: 50 ppm (300 mg/m^3)

DOT Classification: 6.1; Label: KEEP AWAY FROM FOOD

NIOSH REL: (p-Dichlorobenzene): (1.7 ppm LOQ)

SAFETY PROFILE: Confirmed carcinogen. An experimental teratogen. Moderately toxic to humans by ingestion. A human eye irritant. Flammable liquid.

PROP: White crystals or leaflets with strong penetrating odor. Mp: 54°, bp: 174°, flash p: 150°F (CC), d: 1.4581 @ 20.5°/4°, vap press: 10 mm @ 54.8°, vap d: 5.08.

DEQ600 CAS: 91-94-1 $C_{12}H_{10}Cl_2N_2$ HR: 3

3',3'-DICHLOROBENZIDINE

SYNS: C.I. 23060 ♦ DCB ♦ 4,4'-DIAMINO-3,3'-DICHLORODIPHENYL ♦ 3,3'-DICHLORO-4,4'-BIPHENYLDIAMINE

OSHA PEL: Cancer Suspect Agent

ACGIH TLV: Animal Carcinogen

DFG TRK: Animal Carcinogen, Suspected Human Carcinogen

NIOSH REL: (Benzidine-based Dye) Reduce to lowest feasible level

SAFETY PROFILE: Confirmed carcinogen.

PROP: Crystals or needles from alc. Mp: 133°. Insol in water; sol in alc, benzene, and glacial acetic acid.

DEQ800 CAS: 612-83-9 $C_{12}H_{10}Cl_2N_2 \cdot 2ClH$ HR: 3

3,3'-DICHLOROBENZIDINE DIHYDROCHLORIDE

SYN: 3,3'-DICHLORO-(1,1'-BIPHENYL)-4,4'-DIAMINE DIHYDROCHLORIDE

OSHA PEL: Cancer Suspect Agent

SAFETY PROFILE: Confirmed carcinogen. Moderately toxic by ingestion.

DEV000 CAS: 764-41-0 $C_4H_6Cl_2$ HR: 3

1,4-DICHLORO-2-BUTENE

SYNS: DCB ♦ 1,4-DICHLOROBUTENE-2 (MAK)

ACGIH TLV: Animal Carcinogen, Suspected Human Carcinogen

DFG MAK: Animal Carcinogen, Suspected Human Carcinogen

SAFETY PROFILE: Confirmed carcinogen. Poison by ingestion and inhalation. Moderately toxic by skin contact. An experimental teratogen. A severe skin and eye irritant.

PROP: Colorless liquid. Mp: 1–3°, bp: 156°, d: 1.183 @ 25°/4°.

DFA600 CAS: 75-71-8 CCl_2F_2 HR: 1

DICHLORODIFLUOROMETHANE

DOT: UN 1028

SYNS: DIFLUORODICHLOROMETHANE ♦ FREON 12 ♦ REFRIGERANT 12

OSHA PEL: TWA 1000 ppm

ACGIH TLV: TWA 1000 ppm; Not Classifiable as a Human Carcinogen

DFG MAK: 1000 ppm (5000 mg/m^3)

DOT Classification: 2.2; Label: Nonflammable Gas

SAFETY PROFILE: Narcotic in high concentrations. Nonflammable gas.

PROP: Colorless, almost odorless gas. Mp: –158°, bp: –29°, vap press: 5 atm @ 16.1°.

DFE200 CAS: 118-52-5 $C_5H_6Cl_2N_2O_2$ HR: 2

1,3-DICHLORO-5,5-DIMETHYL HYDANTOIN

SYNS: DCA ♦ DICHLORODIMETHYLHYDANTOIN ♦ 1,3-DICHLORO-5,5'-METHYLHYDANTOIN ♦ HALANE

OSHA PEL: TWA 0.2 mg/m^3; STEL 0.4 mg/m^3

ACGIH TLV: TWA 0.2 mg/m^3; STEL 0.4 mg/m^3

SAFETY PROFILE: Moderately toxic by ingestion. Mildly toxic by inhalation. A severe skin irritant. Will react with water or steam to produce toxic and corrosive fumes.

PROP: Crystals, liberates chlorine on contact with hot water; prisms from $CHCl_3$. Mp: 132°. Subl @ 100°; conflagrates @ 212°; d: 1.5 @ 20°, vap d: 6.8. Sol in H_2O; mod in sol AcOH and EtOH.

DFE800 CAS: 28675-08-3 $C_{12}H_8Cl_2O$ HR: 2

DICHLORODIPHENYL OXIDE

SYNS: DICHLOROPHENYL ETHER

OSHA PEL: TWA 0.5 mg/m^3

SAFETY PROFILE: Moderately toxic by ingestion.

PROP: Liquid. Vap d: 8.2.

DFF809 CAS: 75-34-3 $C_2H_4Cl_2$ HR: 3

1,1-DICHLOROETHANE

DOT: UN 2362

SYNS: CHLORINATED HYDROCHLORIC ETHER ♦ ETHYLIDENE CHLORIDE ♦ ETHYLIDENE DICHLORIDE

OSHA PEL: TWA 100 ppm

ACGIH TLV: TWA 100 ppm; Not Classifiable as a Human Carcinogen

DFG MAK: 100 ppm (410 mg/m^3)

NIOSH REL: (1,1-Dichloroethane) Handle with caution

DOT Classification: 3; Label: Flammable Liquid

SAFETY PROFILE: Moderately toxic by ingestion. Experimental teratogenic effects. A very dangerous fire hazard.

PROP: Colorless liquid; aromatic, ethereal odor; hot, saccharine taste. Mp: –97.7°, lel: 5.6%, fp: –98°, bp: 57.3°, flash p: 22°F (TOC), d: 1.174 @ 20°/4°, vap press: 230 mm @ 25°, vap d: 3.44, autoign temp: 856°F.

DFI200 CAS: 156-59-2 $C_2H_2Cl_2$ HR: 1

cis-DICHLOROETHYLENE

SYN: 1,2-DICHLOROETHYLENE

DFG MAK: 200 ppm (800 mg/m^3)

SAFETY PROFILE: Mildly toxic by ingestion and inhalation. Sometimes thought to be nonflammable, however, it is a dangerous fire hazard when exposed to heat or flame. It can react violently with many substances.

PROP: Colorless liquid; pleasant odor. Mp: –80.5°, bp: 59°, lel: 9.7%, uel: 12.8%, flash p: 39°F, d: 1.291 @ 15°/4°, vap press: 400 mm @ 41.0°, vap d: 3.34.

DFI210 CAS: 540-59-0 $C_2H_2Cl_2$ HR: 3

1,2-DICHLOROETHYLENE

SYNS: ACETYLENE DICHLORIDE ♦ sym-DICHLOROETHYLENE

OSHA PEL: TWA 200 ppm

ACGIH TLV: TWA 200 ppm

DFG MAK: 200 ppm (800 mg/m^3)

SAFETY PROFILE: Poison by inhalation. Moderately toxic by ingestion. A skin irritant.

PROP: Liquid with ethereal odor. Bp: 55°.

DFJ050 CAS: 111-44-4 $C_4H_8Cl_2O$ HR: 3

DICHLOROETHYL ETHER

DOT: UN 1916

SYNS: BIS(2-CHLOROETHYL) ETHER ♦ CHLOROETHYL ETHER ♦ DCEE ♦ 2,2'-DICHLOROETHYL ETHER (MAK)

OSHA PEL: TWA 5 ppm; STEL 10 ppm (skin)

ACGIH TLV: TWA 5 ppm; STEL 10 ppm (skin); Not Classifiable as a Human Carcinogen

DFG MAK: 10 ppm (59 mg/m^3)

DOT Classification: 6.1; Label: Poison, Flammable Liquid

SAFETY PROFILE: A poison by ingestion, skin contact, and inhalation. A skin, eye, and mucous membrane irritant. Flammable liquid.

PROP: Colorless, stable liquid. Bp: 178.5°, fp: –51.9°, flash p: 131°F (CC), d: 1.2220 @ 20°/20°, autoign temp: 696°F, vap press: 0.7 mm @ 20°, vap d: 4.93. Misc in Et_2O, MeOH, and C_6H_6.

DFL000 CAS: 75-43-4 $CHCl_2F$ HR: 1

DICHLOROFLUOROMETHANE

DOT: UN 1029

SYNS: FREON 21

OSHA PEL: TWA 10 ppm

ACGIH TLV: TWA 10 ppm

DFG MAK: 10 ppm (43 mg/m^3)

DOT Classification: 2.2; Label: Nonflammable Gas
SAFETY PROFILE: Mildly toxic by inhalation.
PROP: Heavy, colorless gas. Mp: −135°, bp: 8.9°, d: 1.48, vap press: 2 atm @ 28.4°, vap d: 3.82.

DFU000 CAS: 594-72-9 $C_2H_3Cl_2NO_2$ HR: 3
1,1-DICHLORO-1-NITROETHANE
DOT: UN 2650
SYNS: DICHLORONITROETHANE
OSHA PEL: TWA 2 ppm
ACGIH TLV: TWA 2 ppm
DFG MAK: 10 ppm (60 mg/m^3)
DOT Classification: 6.1; Label: Poison
SAFETY PROFILE: Poison by ingestion. Moderately toxic by inhalation. A strong irritant. Flammable.
PROP: Liquid. Bp: 124°, flash p: 168°F(OC), d: 1.4153 @ 20°/20°, vap d: 4.97.

DGG950 CAS: 542-75-6 $C_3H_4Cl_2$ HR: 3
1,3-DICHLOROPROPENE
DOT: UN 2047
SYNS: α-CHLOROALLYL CHLORIDE ♦ 1,3-DICHLOROPROPYENE-1 ♦ 1,3-DICHLOROPROPYLENE
OSHA PEL: TWA 1 ppm (skin)
ACGIH TLV: TWA 1 ppm (skin); Not Classifiable as a Human Carcinogen
DFG MAK: Animal Carcinogen, Suspected Human Carcinogen
DOT Classification: 3; Label: Flammable Liquid
SAFETY PROFILE: Confirmed carcinogen. Poison by ingestion. Moderately toxic by skin contact. Mildly toxic by inhalation. A strong irritant. A flammable liquid.
PROP: Liquid. Bp: 103–110°, flash p: 95°F, d: 1.22, vap d: 3.8.

DGI400 CAS: 75-99-0 $C_3H_4Cl_2O_2$ HR: 2
2,2-DICHLOROPROPIONIC ACID
SYNS: DALAPON (USDA) ♦ α-DICHLOROPROPIONIC ACID
OSHA PEL: TWA 1 ppm
ACGIH TLV: TWA 1 ppm (Proposed: TWA 5 ppm, Not Classifiable as a Human Carcinogen)
DFG MAK: 1 ppm (5.9 mg/m^3)
SAFETY PROFILE: A corrosive with low toxicity by skin contact. A skin irritant.
PROP: White to tan powder. D: 1.39 @ 22.6°/4°, bp: 185–190°. Sol in water.

DGL600 CAS: 1320-37-2 $C_2Cl_2F_4$ HR: 1
DICHLOROTETRAFLUOROETHANE
DOT: UN 1958
SYNS: R114 (DOT) ♦ TETRAFLUORODICHLOROETHANE
OSHA PEL: TWA 1000 ppm

ACGIH TLV: TWA 1000 ppm
DOT Classification: 2.2; Label: Nonflammable Gas
SAFETY PROFILE: A mildly toxic irritant; narcotic in high concentrations.
PROP: Colorless gas. Bp: 3.5°.

DGW000 CAS: 77-73-6 $C_{10}H_{12}$ HR: 3
DICYCLOPENTADIENE
DOT: UN 2048
SYNS: BICYCLOPENTADIENE ♦ 1,3-CYCLOPENTADIENE, DIMER
OSHA PEL: TWA 5 ppm
ACGIH TLV: TWA 5 ppm
DFG MAK: 0.5 ppm (2.7 mg/m^3)
DOT Classification: 3; Label: Flammable Liquid
SAFETY PROFILE: Poison by ingestion. Moderately toxic by inhalation and skin contact. A severe skin and moderate eye irritant. Dangerous fire hazard.
PROP: Colorless crystals. Mp: 32.9°, bp: 166.6°, d: 0.976 @ 35°, vap press: 10 mm @ 47.6°, vap d: 4.55, flash p: 90°F (OC).

DHF000 CAS: 111-42-2 $C_4H_{11}NO_2$ HR: 3
DIETHANOLAMINE
SYNS: BIS(2-HYDROXYETHYL)AMINE ♦ DEA ♦ 2,2'-IMINOBISETHANOL ♦ 2,2'-IMINODIETHANOL
OSHA PEL: TWA 3 ppm
ACGIH TLV: TWA 0.46 ppm (skin)
SAFETY PROFILE: Moderately toxic by ingestion. Mildly toxic by skin contact. A severe eye and mild skin irritant. Combustible.
PROP: A faintly colored, viscous liquid or deliquescent prisms. Mp: 28°, bp: 270° (decomp), flash p: 305°F (OC), d: 1.0919 @ 30°/20°, autoign temp: 1224°F, vap press: 5 mm @ 138°, vap d: 3.65. Very sol in water.

DHJ200 CAS: 109-89-7 $C_4H_{11}N$ HR: 3
DIETHYLAMINE
DOT: UN 1154
SYNS: 2-AMINOPENTANE ♦ N-ETHYL-ETHANAMINE
OSHA PEL: TWA 10 ppm; STEL 25 ppm
ACGIH TLV: TWA 5 ppm; STEL 15 ppm; Not Classifiable as a Carcinogen
DFG MAK: 5 ppm (15 mg/m^3)
DOT Classification: 3; Label: Flammable Liquid
SAFETY PROFILE: Moderately toxic by ingestion, inhalation, and skin contact. A skin and severe eye irritant. A very dangerous fire hazard.
PROP: Colorless liquid; ammonia-like odor. Mp: –38.9°, bp: 55.5°, flash p: –0.4°F, d: 0.711 @ 18°/4°, fp: –50°, autoign temp: 594°F, vap press: 400 mm @ 38.0°, vap d: 2.53, lel: 1.8%, uel: 10.1%.

DHO500 CAS: 100-37-8 $C_6H_{15}NO$ HR: 3
2-DIETHYLAMINOETHANOL

DOT: UN 2686

SYNS: DEAE ♦ β-DIETHYLAMINOETHANOL ♦ β-DIETHYLAMINOETHYL ALCOHOL ♦ 2-HYDROXYTRIETHYLAMINE

OSHA PEL: TWA 10 ppm (skin)

ACGIH TLV: TWA 10 ppm (skin)

DFG MAK: 5 ppm (24 mg/m^3)

DOT Classification: 3; Label: Flammable Liquid

SAFETY PROFILE: Moderately toxic by ingestion and skin contact. A skin and severe eye irritant. Flammable liquid.

PROP: Colorless, hygroscopic liquid. Bp: 162°, flash p: 140°F (OC), d: 0.8851 @ 20°/20°, vap press: 1.4 mm @ 20°, vap d: 4.03. Sol in water.

DIW400 CAS: 88-10-8 $C_5H_{10}ClNO$ HR: 2

DIETHYLCARBAMOYL CHLORIDE

SYNS: DIETHYLCARBAMIDOYL CHLORIDE ♦ N,N-DIETHYLCARBAMOYL CHLORIDE

DFG MAK: Confirmed Animal Carcinogen with Unknown Relevance to Humans

SAFETY PROFILE: Suspected carcinogen. Reacts with water or steam to produce toxic and corrosive fumes.

PROP: Liquid. Mp: –44°, bp: 186°, vap d: 4.1.

DJG600 CAS: 111-40-0 $C_4H_{13}N_3$ HR: 3

DIETHYLENETRIAMINE

DOT: UN 2079

SYNS: AMINOETHYLETHANEDIAMINE ♦ 3-AZAPENTANE-1,5-DIAMINE ♦ BIS(2-AMINOETHYL)AMINE ♦ DETA ♦ 2,2′-DIAMINODIETHYLAMINE ♦ 2,2′-IMINOBISETHYLAMINE

OSHA PEL: TWA 1 ppm

ACGIH TLV: TWA 1 ppm (skin)

DOT Classification: 8; Label: Corrosive

SAFETY PROFILE: Poison by skin contact. Moderately toxic by ingestion. Corrosive. A severe skin and eye irritant. Combustible.

PROP: Yellow, viscous liquid; mild ammonia-like odor. Mp: –39°, bp: 207°, flash p: 215°F (OC), d: 0.9586 @ 20°/20°, autoign temp: 750°F, vap press: 0.22 mm @ 20°, vap d: 3.48. Misc in H_2O and EtOH.

DJN750 CAS: 96-22-0 $C_5H_{10}O$ HR: 3

DIETHYL KETONE

DOT: UN 1156

SYNS: DEK ♦ DIMETHYLACETONE ♦ METACETONE ♦ METHACETONE ♦ 3-PENTANONE ♦ PROPIONE

OSHA PEL: TWA 200 ppm

ACGIH TLV: TWA 200 ppm; STEL 300 ppm

DOT Classification: 3; Label: Flammable Liquid

SAFETY PROFILE: Moderately toxic by ingestion. A skin and eye irritant. Dangerous fire hazard.

PROP: Colorless, mobile liquid; acetone-like odor. Mp: –42°, bp: 101°, flash p: 55°F, d: 0.8159 @ 19°/4°, vap d: 2.96, autoign temp: 842°F, lel: 1.6%. Mod sol in water; misc in alc and ether.

DJX000 CAS: 84-66-2 $C_{12}H_{14}O_4$ HR: 3

DIETHYL PHTHALATE

SYNS: DIETHYL-o-PHTHALATE ♦ ETHYL PHTHALATE ♦ PHTHALIC ACID, DIETHYL ESTER

OSHA PEL: TWA 5 mg/m^3

ACGIH TLV: TWA 5 mg/m^3, Not Classifiable as a Carcinogen

SAFETY PROFILE: Moderately toxic by ingestion. An eye irritant and systemic irritant by inhalation. An experimental teratogen. Combustible.

PROP: Clear, colorless liquid. Mp: –0.3°, bp: 298°, flash p: 325°F (OC), d: 1.110, vap d: 7.66.

DKB110 CAS: 64-67-5 $C_4H_{10}O_4S$ HR: 3

DIETHYL SULFATE

DOT: UN 1594

SYNS: DIETHYL ESTER SULFURIC ACID ♦ DIETHYL TETRAOXOSULFATE ♦ ETHYL SULFATE

DFG TRK: Animal Carcinogen, Human Suspected Carcinogen

DOT Classification: 6.1; Label: Poison

SAFETY PROFILE: Confirmed carcinogen. Poison by inhalation. Moderately toxic by ingestion and skin contact. A severe skin irritant. An experimental teratogen. Combustible. It can react violently with many substances.

PROP: Colorless, oily liquid; faint ethereal odor. Mp: –25°, bp: 209.5° (decomp to ethyl ether), flash p: 220°F (CC), d: 1.18 @ 18°/0°, autoign temp: 817°F, vap press: 1 mm @ 47.0°, vap d: 5.31. Insol in water; decomp by hot water; misc with alc and ether. Insol in water.

DKG850 CAS: 75-61-6 CBr_2F_2 HR: 1

DIFLUORODIBROMOMETHANE

DOT: UN 1941

SYNS: DIBROMODIFLUOROMETHANE ♦ FREON 12-B2 ♦ R12B2 (DOT)

OSHA PEL: TWA 100 ppm

ACGIH TLV: TWA 100 ppm

DFG MAK: 100 ppm (870 mg/m^3)

DOT Classification: 9; Label: None

SAFETY PROFILE: Mildly toxic by inhalation. A non-flammable liquid.

PROP: Colorless, heavy liquid. Fp: –141°, bp: 23° @ 24.5 mm, d: 2.288 @ 15°/4°, mp: –110° (–1°). Insol in water.

DKM200 CAS: 2238-07-5 $C_6H_{10}O_3$ HR: 3

DIGLYCIDYL ETHER

SYNS: BIS(2,3-EPOXYPROPYL)ETHER ♦ DGE ♦ DI(2,3-EPOXYPROPYL) ETHER

OSHA PEL: TWA 0.1 ppm
ACGIH TLV: TWA 0.1 ppm; Not Classifiable as a Human Carcinogen
DFG MAK: 0.1 ppm (0.54 mg/m^3); Confirmed Animal Carcinogen with Unknown Relevance to Humans
NIOSH REL: (Glycidyl Ethers) CL 1 mg/m^3/15M
SAFETY PROFILE: Suspected carcinogen. Poison by ingestion and inhalation. Moderately toxic by skin contact. A severe eye and skin irritant.
PROP: Liquid. D: 1.126 @ 25°/4°, bp: 98–99° @ 11 mm.

DNI800 CAS: 108-83-8 $C_9H_{18}O$ HR: 3
DIISOBUTYL KETONE
DOT: UN 1157
SYNS: s-DIISOPROPYLACETONE ♦ 2,6-DIMETHYL-4-HEPTANONE ♦ ISOBUTYL KETONE ♦ ISOVALERONE ♦ VALERONE
OSHA PEL: TWA 25 ppm
ACGIH TLV: TWA 25 ppm
DFG MAK: 50 ppm (290 mg/m^3)
NIOSH REL: (Ketones) TWA 140 mg/m^3
DOT Classification: 3; Label: Flammable Liquid
SAFETY PROFILE: Moderately toxic by ingestion and inhalation. Mildly toxic by skin contact. An eye and skin irritant. Flammable liquid.
PROP: Liquid. Bp: 166°, flash p: 140°F, d: 0.81, vap d: 4.9, lel: 0.8% @ 212°F, uel: 6.2% @ 212°F.

DNJ800 CAS: 822-06-0 $C_8H_{12}N_2O_2$ HR: 3
1,6-DIISOCYANATOHEXANE
DOT: UN 2281
SYNS: DESMODUR H ♦ HEXAMETHYLENE DIISOCYANATE (DOT) ♦ 1,6-HEXAMETHYLENE DIISOCYANATE ♦ HMDI
ACGIH TLV: TWA 0.005 ppm
DFG MAK: 0.005 ppm (0.035 mg/m^3)
NIOSH REL: (Diisocyanates) TWA 0.005 ppm; CL 0.02 ppm/10M
DOT Classification: 6.1; Label: Poison
SAFETY PROFILE: Poison by inhalation. Moderately toxic by ingestion and skin contact.
PROP: Oil. D: 1.053 @ 20°/4°, bp: 121–122° @ 9 mm.

DNM200 CAS: 108-18-9 $C_6H_{15}N$ HR: 3
DIISOPROPYLAMINE
DOT: UN 1158
SYNS: DIPA ♦ N-(1-METHYLETHYL)-2-PROPANAMINE ♦ 2-PROPANAMINE, N-(1-METHYLETHYL)-
OSHA PEL: TWA 5 ppm (skin)
ACGIH TLV: TWA 5 ppm (skin)
DOT Classification: 3; Label: Flammable Liquid

SAFETY PROFILE: Moderately toxic by ingestion. Mildly toxic by inhalation. A skin and severe eye irritant. A very dangerous fire hazard.
PROP: Colorless liquid. Bp: 83–84°, flash p: 19.4°F, d: 0.722 @ 220.0°, vap d: 3.5.

DOA800 CAS: 20325-40-0 $C_{14}H_{16}N_2O_2 \cdot 2ClH$ HR: 3
3,3′-DIMETHOXYBENZIDINE DIHYDROCHLORIDE
SYNS: o-DIANISIDINE DIHYDROCHLORIDE ♦ 3,3-DIMETHOXY-(1,1′-BIPHENYL)-4,4′-DIAMINE DIHYDROCHLORIDE
NIOSH REL: (Benzidine-Based Dye) Reduce to lowest feasible level
SAFETY PROFILE: Confirmed carcinogen.

DOO800 CAS: 127-19-5 C_4H_9NO HR: 2
N,N-DIMETHYLACETAMIDE
SYNS: ACETDIMETHYLAMIDE ♦ ACETIC ACID DIMETHYLAMIDE ♦ DIMETHYLACETONE AMIDE ♦ DIMETHYLAMIDE ACETATE ♦ DMA ♦ DMAC
OSHA PEL: TWA 10 ppm (skin)
ACGIH TLV: TWA 10 ppm (skin); Not Classifiable as a Human Carcinogen; BEI: 30 mg/g creatinine of N-methylacetamide in urine at end of shift
DFG MAK: 10 ppm (36 mg/m^3)
SAFETY PROFILE: Moderately toxic by skin contact and inhalation. Mildly toxic by ingestion. Experimental teratogenic effects. A skin and eye irritant. Combustible
PROP: Colorless oily liquid; weak fishy odor. Mp: –20°, bp: 165°, d: 0.943 @ 20°/4°, vap d: 3.01, vap press: 1.3 mm @ 25°, flash p: 171°F (TOC), lel: 1.8%, uel: 11.5% @ 740 mm and 160°. Misc in water.

DOQ800 CAS: 124-40-3 C_2H_7N HR: 3
DIMETHYLAMINE
DOT: UN 1032/UN 1160
SYNS: DMA ♦ N-METHYLMETHANAMINE
OSHA PEL: TWA 10 ppm
ACGIH TLV: TWA 5 ppm; STEL 15 ppm; Not Classifiable as a Human Carcinogen
DFG MAK: 2 ppm (3.7 mg/m^3)
DOT Classification: 2.1; Label: Flammable Gas (UN 1032); DOT Class: 3; Label: Flammable Liquid (UN 1160)
SAFETY PROFILE: Poison by ingestion. Moderately toxic by inhalation. An eye irritant. Corrosive to the eyes, skin, and mucous membranes. A flammable gas.
PROP: Gas. D: 0.680 @ 0°/4°, mp: –96°, bp: 7°. Very sol in water.

DOT300 CAS: 60-11-7 $C_{14}H_{15}N_3$ HR: 3
4-DIMETHYLAMINOAZOBENZENE
SYNS: C.I. 11020 ♦ DAB ♦ N,N-DIMETHYL-p-PHENYLAZOANILINE
OSHA PEL: Cancer Suspect Agent
NIOSH REL: (4-Dimethylaminoazobenzene) TWA use 29 CFR 1910.1015
SAFETY PROFILE: Confirmed carcinogen. Poison by ingestion. Experimental teratogenic effects.

PROP: Yellow, crystalline tablets; yellow leaflets from EtOH. Mp: 115°. Sol in EtOH, Me_2CO, and C_6H_6; insol in H_2O.

DQF800 CAS: 121-69-7 $C_8H_{11}N$ HR: 3

N,N-DIMETHYLANILINE

DOT: UN 2253

SYNS: N-DIMETHYL-ANILINE (OSHA) ♦ N,N-DIMETHYLBENZENEAMINE ♦ DIMETHYLPHENYLAMINE ♦ N,N-DIMETHYLPHENYLAMINE

OSHA PEL: TWA 5 ppm; STEL 10 ppm (skin)

ACGIH TLV: TWA 5 ppm; STEL 10 ppm (skin); Not Classifiable as a Human Carcinogen

DFG MAK: 5 ppm (25 mg/m^3); Confirmed Animal Carcinogen with Unknown Relevance to Humans

DOT Classification: 6.1; Label: Poison

SAFETY PROFILE: Suspected carcinogen. Human poison by ingestion. Moderately toxic by inhalation and skin contact. A skin irritant. Flammable liquid.

PROP: Yellowish-brown oily liquid. Mp: 2.5°, bp: 193.1°, flash p: 145°F (CC), d: 0.9557 @ 20°/4°, ULC: 20–25, autoign temp: 700°F, vap press: 1 mm @ 29.5°, vap d: 4.17.

DQT200 CAS: 75-83-2 C_6H_{14} HR: 3

2,2-DIMETHYLBUTANE

SYN: NEOHEXANE (DOT)

OSHA PEL: TWA 500 ppm; STEL 1000 ppm

ACGIH TLV: TWA 500 ppm; STEL 1000 ppm

DFG MAK: 200 ppm (720 mg/m^3)

NIOSH REL: (Alkanes) TWA 350 mg/m^3

SAFETY PROFILE: Probably an irritant and narcotic in high concentration. A very dangerous fire and explosion hazard when exposed to heat or flame.

PROP: Liquid. Bp: 49.7°, mp: –98.2°, flash p: –54°F, fp: –101.9°, d: 0.649, autoign temp: 797°F, vap press: 400 mm @ 31.0°, vap d: 3.00, lel: 1.2%, uel: 7.0%.

DQT400 CAS: 79-29-8 C_6H_{14} HR: 3

2,3-DIMETHYLBUTANE

DOT: UN 2457

OSHA PEL: TWA 500 ppm; STEL 1000 ppm

ACGIH TLV: TWA 500 ppm; STEL 1000 ppm

DFG MAK: 200 ppm (720 mg/m^3)

NIOSH REL: TWA (Alkanes) 350 mg/m^3

DOT Classification: 3; Label: Flammable Liquid

SAFETY PROFILE: Probably an irritant and narcotic in high concentration. A very dangerous fire and explosion hazard.

PROP: Liquid. Mp: –135°, bp: 58.0°, flash p: –20°F, d: 0.662 @ 20°/4°, autoign temp: 788°F, vap press: 400 mm @ 39.0°, vap d: 3.0, lel: 1.2%, uel: 7.0%.

DQY950 CAS: 79-44-7 C_3H_6ClNO HR: 3

DIMETHYLCARBAMOYL CHLORIDE

DOT: UN 2262

SYNS: CHLOROFORMIC ACID DIMETHYLAMIDE ♦ DDC ♦ (DIMETHYLAMINO)CARBONYL CHLORIDE ♦ N,N-DIMETHYLAMINOCARBONYL CHLORIDE ♦ N,N-DIMETHYLCARBAMIDOYL CHLORIDE ♦ DIMETHYL CARBAMOYL CHLORIDE (ACGIH,DOT) ♦ DIMETHYLCHLOROFORMAMIDE ♦ DMCC

ACGIH TLV: Suspected Human Carcinogen

DFG MAK: Animal Carcinogen, Suspected Human Carcinogen

DOT Classification: 8; Label: Corrosive

SAFETY PROFILE: Confirmed carcinogen. Moderately toxic by inhalation and ingestion. Will react with water or steam to produce toxic and corrosive fumes.

PROP: Liquid. Mp: –33°, bp: 165–167°, d: 1.678 @ 20°/4°, vap d: 3.73.

DSB000 CAS: 68-12-2 C_3H_7NO HR: 3

DIMETHYLFORMAMIDE

DOT: UN 2265

SYNS: N,N-DIMETHYLFORMAMIDE (DOT) ♦ DMF ♦ DMFA ♦ N-FORMYLDIMETHYLAMINE

OSHA PEL: TWA 10 ppm (skin)

ACGIH TLV: TWA 10 ppm (skin); Not Classifiable as a Human Carcinogen; BEI: 40 mg/L N-methylforamide in urine at end of shift

DFG MAK: 10 ppm (30 mg/m^3)

DOT Classification: 3; Label: Flammable Liquid

SAFETY PROFILE: Suspected carcinogen. Moderately toxic by ingestion. Mildly toxic by skin contact and inhalation. Experimental teratogenic effects. A skin and severe eye irritant. Flammable liquid. It can react violently with many substances.

PROP: Colorless, mobile liquid; fishy or faint amine odor. Mp: –61°, bp: 152.8°, lel: 2.2% @ 100°, uel: 15.2% @ 100°, flash p: 136°, d: 0.945 @ 22.4°/4°, autoign temp: 833°F, vap press: 3.7 mm @ 25°, vap d: 2.51. Misc in H_2O, EtOH, Et_2O, C_6H_6, and $CHCl_3$.

DSF400 CAS: 57-14-7 $C_2H_8N_2$ HR: 3

1,1-DIMETHYLHYDRAZINE

DOT: UN 1163

SYNS: DIMAZINE ♦ DIMETHYLHYDRAZINE, unsymmetrical (DOT) ♦ UDMH (DOT)

OSHA PEL: TWA 0.5 ppm (skin)

ACGIH TLV: TWA 0.5 ppm (skin); Suspected Human Carcinogen (Proposed: TWA 0.01 ppm (skin); Suspected Human Carcinogen)

DFG MAK: Animal Carcinogen, Suspected Human Carcinogen

NIOSH REL: (Hydrazines) CL 0.15 mg/m^3/2H

DOT Classification: 6.1; Label: Poison, Flammable Liquid, Corrosive

SAFETY PROFILE: Confirmed carcinogen. Poison by ingestion. Moderately toxic by inhalation and skin contact. Corrosive. A dangerous fire hazard. It can react violently with many substances.
PROP: Colorless liquid; ammonia-like odor. Hygroscopic, water-misc. Mp: –58°, bp: 63.3°, flash p: 5°F, d: 0.791 @ 22°, vap press: 157 mm @ 25°, vap d: 1.94, autoign temp: 480°F, lel: 2%, uel: 95%. Sol in H_2O and EtOH.

DSF600 CAS: 540-73-8 $C_2H_8N_2$ HR: 3
1,2-DIMETHYLHYDRAZINE
DOT: UN 2382
SYNS: DIMETHYLHYDRAZINE, symmetrical (DOT) ♦ DMH ♦ HYDRAZOMETHANE ♦ SDMH
DFG MAK: Animal Carcinogen, Suspected Human Carcinogen
DOT Classification: 3; Label: Flammable Liquid, Poison
SAFETY PROFILE: Confirmed carcinogen An experimental teratogen. Poison by ingestion. Moderately toxic by inhalation. A very dangerous fire hazard.
PROP: Clear, colorless, flammable, hygroscopic, fuming liquid; fishy ammonia odor. Flash p: <73.4°F, bp: 81°, mp: –9°, d: 0.8274 @ 20°/4°. Sol in H_2O, EtOH, etc.

DTR200 CAS: 131-11-3 $C_{10}H_{10}O_4$ HR: 2
DIMETHYL PHTHALATE
SYNS: 1,2-BENZENEDICARBOXYLIC ACID DIMETHYL ESTER ♦ DMP ♦ METHYL PHTHALATE ♦ PHTHALIC ACID METHYL ESTER
OSHA PEL: TWA 5 mg/m^3
ACGIH TLV: TWA 5 mg/m^3
SAFETY PROFILE: Moderately toxic by ingestion. Mildly toxic by inhalation. Experimental teratogenic effects. An eye irritant. Combustible.
PROP: Colorless, odorless liquid. Mp: 0°, bp: 282.4°, flash p: 295°F (CC), d: 1.189 @ 25°/25°, autoign temp: 1032°F, vap d: 6.69, vap press: 1 mm @ 100.3°.

DUD100 CAS: 77-78-1 $C_2H_6O_4S$ HR: 3
DIMETHYL SULFATE
DOT: UN 1595
SYNS: DIMETHYL MONOSULFATE ♦ DMS ♦ METHYL SULFATE (DOT) ♦ SULFURIC ACID, DIMETHYL ESTER
OSHA PEL: TWA 0.1 ppm (skin)
ACGIH TLV: TWA 0.1 ppm (skin); Animal Carcinogen
DFG TRK: Production: 0.02 ppm; Use: 0.04 ppm; Animal Carcinogen, Suspected Human Carcinogen
DOT Classification: 3; Label: Poison, Corrosive
SAFETY PROFILE: Confirmed carcinogen An experimental teratogen. A poison by ingestion and inhalation. A corrosive irritant to skin, eyes, and mucous membranes. Flammable liquid.

PROP: Colorless, odorless liquid. Mp: –31.8°, fp: –27°, bp: 188° (decomp), flash p: 182°F (OC), d: 1.332 @ 15°, vap d: 4.35, autoign temp: 370°F. Sltly sol in H_2O, hexane, EtOH, C_6H_6; sol in Et_2O and Me_2CO.

DUQ200 CAS: 99-65-0 $C_6H_4N_2O_4$ HR: 3
m-DINITROBENZENE
DOT: UN 1597
SYNS: BINITROBENZENE ♦ 1,3-DINITROBENZENE
OSHA PEL: TWA 1 mg/m^3 (skin)
ACGIH TLV: TWA 0.15 ppm (skin)
DFG MAK: Confirmed Animal Carcinogen with Unknown Relevance to Humans
DOT Classification: 6.1; Label: Poison
SAFETY PROFILE: Suspected carcinogen. Human poison by ingestion. An eye irritant.
PROP: Yellowish crystals from alc. Mp: 89°, bp: 291°.

DUQ400 CAS: 528-29-0 $C_6H_4N_2O_4$ HR: 3
o-DINITROBENZENE
DOT: UN 1597
SYN: 1,2-DINITROBENZENE
OSHA PEL: TWA 1 mg/m^3 (skin)
ACGIH TLV: TWA 0.15 ppm (skin)
DFG MAK: Confirmed Animal Carcinogen with Unknown Relevance to Humans
DOT Classification: 6.1; Label: Poison
SAFETY PROFILE: Suspected carcinogen. Poison by inhalation and ingestion. Moderately toxic by skin contact. Combustible. A severe explosion hazard when shocked.
PROP: Colorless needles or plates from alc. Mp: 118°, bp: 319°, flash p: 302°F (CC), d: 1.571 @ 0°/4°, vap d: 5.79. Sol in EtOH and $CHCl_3$; sltly sol in H_2O.

DUQ600 CAS: 100-25-4 $C_6H_4N_2O_4$ HR: 3
p-DINITROBENZENE
DOT: UN 1597
OSHA PEL: TWA 1 mg/m^3 (skin)
ACGIH TLV: TWA 0.15 ppm (skin)
DFG MAK: Confirmed Animal Carcinogen with Unknown Relevance to Humans
DOT Classification: 6.1; Label: Poison
SAFETY PROFILE: Suspected carcinogen. Poison by ingestion.
PROP: White crystals, needles or prisms from alc. Mp: 173°, bp: 299°. Volatile with steam.

DUS700 CAS: 534-52-1 $C_7H_6N_2O_5$ HR: 3
DINITRO-o-CRESOL
SYNS: 2,4-DINITRO-o-CRESOL ♦ 2,4-DINITRO-6-METHYLPHENOL ♦ DINOC ♦ 2-METHYL-4,6-DINITROPHENOL

OSHA PEL: TWA 0.2 mg/m^3 (skin)
ACGIH TLV: TWA 0.2 mg/m^3 (skin)
DFG MAK: 0.2 mg/m^3
NIOSH REL: (Dinitro-o-Cresol) TWA 0.2 mg/m^3
DOT Classification: 6.1; Label: Poison
SAFETY PROFILE: A poison by ingestion, inhalation, skin contact. An eye and skin irritant.
PROP: Yellow, prismatic crystals from alc. Mp: 85.8°, vap d: 6.82.

DVG600 CAS: 25321-14-6 $C_7H_6N_2O_4$ HR: 3
DINITROTOLUENE
DOT: UN 2038
SYNS: DINITROPHENYLMETHANE ♦ METHYLDINITROBENZENE
OSHA PEL: TWA 1.5 mg/m^3 (skin)
ACGIH TLV: TWA 0.2 mg/m^3 (skin); Animal Carcinogen
DFG MAK: Animal Carcinogen, Suspected Human Carcinogen
NIOSH REL: (Dinitrotoluene) Reduce to lowest level
DOT Classification: 6.1; Label: Poison
SAFETY PROFILE: Confirmed carcinogen. An experimental teratogen. A poison. Flammable.

DVL700 CAS: 117-81-7 $C_{24}H_{38}O_4$ HR: 3
DI-sec-OCTYL PHTHALATE
SYNS: BEHP ♦ BIS(2-ETHYLHEXYL)PHTHALATE ♦ DEHP ♦ DI(2-ETHYLHEXYL)PHTHALATE ♦ DIOCTYL PHTHALATE ♦ DOP ♦ 2-ETHYLHEXYL PHTHALATE ♦ OCTYL PHTHALATE ♦ PHTHALIC ACID DIOCTYL ESTER
OSHA PEL: TWA 5 mg/m^3; STEL 10 mg/m^3
ACGIH TLV: TWA 5 mg/m^3; Confirmed Animal Carcinogen with Unknown Revelance to Humans
DFG MAK: 10 mg/m^3
NIOSH REL: (DEHP) Reduce to lowest feasible level
SAFETY PROFILE: Confirmed carcinogen. Experimental teratogenic data. A mild skin and eye irritant.
PROP: A liquid. D: 0.986 @ 20°, mp: –46°, bp: 231° @ 5 mm.

DVQ000 CAS: 123-91-1 $C_4H_8O_2$ HR: 3
DIOXANE
DOT: UN 1165
SYNS: 1,4-DIETHYLENE DIOXIDE ♦ 1,4-DIOXANE (MAK) ♦ GLYCOL ETHYLENE ETHER ♦ TETRAHYDRO-p-DIOXIN
OSHA PEL: TWA 25 ppm (skin)
ACGIH TLV: TWA 20 ppm (skin); Confirmed Animal Carcinogen with Unknown Revelance to Humans

DFG MAK: 20 ppm (73 mg/m^3); Not Classifiable as a Human Carcinogen
NIOSH REL: CL (Dioxane) 1 ppm/30M
DOT Classification: 3; Label: Flammable Liquid
SAFETY PROFILE: Confirmed carcinogen. An experimental teratogen. Moderately toxic by ingestion and inhalation. Mildly toxic by skin contact. An eye and skin irritant.A very dangerous fire and explosion hazard. It can react violently with many substances.
PROP: Colorless liquid with pleasant odor. Mp: 12°, fp: 11°, bp: 101.1°, lel: 2.0%, uel: 22.2%, flash p: 54°F (CC), d: 1.0353 @ 20°/4°, autoign temp: 356°F, vap press: 40 mm @ 25.2°, vap d: 3.03. Sol in EtOH and C_6H_6.

DVQ709 CAS: 78-34-2 $C_{12}H_{26}O_6P_2S_4$ HR: 3
DIOXATHION
SYNS: 2,3-p-DIOXANDITHIOL S,S-BIS(O,O-DIETHYL PHOSPHORODITHIOATE) ♦ p-DIOXANE-2,3-DIYL ETHYL PHOSPHORODITHIOATE ♦ PHOSPHORODITHIOIC ACID-S,S'-1,4-DIOXANE-2,3-DIYL-O,O,O',O'-TETRAETHYL ESTER
OSHA PEL: TWA 0.2 mg/m^3 (skin)
ACGIH TLV: TWA 0.2 mg/m^3 (skin); Not Classifiable as a Human Carcinogen
SAFETY PROFILE: Poison by ingestion, inhalation, and skin contact.
PROP: Nonvolatile, stable solid or brown liquid (tech grade). D: 1.257 @ 26°/4°, mp: –20°, bp: 60–68° @ 0.5 mm. Nonflammable. Insol in water.

DVX800 CAS: 122-39-4 $C_{12}H_{11}N$ HR: 3
DIPHENYLAMINE
SYNS: ANILINOBENZENE ♦ C.I. 10355 ♦ DFA ♦ N-PHENYLANILINE ♦ N-PHENYLBENEZENAMINE
OSHA PEL: TWA 10 mg/m^3
ACGIH TLV: TWA 10 mg/m^3; Not Classifiable as a Human Carcinogen
SAFETY PROFILE: Poison by ingestion. Experimental teratogenic effects. Combustible.
PROP: Crystals; floral odor. Mp: 52.9°, bp: 302.0°, flash p: 307°F (CC), d: 1.16, autoign temp: 1173°F, vap press: 1 mm @ 108.3°, vap d: 5.82. Sol in benzene, ether, and carbon disulfide; insol in water.

DWQ000 CAS: 7727-21-1 $H_2O_8S_2{\bullet}2K$ HR: 3
DIPOTASSIUM PERSULFATE
DOT: UN 1492
SYNS: PEROXYDISULFURIC ACID DIPOTASSIUM SALT ♦ POTASSIUM PEROXYDISULFATE ♦ POTASSIUM PERSULFATE (DOT)
ACGIH TLV: TWA 0.1 mg(S_2O_8)/m^3 (Proposed: TWA 0.1 mg/m^3)
DOT Classification: 5.1; Label: Oxidizer
SAFETY PROFILE: Moderately toxic by ingestion. An irritant and allergen. Flammable.

PROP: White, odorless, colorless, triclinic crystals. Mp: decomp @ 100°, d: 2.477. Decomp on heating to $K_2S_2O_7$ with loss of O_2. Mod sol in H_2O.

DWT200 CAS: 34590-94-8 $C_7H_{16}O_3$ HR: 2

DIPROPYLENE GLYCOL METHYL ETHER

SYNS: DIPROPYLENE GLYCOL MONOMETHYL ETHER
OSHA PEL: TWA 100 ppm; STEL 150 ppm (skin)
ACGIH TLV: TWA 100 ppm; STEL 150 ppm (skin)
DFG MAK: 50 ppm (310 mg/m^3)
SAFETY PROFILE: Mildly toxic by ingestion and skin contact. A skin and eye irritant. A mild allergen. Combustible.
PROP: Liquid. Bp: 190°, d: 0.951, vap d: 5.11, flash p: 185°F.

DWT600 CAS: 123-19-3 $C_7H_{14}O$ HR: 3

DIPROPYL KETONE

DOT: UN 2710
SYNS: BUTYRONE (DOT) ♦ 4-HEPTANONE ♦ PROPYL KETONE
OSHA PEL: TWA 50 ppm
ACGIH TLV: TWA 50 ppm
DOT Classification: 3; Label: Flammable Liquid
SAFETY PROFILE: Moderately toxic by ingestion, inhalation, and skin contact. A skin and eye irritant. Flammable liquid.
PROP: Colorless, refractive liquid. Bp: 144°, mp: –32.6°, vap press: 5.2 mm @ 20°, flash p: 120°F (CC), d: 0.815, vap d: 3.93.

DXQ745 CAS: 108-57-6 $C_{10}H_{10}$ HR: 2

DIVINYLBENZENE

SYNS: m-DIVINYLBENZENE ♦ m-VINYLSTYRENE
OSHA PEL: 10 ppm
ACGIH TLV: 10 ppm
SAFETY PROFILE: An eye irritant. Combustible.
PROP: Pale straw-colored liquid. Bp: 195–200°, mp: –87°, d: 0.918, flash p: 165F°. Not misc in water; sol in ether and methanol.

EAT900 CAS: 13838-16-9 $C_3H_2ClF_5O$ HR: 2

ENFLURANE

SYNS: 2-CHLORO-1,1,2-TRIFLUOROETHYL DIFLUOROMETHYL ETHER ♦ METHYLFLURETHER
ACGIH TLV: TWA 75 ppm; Not Classifiable as a Human Carcinogen
DFG MAK: 20 ppm
NIOSH REL: (Waste Anesthetic Gases and Vapors) CL 2 ppm/1H
SAFETY PROFILE: Mildly toxic by inhalation, ingestion. An experimental teratogen. An eye irritant. An anesthetic.

EAZ500 CAS: 106-89-8 C_3H_5ClO HR: 3

EPICHLOROHYDRIN
DOT: UN 2023
SYNS: 1-CHLORO-2,3-EPOXYPROPANE ♦ (CHLOROMETHYL)ETHYLENE OXIDE ♦ CHLOROMETHYLOXIRANE ♦ ECH ♦ 2,3-EPOXYPROPYL CHLORIDE ♦ GLYCEROL EPICHLORHYDRIN
OSHA PEL: TWA 2 ppm (skin)
ACGIH TLV: TWA 0.5 ppm (skin); Animal Carcinogen
DFG TRK: Animal Carcinogen, Suspected Human Carcinogen
NIOSH REL: Minimize exposure
DOT Classification: 6.1; Label: Poison
SAFETY PROFILE: Confirmed carcinogen. Poison by ingestion and skin contact. Moderately toxic by inhalation. An experimental teratogen. A skin and eye irritant. Flammable liquid. It can react violently with many substances.
PROP: Colorless, mobile liquid; irritating chloroform-like odor. Bp: 117.9°, fp: – 57.1°, flash p: 105.1°F (OC) (40°C), mp: –25.6°C, d: 1.1761 @ 20°/20°, vap press: 10 mm @ 16.6°, vap d: 3.29.

EEA500 CAS: 107-15-3 $C_2H_8N_2$ HR: 3
1,2-ETHANEDIAMINE
DOT: UN 1604
SYNS: 1,2-DIAMINOETHANE ♦ DIMETHYLENEDIAMINE ♦ ETHYLENEDIAMINE (OSHA)
OSHA PEL: TWA 10 ppm
ACGIH TLV: TWA 10 ppm; Not Classifiable as a Human Carcinogen
DFG MAK: 10 ppm (25 mg/m^3)
DOT Classification: 8; Label: Corrosive, Flammable Liquid
SAFETY PROFILE: A human poison by inhalation. Moderately toxic by ingestion and skin contact. Corrosive. A severe skin and eye irritant. An allergen and sensitizer. Flammable liquid. It can react violently with many substances.
PROP: Volatile, colorless, clear, thick, strongly alkaline, hygroscopic liquid; ammonia-like odor. Mp: 8.5°, bp: 117.2°, flash p: 110°F (CC), d: 0.8994 @ 20°/4°, vap press: 10.7 mm @ 20°, vap d: 2.07, autoign temp: 725°F. Sol in EtOH and H_2O (with hydration); insol in C_6H_6; sltly sol in Et_2O.

EEC600 CAS: 141-43-5 C_2H_7NO HR: 3
ETHANOLAMINE
DOT: UN 2491
SYNS: 2-AMINOETHANOL (MAK) ♦ β-AMINOETHYL ALCOHOL ♦ GLYCINOL ♦ 2-HYDROXYETHYLAMINE ♦ MONOETHANOLAMINE
OSHA PEL: TWA 3 ppm; STEL 6 ppm
ACGIH TLV: TWA 3 ppm; STEL 6 ppm
DFG MAK: 2 ppm (5.1 mg/m^3)
DOT Classification: 8; Label: Corrosive
SAFETY PROFILE: Moderately toxic by ingestion and skin contact. A corrosive irritant to skin, eyes, and mucous membranes. Flammable. It can react violently with many substances.

PROP: Colorless, viscous, hygroscopic liquid with ammonia-like odor. Bp: 170.5°, fp: 10.5°, flash p: 200°F (OC), d: 1.012 @ 25°/4°, vap press: 6 mm @ 60°, vap d: 2.11. Misc in water and alc; sltly sol in benzene; sol in chloroform.

EES350 CAS: 110-80-5 $C_4H_{10}O_2$ HR: 3

2-ETHOXYETHANOL

DOT: UN 1171

SYNS: CELLOSOLVE (DOT) ♦ ETHYLENE GLYCOL ETHYL ETHER ♦ ETHYLENE GLYCOL MONOETHYL ETHER (DOT) ♦ GLYCOL ETHYL ETHER ♦ GLYCOL MONOETHYL ETHER

OSHA PEL: TWA 200 ppm (skin)

ACGIH TLV: TWA 5 ppm (skin); BEI: 100 mg/g creatinine of 2-ethoxyacetic acid in urine end of shift at end of workweek

DFG MAK: 5 ppm (19 mg/m^3)

NIOSH REL: (Glycol Ethers) Reduce to lowest level

DOT Classification: 3; Label: Flammable Liquid

SAFETY PROFILE: Moderately toxic by ingestion and skin contact. Mildly toxic by inhalation. An experimental teratogen. A mild eye and skin irritant. Combustible

PROP: Colorless liquid; practically odorless. Bp: 135.1°, lel: 1.8%, uel: 14%, fp: –70°, flash p: 202°F (CC), d: 0.9360 @ 15°/15°, autoign temp: 455°F, vap press: 3.8 mm @ 20°, vap d: 3.10. Misc in H_2O, EtOH, Et_2O, and Me_2CO.

EES400 CAS: 111-15-9 $C_6H_{12}O_3$ HR: 3

2-ETHOXYETHYL ACETATE

DOT: UN 1172

SYNS: CELLOSOLVE ACETATE (DOT) ♦ CSAC ♦ 2-ETHOXYETHANOL ACETATE ♦ ETHYLENE GLYCOL MONOETHYL ETHER ACETATE (MAK, DOT) ♦ GLYCOL MONOETHYL ETHER ACETATE

OSHA PEL: TWA 100 ppm (skin)

ACGIH TLV: TWA 5 ppm (skin); BEI: 100 mg/g creatinine of 2-ethoxyacetic acid in urine end of shift at end of workweek

DFG MAK: 5 ppm (27 mg/m^3)

DOT Classification: 3; Label: Flammable Liquid

SAFETY PROFILE: Moderately toxic by ingestion. A skin and eye irritant. An experimental teratogen. Flammable.

PROP: Colorless liquid with a mild, pleasant, ester-like odor. Mp: –61°, bp: 156.4°, flash p: 117°F (COC), lel: 1.7%, fp: –61.7°, d: 0.9748 @ 20°/20°, autoign temp: 715°F, vap press: 1.2 mm @ 20°, vap d: 4.72.

EFR000 CAS: 141-78-6 $C_4H_8O_2$ HR: 3

ETHYL ACETATE

DOT: UN 1173

SYNS: ACETOXYETHANE ♦ ETHYL ACETIC ESTER ♦ ETHYL ETHANOATE ♦ VINEGAR NAPHTHA

OSHA PEL: TWA 400 ppm

ACGIH TLV: TWA 400 ppm; Not Classifiable as a Human Carcinogen

DFG MAK: 400 ppm (1500 mg/m^3)
DOT Classification: 3; Label: Flammable Liquid
SAFETY PROFILE: Poison by inhalation. Mildly toxic by ingestion. Human eye irritant. It can cause dermatitis. Highly flammable liquid. It can react violently with many substances.
PROP: A volatile, flammable, colorless liquid with fragrant fruity odor. Mp: –83.6°, bp: 77.15°, ULC: 85–90, lel: 2.2%, uel: 11%, flash p: 24°F, d: 0.8946 @ 25°, autoign temp: 800°F, vap press: 100 mm @ 27.0°, vap d: 3.04. Misc with alc, ether, glycerin, volatile oils, water @ 54°, and most org solvs.

EFT000 CAS: 140-88-5 $C_5H_8O_2$ HR: 3

ETHYL ACRYLATE

DOT: UN 1917
SYNS: ACRYLIC ACID ETHYL ESTER ♦ ETHYL PROPENOATE ♦ 2-PROPENOIC ACID, ETHYL ESTER (MAK)
OSHA PEL: TWA 5 ppm; STEL 25 ppm (skin)
ACGIH TLV: TWA 5 ppm; STEL 15 ppm; Suspected Human Carcinogen
DFG MAK: 5 ppm (21 mg/m^3)
DOT Classification: 3; Label: Flammable Liquid
SAFETY PROFILE: Confirmed carcinogen. Poison by ingestion and inhalation. Moderately toxic by skin contact. A skin and eye irritant. Flammable liquid.
PROP: Colorless liquid; acrid, penetrating odor. Mp: –71.2°, bp: 99.8°, fp: <–72°, lel: 1.8%, flash p: 60°F (OC), d: 0.916–0.919, vap press: 29.3 mm @ 20°, vap d: 3.45. Misc with alc, ether; sltly sol in water.

EFU000 CAS: 64-17-5 C_2H_6O HR: 3

ETHYL ALCOHOL

DOT: UN 1170/UN 1986/UN 1987
SYNS: ALCOHOLS, n.o.s. (UN 1987) (DOT) ♦ ETHANOL (MAK) ♦ GRAIN ALCOHOL
OSHA PEL: TWA 1000 ppm
ACGIH TLV: TWA 1000 ppm; Not Classifiable as a Human Carcinogen
DFG MAK: 500 ppm (960 mg/m^3)
DOT Classification: 3; Label: Flammable Liquid (UN 1987, UN 1170); DOT Class: 3; Label: Flammable Liquid, Poison (UN 1986)
SAFETY PROFILE: Confirmed human carcinogen for ingestion of beverage alcohol. Experimental teratogenic data. Moderately toxic to humans by ingestion. An eye and skin irritant. Flammable liquid. It can react violently with many substances.
PROP: Clear, colorless, very mobile liquid; fragrant odor and burning taste. Bp: 78.32°, ULC: 70, lel: 3.3%, uel: 19% @ 60°, fp: –117°, flash p: 55.6°F, d: 0.7893 @ 20°/4°, autoign temp: 793°F, vap press: 40 mm @ 19°, vap d: 1.59, refr index: 1.364. Misc in water, alc, chloroform, ether, and most org solvs.

EFU400 CAS: 75-04-7 C_2H_7N HR: 3

ETHYLAMINE

DOT: UN 1036/UN 2270
SYNS: 1-AMINOETHANE ♦ MONOETHYLAMINE (DOT)

OSHA PEL: TWA 10 ppm
ACGIH TLV: TWA 5 ppm; 15 ppm STEL (skin)
DFG MAK: 5 ppm (9.4 mg/m^3)
DOT Classification: 2.1; Label: Flammable Gas (UN 1036); DOT Class: 3; Label: Flammable Liquid (UN 2270)
SAFETY PROFILE: A poison by ingestion and skin contact. Moderately toxic by inhalation. A severe eye irritant. A very dangerous fire hazard.
PROP: Colorless gas or liquid; strong ammonia-like odor. Bp: 16.6°, flammable, lel: 4.95%, uel: 20.75%, fp: –80.6°, flash p: –0.4°F, d: 0.662 @ 20°/4°, autoign temp: 725°F, vap d: 1.56, vap press: 400 mm @ 20°. Misc with water, alc, and ether; salted out by NaOH.

EGI750 CAS: 541-85-5 $C_8H_{16}O$ HR: 3

ETHYL AMYL KETONE

SYNS: 3-METHYL-5-HEPTANONE ♦ 5-METHYL-3-HEPTANONE
OSHA PEL: TWA 25 ppm
ACGIH TLV: TWA 25 ppm
DOT Classification: 3; Label: Flammable Liquid
SAFETY PROFILE: Moderately irritating to skin, eyes, and mucous membranes by inhalation and ingestion. Flammable liquid.
PROP: Liquid; mild, fruity odor. Bp: 157–162°, d: 0.822 @ 20°/20°, flash p: 138°F. Sol in many org solvs.

EGP500 CAS: 100-41-4 C_8H_{10} HR: 3

ETHYL BENZENE

DOT: UN 1175
SYNS: ETHYLBENZOL ♦ PHENYLETHANE
OSHA PEL: TWA 100 ppm; STEL 125 ppm
ACGIH TLV: TWA 100 ppm; STEL 125 ppm; (Proposed: TWA 100 ppm; STEL 125 ppm; Confirmed Animal Carcinogen with Unknown Revelance to Humans) BEI: 1.5 g/g creatinine of manelic acid) in urine at end of shift at end of workweek
DFG MAK: 100 ppm (440 mg/m^3)
NIOSH REL: (Ethyl Benzene) TWA 100 ppm; STEL 125 ppm
DOT Classification: 3; Label: Flammable Liquid
SAFETY PROFILE: Moderately toxic by ingestion. Mildly toxic by inhalation and skin contact. An experimental teratogen. An eye and skin irritant. A very dangerous fire and explosion hazard.
PROP: Colorless liquid; aromatic odor. Bp: 136.2°, fp: –94.9°, flash p: 59°F, d: 0.8669 @ 20°/4°, autoign temp: 810°F, vap press: 10 mm @ 25.9°, vap d: 3.66, lel: 1.2%, uel: 6.8%. Misc in alc and ether; insol in NH_3; sol in SO_2.

EGV400 CAS: 74-96-4 C_2H_5Br HR: 3

ETHYL BROMIDE

DOT: UN 1891
SYNS: BROMIC ETHER ♦ BROMOETHANE ♦ HYDROBROMIC ETHER ♦ MONOBROMOETHANE

OSHA PEL: TWA 200 ppm; STEL 250 ppm
ACGIH TLV: TWA 5 ppm (skin); Animal Carcinogen
DFG MAK: Animal Carcinogen, Suspected Human Carcinogen
DOT Classification: 6.1; Label: Poison
SAFETY PROFILE: Confirmed carcinogen. Moderately toxic by ingestion. Mildly toxic by inhalation. An eye and skin irritant. Dangerously flammable.
PROP: Colorless, volatile liquid. Mp: –119°, bp: 38.4°, fp: –125.5°, lel: 6.7%, uel: 11.3%, flash p: <–4°F, d: 1.451 @ 20°/4°, autoign temp: 952°F, vap press: 400 mm @ 21°, vap d: 3.76.

EHA600 CAS: 106-35-4 $C_7H_{14}O$ HR: 3
ETHYL BUTYL KETONE
SYNS: n-BUTYL ETHYL KETONE ♦ 3-HEPTANONE
OSHA PEL: TWA 50 ppm
ACGIH TLV: TWA 50 ppm; STEL 75 ppm
DOT Classification: 3; Label: Flammable Liquid
SAFETY PROFILE: Moderately toxic by ingestion and inhalation. A skin and eye irritant. A flammable liquid.
PROP: Clear mobile liquid; fatty odor. Mp: –36.7°, bp: 149–152°, flash p: 115°F (OC), d: 0.8198 @ 20°/20°, vap d: 3.93. Misc with alc, ether, water @ 149°.

EHH000 CAS: 75-00-3 C_2H_5Cl HR: 3
ETHYL CHLORIDE
DOT: UN 1037
SYNS: HYDROCHLORIC ETHER ♦ MONOCHLORETHANE ♦ MURIATIC ETHER
OSHA PEL: TWA 1000 ppm
ACGIH TLV: TWA 1000 ppm
DFG MAK: Confirmed Animal Carcinogen with Unknown Relevance to Humans
NIOSH REL: (Chloroethane) Handle with caution
DOT Classification: 2.1; Label: Flammable Gas
SAFETY PROFILE: Suspected carcinogen. Mildly toxic by inhalation. An irritant to skin, eyes, and mucous membranes. A very dangerous fire hazard.
PROP: Colorless liquid or gas which is volatile at room temp; ether-like odor, burning taste. Bp: 12.3°, lel: 3.8%, uel: 15.4%, fp: –142.5°, flash p: –58°F (CC), d: 0.917 @ 6°/6°, autoign temp: 966°F, vap press: 1000 mm @ 20°, vap d: 2.22; misc in alc and ether. Sltly sol in water.

EIK000 CAS: 78-78-4 C_5H_{12} HR: 3
ETHYLDIMETHYLMETHANE
DOT: UN 1265
SYNS: ISOAMYLHYDRIDE ♦ ISOPENTANE (DOT) ♦ 2-METHYLBUTANE
ACGIH TLV: TWA 600 ppm
DFG MAK: 1000 ppm (3000 mg/m^3)
NIOSH REL: (Alkanes) TWA 350 mg/m^3
DOT Classification: 3; Label: Flammable Liquid
SAFETY PROFILE: Mildly toxic and narcotic by inhalation. Flammable liquid.

PROP: Colorless liquid with pleasant odor. Fp: –160.5°, bp: 30–30.2°, flash p: <–60°F (CC), d: 0.620 @ 20°/4°, vap press: 595 mm @ 21.1°, vap d: 2.48, lel: 1.4%, uel: 7.6%.

EIU800 CAS: 107-07-3 C_2H_5ClO HR: 3

ETHYLENE CHLOROHYDRIN

DOT: UN 1135

SYNS: 2-CHLOROETHANOL (MAK) ♦ 2-CHLOROETHYL ALCOHOL ♦ GLYCOL MONOCHLOROHYDRIN ♦ 2-MONOCHLOROETHANOL

OSHA PEL: CL 1 ppm (skin)

ACGIH TLV: CL 1 ppm (skin); Not Classifiable as a Human Carcinogen

DFG MAK: 1 ppm (3.3 mg/m^2)

DOT Classification: 6.1; Label: Poison

SAFETY PROFILE: A poison by ingestion, inhalation, and skin contact. An experimental teratogen. A severe eye and mild skin irritant. Flammable liquid.

PROP: Colorless liquid; faint, ethereal odor. Mp: –69°, fp: –67.5°, bp: 128.8°, flash p: 140°F (OC), d: 1.197 @ 20°/4°, autoign temp: 797°F, vap press: 10 mm @ 30.3°, vap d: 2.78, lel: 4.9%, uel: 15.9%. Misc in water.

EIY500 CAS: 106-93-4 $C_2H_4Br_2$ HR: 3

1,2-ETHYLENE DIBROMIDE

DOT: UN 1605

SYNS: DBE ♦ 1,2-DIBROMOETHANE (MAK) ♦ GLYCOL DIBROMIDE

OSHA PEL: TWA 20 ppm; CL 30 ppm; Pk 50 ppm/5M/8H

ACGIH TLV: Animal Carcinogen

DFG TRK: Animal Carcinogen, Suspected Human Carcinogen

NIOSH REL: (EDB) 0.045 ppm; CL 1 mg/m^3/15M

DOT Classification: 6.1; Label: Poison

SAFETY PROFILE: Confirmed carcinogen. An experimental teratogen. Human poison by ingestion. Moderately toxic by inhalation. A severe skin and eye irritant.

PROP: Colorless, heavy liquid; sweet odor. Bp: 131.4°, fp: 9.3°, flash p: none, d: 2.178 @ 20°/4, mp: 10°, vap press: 17.4 mm @ 30°, vap d: 6.48.

EIY600 CAS: 107-06-2 $C_2H_4Cl_2$ HR: 3

ETHYLENE DICHLORIDE

DOT: UN 1184

SYNS: 1,2-DCE ♦ 1,2-DICHLOROETHANE ♦ EDC ♦ GLYCOL DICHLORIDE

OSHA PEL: TWA 1 ppm; STEL 2 ppm

ACGIH TLV: TWA 10 ppm; Not Classifiable as a Human Carcinogen

DFG MAK: Confirmed Animal Carcinogen, Suspected Human Carcinogen

NIOSH REL: (Ethylene Dichloride) TWA 1 ppm; CL 2 ppm/15M

DOT Classification: 3; Label: Flammable Liquid, Poison

SAFETY PROFILE: Confirmed carcinogen. A human poison by ingestion. Moderately toxic by inhalation and skin contact. Experimental teratogenic effects. A skin and severe eye irritant. Flammable liquid.

PROP: Colorless, clear liquid; pleasant odor, sweet taste. Bp: 83.5°, ULC: 60–70, lel: 6.2%, uel: 15.9%, fp: –35.7°, flash p: 56°F, d: 1.257 @ 20°/4°, autoign temp: 775°F, vap press: 100 mm @ 29.4°, vap d: 3.35, refr index: 1.445 @ 20°. Sol in alc, ether, acetone, carbon tetrachloride; sltly sol in water.

EJC500 CAS: 107-21-1 $C_2H_6O_2$ HR: 3

ETHYLENE GLYCOL

SYNS: 1,2-DIHYDROXYETHANE ♦ 1,2-ETHANEDIOL ♦ ETHYLENE ALCOHOL ♦ MONOETHYLENE GLYCOL

OSHA PEL: CL 50 ppm

ACGIH TLV: CL 50 ppm (vapor)

DFG MAK: 10 ppm (26 mg/m^3)

SAFETY PROFILE: Human poison by ingestion. An experimental teratogen. A skin, eye, and mucous membrane irritant. Combustible. It can react violently with many substances.

PROP: Colorless, sweet-tasting, hygroscopic, viscid, poisonous liquid. Fp: –13°, mp: –15.6°, bp: 197.5°, lel: 3.2%, flash p: 232°F (CC), d: 1.113 @ 25°/25°, autoign temp: 752°F, vap d: 2.14, vap press: 0.05 mm @ 20°. Misc in H_2O, EtOH, MeOH, Me_2CO, AcOH, and Py. Immisc in $CHCl_3$, CCl_4, Et_2O, C_6H_6, CS_2, and ligroin.

EJG000 CAS: 628-96-6 $C_2H_4N_2O_6$ HR: 3

ETHYLENE GLYCOL DINITRATE

SYNS: DINITROGLYCOL ♦ GLYCOL DINITRATE

OSHA PEL: STEL 0.1 mg/m^3 (skin)

ACGIH TLV: TWA 0.05 ppm (skin)

DFG MAK: 0.05 ppm (0.32 mg/m^3)

NIOSH REL: (Nitroglycerin) CL 0.1 mg/m^3/20M

DOT Classification: Forbidden

SAFETY PROFILE: Can cause lowered blood pressure leading to headache, dizziness, and weakness. Used as an explosive.

PROP: Yellow liquid. Mp: –22.3°, bp: 105.5° @ 19 mm, explodes @ 114°, d: 1.483 @ 8°, vap d: 5.25.

EJH500 CAS: 109-86-4 $C_3H_8O_2$ HR: 3

ETHYLENE GLYCOL METHYL ETHER

DOT: UN 1188

SYNS: EGM ♦ EGME ♦ ETHYLENE GLYCOL MONOMETHYL ETHER (MAK, DOT) ♦ GLYCOL MONOMETHYL ETHER ♦ 2-METHOXYETHANOL (ACGIH) ♦ METHYL CELLOSOLVE (OSHA, DOT) ♦ MONOMETHYL ETHER of ETHYLENE GLYCOL

OSHA PEL: TWA 25 ppm (skin)

ACGIH TLV: TWA 5 ppm (skin)

DFG MAK: 5 ppm (16 mg/m^3)

NIOSH REL: TWA (Glycol Ethers) Reduce to lowest level

DOT Classification: 3; Label: Flammable Liquid

SAFETY PROFILE: Moderately toxic to humans by ingestion. Experimental teratogenic effects. A skin and eye irritant. Flammable liquid when exposed to heat or flame. A moderate explosion hazard.
PROP: Colorless liquid; mild, agreeable odor. Misc in water, alc, ether, benzene. Bp: 124.5°, fp: –86.5°, flash p: 115°F (OC), lel: 2.5%, uel: 14%, d: 0.9660 @ 20°/4°, autoign temp: 545°F, vap press: 6.2 mm @ 20°, vap d: 2.62.

EJJ500 CAS: 110-49-6 $C_5H_{10}O_3$ HR: 3
ETHYLENE GLYCOL MONOMETHYL ETHER ACETATE
DOT: UN 1189
SYNS: ETHYLENE GLYCOL METHYL ETHER ACETATE ♦ GLYCOL MONOMETHYL ETHER ACETATE ♦ 2-METHOXYETHYL ACETATE (ACGIH) ♦ METHYL CELLOSOLVE ACETATE (OSHA, DOT) ♦ METHYL GLYCOL ACETATE ♦ METHYL GLYCOL MONOACETATE
OSHA PEL: TWA 25 ppm (skin)
ACGIH TLV: TWA 5 ppm (skin)
DFG MAK: 5 ppm (25 mg/m^3)
DOT Classification: 3; Label: Flammable Liquid
SAFETY PROFILE: Moderately toxic by ingestion. Mildly toxic by inhalation and skin contact. An inhalation irritant in humans. An eye irritant. Flammable liquid.
PROP: Colorless liquid; pleasant, sweet, ether odor. Bp: 143°, fp: –70°, flash p: 111°F (CC), d: 1.005 @ 20°/20°, vap d: 4.07, lel: 1.7%, uel: 8.2%. Sol in water.

EJM900 CAS: 151-56-4 C_2H_5N HR: 3
ETHYLENEIMINE
DOT: UN 1185
SYNS: AZIRANE ♦ DIHYDROAZIRENE ♦ DIMETHYLENEIMINE ♦ ETHYLIMINE
OSHA PEL: TWA 1 mg/m^3 (skin); Cancer Suspect Agent
ACGIH TLV: TWA 0.5 ppm (skin); Animal Carcinogen
DFG TRK: Animal Carcinogen, Suspected Human Carcinogen
NIOSH REL: (Ethyleneimine) TWA use 29 CFR 1910.1012
DOT Classification: 6.1; Label: Poison, Flammable Liquid
SAFETY PROFILE: Confirmed carcinogen. An experimental teratogen. Poison by ingestion, skin contact, and inhalation. A skin, mucous membrane, and severe eye irritant. An allergic sensitizer of skin. A very dangerous fire hazard. It can react violently with many substances.
PROP: Oily, water-white liquid. Pungent ammonia-like odor. Bp: 55–56°, fp: –71.5°, flash p: 12°F, d: 0.832 @ 20°/4°, autoign temp: 608°F, vap press: 160 mm @ 20°, vap d: 1.48, lel: 3.6%, uel: 46%. Misc in water.

EJN500 CAS: 75-21-8 C_2H_4O HR: 3
ETHYLENE OXIDE
DOT: UN 1040
SYNS: AMPROLENE ♦ DIMETHYLENE OXIDE ♦ 1,2-EPOXYETHANE ♦ ETHENE OXIDE ♦ ETO ♦ OXACYCLOPROPANE ♦ OXYFUME
OSHA PEL: TWA 1 ppm; Cancer Hazard

ACGIH TLV: TWA 1 ppm; Suspected Human Carcinogen
DFG TRK: Animal Carcinogen, Suspected Human Carcinogen
NIOSH REL: (Ethylene Oxide) TWA 0.1 ppm; CL 5 ppm/10M/D
DOT Classification: 2.3; Label: Poison Gas, Flammable Gas
SAFETY PROFILE: Confirmed human carcinogen. An experimental teratogen. Poison by ingestion. Moderately toxic by inhalation. A skin and eye irritant. Highly flammable liquid or gas. It can react violently with many substances.
PROP: Colorless gas at room temperature. Mp: –111.3°, bp: 10.7°, ULC: 100, lel: 3.0%, uel: 100%, flash p: –4°F, d: 0.8711 @ 20°/20°, autoign temp: 804°F, vap press: 1095 mm @ 20°, vap d: 1.52. Misc in water and alc; very sol in ether.

EJU000 CAS: 60-29-7 $C_4H_{10}O$ HR: 3
ETHYL ETHER
DOT: UN 1155
SYNS: ANESTHETIC ETHER ♦ DIETHYL ETHER (DOT) ♦ DIETHYL OXIDE
OSHA PEL: TWA 400 ppm; STEL 500 ppm
ACGIH TLV: TWA 400 ppm; STEL 500 ppm
DFG MAK: 400 ppm (1200 mg/m^3)
DOT Classification: 3; Label: Flammable Liquid
SAFETY PROFILE: Moderately toxic by ingestion. Mildly toxic by inhalation. A severe eye and moderate skin irritant. A very dangerous fire hazard. It can react violently with many substances.
PROP: A clear, volatile liquid; sweet, pungent odor. Mp: –116.2°, bp: 34.6°, ULC: 100, lel: 1.85%, uel: 36%, flash p: –49°F, d: 0.7135 @ 20°/4°, autoign temp: 320°F, vap press: 442 mm @ 20°, vap d: 2.56. Sol in H_2SO_4; sltly sol in H_2O; misc in most org solvs.

EKL000 CAS: 109-94-4 $C_3H_6O_2$ HR: 3
ETHYL FORMATE
DOT: UN 1190
SYNS: ETHYL FORMIC ESTER ♦ ETHYL METHANOATE ♦ FORMIC ACID, ETHYL ESTER ♦ FORMIC ETHER
OSHA PEL: TWA 100 ppm
ACGIH TLV: TWA 100 ppm
DFG MAK: 100 ppm (310 mg/m^3)
DOT Classification: 3; Label: Flammable Liquid
SAFETY PROFILE: Moderately toxic by ingestion. Mildly toxic by skin contact and inhalation. A skin and eye irritant. Highly flammable liquid.
PROP: Colorless, mobile flammable liquid; sharp, pleasant, rum-like odor. Mp: –79°, bp: 54.3°, lel: 2.7%, uel: 13.5%, flash p: –4°F (CC), d: 0.9236 @ 20°/20°, refr index: 1.359, autoign temp: 851°F, vap press: 100 mm @ 5.4°, vap d: 2.55. Misc in EtOH, Et_2O, C_6H_6; sltly sol in and gradually hydrated by H_2O.

ELO500 CAS: 16219-75-3 C_9H_{12} HR: 2
ETHYLIDENE NORBORNENE
SYNS: 5-ETHYLIDENE-2-NORBORNENE

OSHA PEL: CL 5 ppm
ACGIH TLV: CL 5 ppm
SAFETY PROFILE: Moderately toxic by ingestion. Mildly toxic by inhalation and skin contact. A skin irritant.
PROP: Bp: 70.2–70.4° @ 58 mm.

ENL000 CAS: 100-74-3 $C_6H_{13}NO$ HR: 3
N-ETHYLMORPHOLINE
SYNS: 4-ETHYLMORPHOLINE ♦ NEM
OSHA PEL: TWA 5 ppm (skin)
ACGIH TLV: TWA 5 ppm (skin)
SAFETY PROFILE: Moderately toxic by ingestion. Mildly toxic by inhalation. A skin and severe eye irritant. A very dangerous fire hazard.
PROP: Colorless liquid. Bp: 138°, flash p: 89.6°F (OC), d: 0.916 @ 20°/20°, vap d: 4.00.

EPF550 CAS: 78-10-4 $C_8H_{20}O_4Si$ HR: 3
ETHYL SILICATE
DOT: UN 1292
SYNS: ETHYL ORTHOSILICATE ♦ TETRAETHYL ORTHOSILICATE (DOT) ♦ TETRAETHYL SILICATE (DOT)
OSHA PEL: TWA 10 ppm
ACGIH TLV: TWA 10 ppm
DFG MAK: 10 ppm (86 mg/m^3)
DOT Classification: 3; Label: Flammable Liquid
SAFETY PROFILE: A skin, mucous membrane, and severe eye irritant. Flammable liquid.
PROP: Colorless liquid. Mp: –77°, bp: 165–166°, flash p: 125°F (52°C), d: 0.933 @ 20°/4°, n: (25/D) 1.3818. Viscosity 0.6 cps. Practically insol in water with slow decomp. Miscible with alc.

FAU000 CAS: 7705-08-0 Cl_3Fe HR: 3
FERRIC CHLORIDE
DOT: UN 1773/UN 2582
SYNS: IRON CHLORIDE ♦ IRON TRICHLORIDE
OSHA PEL: TWA 1 $mg(Fe)/m^3$
ACGIH TLV: TWA 1 $mg(Fe)/m^3$
DOT Classification: 8; Label: Corrosive
SAFETY PROFILE: Poison by ingestion. Corrosive. Probably an eye, skin, and mucous membrane irritant.
PROP: Black-brown solid or hygroscopic dark-green or black crystals. Mp: 303°, bp: 315°, d: 2.90 @ 25°, vap press: 1 mm @ 194.0°. Aq solns are strongly acidic. Sol in H_2O to give hydrates; sol in MeOH and Et_2O.

FBC000 CAS: 102-54-5 $C_{10}H_{10}Fe$ HR: 3

FERROCENE

SYNS: BISCYCLOPENTADIENYLIRON ♦ DICYCLOPENTADIENYL IRON (OSHA, ACGIH) ♦ IRON BIS(CYCLOPENTADIENE)

OSHA PEL: TWA Total Dust: 10 mg/m^3; Respirable Fraction: 5 mg/m^3

ACGIH TLV: TWA 10 mg/m^3

SAFETY PROFILE: Moderately toxic by ingestion. Flammable.

PROP: Orange crystals from alc (aq); camphor odor. Mp: 172.5–173°, bp: 249°, subl @ >100°, volatile in steam. Insol in water; sol in alcohol and ether.

FBI000 CAS: 7758-94-3 Cl_2Fe HR: 3

FERROUS CHLORIDE

DOT: UN 1759/UN 1760

SYNS: IRON DICHLORIDE ♦ IRON PROTOCHLORIDE

OSHA PEL: TWA 1 $mg(Fe)/m^3$

ACGIH TLV: TWA 1 $mg(Fe)/m^3$

DOT Classification: 8; Label: Corrosive

SAFETY PROFILE: Poison by ingestion. Corrosive. Probably an irritant to the eyes, skin, and mucous membranes.

PROP: White crystals when pure; hygroscopic. Green to yellow, deliquescent crystals. Mp: 676°, bp: 1012°, d: 3.16, vap press: 10 mm @ 700°. Sol in H_2O; insol in Et_2O; sltly sol in C_6H_6.

FBK000 CAS: 299-29-6 $C_{12}H_{22}O_{14}{\bullet}Fe$ HR: 3

FERROUS GLUCONATE

SYNS: IRON GLUCONATE

OSHA PEL: TWA 1 $mg(Fe)/m^3$

ACGIH TLV: TWA 1 $mg(Fe)/m^3$

SAFETY PROFILE: Moderately toxic by ingestion.

PROP: Yellowish-gray or pale-greenish-yellow, fine powder or granules with slt odor of burned sugar. Sol in water and glycerin; insol in alc.

FBN100 CAS: 7720-78-7 $O_4S{\bullet}Fe$ HR: 3

FERROUS SULFATE

SYNS: FERROSULFATE ♦ IRON MONOSULFATE ♦ IRON(II) SULFATE (1:1)

OSHA PEL: TWA 1 $mg(Fe)/m^3$

ACGIH TLV: TWA 1 $mg(Fe)/m^3$

SAFETY PROFILE: A human poison by ingestion. Experimental teratogenic effects.

PROP: Grayish white to buff powder. Slowly sol in water; insol in alc.

FBP000 CAS: 12604-58-9 HR: 2

FERROVANADIUM DUST

OSHA PEL: TWA 1 mg/m^3; STEL 3 mg/m^3

ACGIH TLV: TWA 1 mg/m^3; STEL 3 mg/m^3

DFG MAK: 1 mg/m^3

NIOSH REL: (Vanadium) TWA 1.0 $mg(V)/m^3$

SAFETY PROFILE: Can cause pulmonary damage. Combustible.

PROP: A gray to black dust.

FBQ000 HR: 2

FIBROUS GLASS

SYNS: FIBERGLASS ♦ FIBROUS GLASS DUST (ACGIH) ♦ GLASS FIBERS

OSHA PEL: TWA 15 mg/m^3 (total dust); 5 mg/m^3 (nuisance dust)

ACGIH TLV: TWA 10 mg/m^3 (dust)

NIOSH REL: TWA 5 mg/m^3 (total fibrous glass)

SAFETY PROFILE: Suspected carcinogen. Exposure to glass fibers sometimes causes irritation of the skin and, less frequently, irritation of the eyes, nose, or throat.

PROP: Is of a borosilicate variety, of low alkalinity, and consists of calcia-alumina-silicate (85INA8 5,270,86).

FDD125 CAS: 16872-11-0 $BF_4 \cdot H$ HR: 2

FLUOBORIC ACID

DOT: UN 1775SYNS: BOROFLUORIC ACID ♦ FLUOBORIC ACID (DOT) ♦ HYDROFLUOBORIC ACID ♦ HYDROGEN TETRAFLUOROBORATE ♦ TETRAFLUOROBORIC ACID

OSHA PEL: TWA 2.5 $mg(F)/m^3$

ACGIH TLV: TWA 2.5 $mg(F)/m^3$; BEI: 3 mg/g creatinine of fluorides in urine prior to shift; 10 mg/g creatinine of fluorides in urine at end of shift.

NIOSH REL: (Fluorides, inorganic) TWA 2.5 $mg(F)/m^3$

DOT Classification: 8; Label: Corrosive

SAFETY PROFILE: A corrosive acid.

FDR000 CAS: 53-96-3 $C_{15}H_{13}NO$ HR: 3

N-FLUOREN-2-YL ACETAMIDE

SYNS: 2-AAF ♦ 2-ACETAMINOFLUORENE ♦ 2-ACETYLAMINOFLUORENE (OSHA) ♦ FAA ♦ 2-FLUORENYLACETAMIDE ♦ N-2-FLUORENYLACETAMIDE

OSHA PEL: Cancer Suspect Agent

NIOSH REL: (2-Acetylaminofluorene) TWA use 29 CFR 1910.1014

SAFETY PROFILE: Confirmed human carcinogen. An experimental teratogen. Moderately toxic by ingestion and routes.

FEY000 HR: 2

FLUORIDES

OSHA PEL: TWA 2.5 $mg(F)/m^3$

ACGIH TLV: TWA 2.5 $mg(F)/m^3$; BEI: 3 mg/g creatinine of fluorides in urine prior to shift; 10 mg/g creatinine of fluorides in urine at end of shift.

DFG MAK: 2.5 mg/m^3; BAT: 7 mg/kg creatinine in urine at end of exposure; 4 mg/kg creatinine in urine about 16 hours after end of exposure
NIOSH REL: TWA 2.5 mg(F)/m^3
SAFETY PROFILE: Inorganic fluorides are generally highly irritating and toxic. Irritants to the eyes, skin, and mucous membranes. Organic fluorides are generally less toxic than other halogenated hydrocarbons. Fluorocarbons are chemically inert to most materials.

FEZ000 CAS: 7782-41-4 F_2 HR: 3
FLUORINE
DOT: UN 1045
OSHA PEL: TWA 0.1 ppm
ACGIH TLV: TWA 1 ppm; STEL 2 ppm
DFG MAK: 0.1 ppm (0.16 mg/m^3)
DOT Classification: 2.3; Label: Poison Gas, Oxidizer
SAFETY PROFILE: A poison gas. A most powerful caustic irritant to tissue. A very dangerous fire and explosion hazard. A powerful oxidizer. It can react violently with many substances.
PROP: Pale-yellow gas (turning white at –2°) which reacts with most organic and inorganic materials. Powerful oxidant. Mp: –218°, bp: –187°, d: 1.14 @ –200°, 1.108 @ –188°, vap d: 1.695.

FMV000 CAS: 50-00-0 CH_2O HR: 3
FORMALDEHYDE
DOT: UN 1198/UN 2209
SYNS: FORMALIN (DOT) ♦ METHYL ALDEHYDE
OSHA PEL: TWA 0.75 ppm; STEL 2 ppm
ACGIH TLV: TWA 1 ppm; Suspected Human Carcinogen (Proposed: CL 0.3 ppm (skin, sensitizer); Suspected Human Carcinogen)
DFG MAK: 0.5 ppm (0.6 mg/m^3); Confirmed Animal Carcinogen with Unknown Relevance to Humans
NIOSH REL: (Formaldehyde) Limit to lowest feasible level
DOT Classification: 9; Label: None (UN 2209); DOT Class: 3; Label: Flammable Liquid (UN 1198)
SAFETY PROFILE: Confirmed carcinogen. An experimental teratogen. A poison by ingestion, skin contact, inhalation. A human skin and eye irritant. Flammable liquid.
PROP: Clear, water-white, very sltly acid gas or liquid; pungent odor. Pure formaldehyde is not available commercially because of its tendency to polymerize. It is sold as aqueous solns containing from 37 to 50% formaldehyde by weight and varying amounts of methanol. Some alcoholic solns are used industrially, and the physical properties and hazards may be greatly influenced by the solvent. Lel: 7.0%, uel: 73.0%, autoign temp: 806°F, mp: –92°, d: 1.083, bp: –21°, flash p: (37%, methanol-free) 185°F, flash p: (15%, methanol-free) 122°F. Sol in H_2O and most org solvs except pet ether.

FMY000 CAS: 75-12-7 CH_3NO HR: 3

FORMAMIDE

SYNS: CARBAMALDEHYDE ♦ METHANAMIDE
OSHA PEL: TWA 20 ppm; STEL 30 ppm
ACGIH TLV: TWA 10 ppm (skin)
SAFETY PROFILE: Poison by skin contact. Moderately toxic by ingestion. An irritant to skin, eyes, and mucous membranes. Experimental teratogenic effects. Combustible.
PROP: Colorless, odorless, hygroscopic, sltly viscous, oily liquid. Mp: 2.5°, fp: 2.6°, vap press: 29.7 mm @ 129.4°, flash p: 310°F (COC), bp: 70.5° @ 1 mm, d: 1.134 @ 20°/40°, 1.1292 @ 25°/4°. Misc in H_2O, MeOH; very sltly sol in Et_2O, C_6H_6; insol in $CHCl_3$, hexane.

FNA000 CAS: 64-18-6 CH_2O_2 HR: 3

FORMIC ACID

DOT: UN 1779
SYNS: FORMYLIC ACID ♦ METHANOIC ACID
OSHA PEL: TWA 5 ppm
ACGIH TLV: 5 ppm; STEL 10 ppm
DFG MAK: 5 ppm (9.5 mg/m^3)
DOT Classification: 8; Label: Corrosive
SAFETY PROFILE: Poison by inhalation. Moderately toxic by ingestion. Corrosive. A skin and severe eye irritant. Combustible liquid.
PROP: Colorless, fuming liquid; pungent, penetrating odor. Bp: 100.8°, fp: 8.2°, flash p: 156°F (OC), d: 1.220 @ 20°/4°, 1.220 @ 20°/4°, mp: 8.4°, autoign temp: 1114°F, vap press: 40 mm @ 24.0°, vap d: 1.59, flash p: (90% soln) 122°F, autoign temp: (90% soln) 813°F, lel: (90% soln) 18%, uel: (90% soln) 57%. Misc in H_2O, EtOH, Et_2O; mod sol in C_6H_6.

FOO000 CAS: 76-13-1 $C_2Cl_3F_3$ HR: 2

FREON 113

SYNS: REFRIGERANT 113 ♦ TRICHLOROTRIFLUOROETHANE ♦ 1,1,2-TRICHLORO-1,2,2-TRIFLUOROETHANE (OSHA, ACGIH, MAK)
OSHA PEL: TWA 1000 ppm; STEL 1250 ppm
ACGIH TLV: TWA 1000 ppm; STEL 1250 ppm; Not Classifiable as a Human Carcinogen
DFG MAK: 500 ppm (3900 mg/m^3)
SAFETY PROFILE: Mildly toxic by ingestion and inhalation. A skin irritant. Combustible.
PROP: Colorless gas. Mp: –36.4°, bp: 45.8°, d: 1.5702, autoign temp: 1256°F.

FOO509 CAS: 76-14-2 $C_2Cl_2F_4$ HR: 1

FREON 114

SYNS: 1,2-DICHLORO-1,1,2,2-TETRAFLUOROETHANE (MAK) ♦ DICHLOROTETRAFLUOROETHANE (OSHA, ACGIH) ♦ FLUOROCARBON 114 ♦ 1,1,2,2-TETRAFLUORO-1,2-DICHLOROETHANE
OSHA PEL: TWA 1000 ppm

ACGIH TLV: TWA 1000 ppm; Not Classifiable as a Human Carcinogen
DFG MAK: 1000 ppm (7100 mg/m^3)
SAFETY PROFILE: An asphyxiant.
PROP: Colorless, practically odorless, noncorrosive, nonirritating, nonflammable gas. Faint, ether-like odor in high concentrations. D: 1.5312, mp: –94°, bp: 4.1°, n: (0/D) 1.3092. Insol in water; sol in alc and ether.

FPQ875 CAS: 98-01-1 $C_5H_4O_2$ HR: 3
FURFURAL
DOT: UN 1199
SYNS: 2-FURALDEHYDE ♦ FURFURALDEHYDE ♦ 2-FURYL-METHANAL
OSHA PEL: TWA 2 ppm (skin)
ACGIH TLV: TWA 2 ppm (skin); Animal Carcinogen: BEI: 200 mg/g creatinine of total furoic acid in urine at end of shift
DFG MAK: Confirmed Animal Carcinogen with Unknown Relevance to Humans
DOT Classification: 3; Label: Flammable Liquid
SAFETY PROFILE: Confirmed carcinogen. Poison by ingestion. Moderately toxic by inhalation and skin contact. A skin and eye irritant. Flammable liquid.
PROP: Colorless–yellowish liquid; almond-like odor. Bp: 161.7° @ 764 mm, lel: 2.1%, uel: 19.3%, flash p: 140°F (CC), d: 1.154–1.158, refr index: 1.522–1.528, autoign temp: 600°F, vap d: 3.31. Sol in water; misc with alc.

FPU000 CAS: 98-00-0 $C_5H_6O_2$ HR: 3
FURFURYL ALCOHOL
DOT: UN 2874
SYNS: 2-FURANMETHANOL ♦ FURYL ALCOHOL ♦ 2-HYDROXYMETHYLFURAN
OSHA PEL: TWA 10 ppm; STEL 15 ppm (skin)
ACGIH TLV: TWA 10 ppm; STEL 15 ppm (skin)
DFG MAK: 10 ppm (41 mg/m^3)
NIOSH REL: (Furfuryl Alcohol) TWA 200 mg/m^3
DOT Classification: 6.1; Label: KEEP AWAY FROM FOOD
SAFETY PROFILE: Poison by ingestion, skin contact. Moderately toxic by inhalation. An eye irritant. Flammable.
PROP: Clear, colorless, mobile liquid. Mp: –31°, lel: 1.8%, uel: 16.3% (between 72 and 122°), bp: 171° @ 750 mm, flash p: 167°F (OC), d: 1.129 @ 20°/4°, autoign temp: 915°F, vap press: 1 mm @ 31.8°, vap d: 3.37. Misc in H_2O; very sol in EtOH and Et_2O.

GBY000 CAS: 8006-61-9 HR: 3
GASOLINE
DOT: UN 1203/UN 1257
OSHA PEL: TWA 300 ppm; STEL 500 ppm
ACGIH TLV: TWA 300 ppm; STEL 500 ppm; Animal Carcinogen
DOT Classification: 3; Label: Flammable Liquid

SAFETY PROFILE: Confirmed carcinogen. Mildly toxic by inhalation. Pulmonary aspiration can cause severe pneumonitis. A human eye irritant. A very dangerous fire and explosion hazard.
PROP: Clear, aromatic, volatile liquid; a mixture of aliphatic hydrocarbons. Flash p: –50°F, d: <1.0, vap d: 3.0–4.0, ULC: 95–100, lel: 1.3%, uel: 6.0%, autoign temp: 536–853°F, bp: initially 39°, after 10% distilled = 60°, after 50% = 110°, after 90% = 170°, final bp: = 204°. Insol in water; freely sol in abs alc, ether, chloroform, and benzene.

GEI100 CAS: 7782-65-2 GeH_4 HR: 3
GERMANIUM TETRAHYDRIDE
DOT: UN 2192
SYNS: GERMANE (DOT) ♦ GERMANIUM HYDRIDE
ACGIH TLV: TWA 0.2 ppm
DOT Classification: 2.3; Label: Poison Gas, Flammable Gas
SAFETY PROFILE: Poison by inhalation. Moderately toxic by ingestion. Ignites spontaneously in air.
PROP: Colorless gas. Mp: –165°, bp: –90°, d: 1.523 @ –142°/4°. Insol in H_2O; sol in aq NH_3, aq NaOCl; sltly sol in hot HCl.

GFQ000 CAS: 111-30-8 $C_5H_8O_2$ HR: 3
GLUTARALDEHYDE
SYNS: GLUTARIC DIALDEHYDE ♦ 1,5-PENTANEDIONE
OSHA PEL: CL 0.2 ppm
ACGIH TLV: CL 0.05 ppm (skin, sensitizer); Not Classifiable as a Human Carcinogen
DFG MAK: 0.1 ppm (0.42 mg/m^3)
SAFETY PROFILE: Poison by ingestion. Moderately toxic by inhalation, skin contact. Experimental teratogenic effects. A severe eye and human skin irritant.
PROP: Oil. Bp: 71–72° @ 10 mm.

GGA000 CAS: 56-81-5 $C_3H_8O_3$ HR: 3
GLYCERIN
SYNS: GLYCERITOL ♦ GLYCEROL ♦ GLYCYL ALCOHOL ♦ 1,2,3-PROPANETRIOL ♦ 1,2,3-TRIHYDROXYPROPANE
OSHA PEL: TWA Total Mist: 10 mg/m^3; Respirable Fraction: 5 mg/m^3
ACGIH TLV: TWA 10 mg/m^3 (vapor)
SAFETY PROFILE: Mildly toxic by ingestion. A skin and eye irritant. Combustible liquid. It can react violently with many substances.
PROP: Colorless or pale-yellow liquid; odorless, syrupy, sweet and warm taste. Mp: 17.9 (solidifies at a much lower temp), bp: 290° (part decomp), ULC: 10–20, flash p: 320°F, d: 1.260 @ 20°/4°, autoign temp: 698°F, vap press: 0.0025 mm @ 50°, vap d: 3.17. Misc in H_2O, EtOH; insol in C_6H_6, $CHCl_3$, and CCl_4.

GGW500 CAS: 556-52-5 $C_3H_6O_2$ HR: 3

GLYCIDOL
SYNS: 2,3-EPOXY-1-PROPANOL (OSHA) ♦ GLYCIDYL ALCOHOL ♦ 3-HYDROXY-1,2-EPOXYPROPANE
OSHA PEL: TWA 25 ppm
ACGIH TLV: TWA 2 ppm; Animal Carcinogen
DFG MAK: 50 ppm (150 mg/m^3)
SAFETY PROFILE: Confirmed carcinogen. Moderately toxic by ingestion, inhalation, and skin contact. Experimental teratogenic effects. A skin irritant. Explodes when heated or in the presence of strong acids, bases, metals.
PROP: Colorless liquid. D: 1.165 @ 0°/4°, bp: 167° (decomp). Entirely sol in water, alc, and ether.

HAR000 CAS: 76-44-8 $C_{10}H_5Cl_7$ HR: 3
HEPTACHLOR
SYNS: 3-CHLOROCHLORDENE ♦ 3,4,5,6,7,8,8-HEPTACHLORODICYCLOPENTADIENE
OSHA PEL: TWA 0.5 mg/m^3 (skin)
ACGIH TLV: 0.05 mg/m^3 (skin); Animal Carcinogen
DFG MAK: 0.5 mg/m^3, Confirmed Animal Carcinogen with Unknown Relevance to Humans
SAFETY PROFILE: Confirmed carcinogen. A poison by ingestion, skin contact.
PROP: Crystals. Mp: 96°. Nearly insol in water; sol in org solvs.

HBC500 CAS: 142-82-5 C_7H_{16} HR: 3
HEPTANE
DOT: UN 1206
SYNS: DIPROPYL METHANE ♦ HEPTYL HYDRIDE
OSHA PEL: TWA 400 ppm; STEL 500 ppm
ACGIH TLV: TWA 400 ppm; STEL 500 ppm
DFG MAK: 500 ppm (2100 mg/m^3)
NIOSH REL: TWA (Alkanes) 350 mg/m^3
DOT Classification: 3; Label: Flammable Liquid
SAFETY PROFILE: Mildly toxic by inhalation. A flammable liquid.
PROP: Colorless liquid. Bp: 98.52°, lel: 1.05%, uel: 6.7%, mp: –91.61°, flash p: 25°F (CC), d: 0.684 @ 20°/4°, autoign temp: 433.4°F, vap press: 40 mm @ 22.3°, vap d: 3.45. Sltly sol in alc; misc in ether and chloroform; insol in water.

HBD500 CAS: 1639-09-4 $C_7H_{16}S$ HR: 3
1-HEPTANETHIOL
DOT: UN 1228/UN 3071
SYNS: HEPTYL MERCAPTAN ♦ n-HEPTYLMERCAPTAN
NIOSH REL: (n-Alkane Mono Thiols) CL 0.5 ppm/15M
DOT Classification: 3; Label: Flammable Liquid, Poison (UN 1228); DOT Class: 6.1; Label: Poison, Flammable Liquid (UN 3071)
SAFETY PROFILE: A poison. Toxic by inhalation. A flammable liquid.
PROP: A liquid with powerful odor. Bp: 173–176°.

HCB000 CAS: 13007-92-6 C_6CrO_6 HR: 3

HEXACARBONYLCHROMIUM

SYNS: CHROMIUM CARBONYL (MAK) ♦ CHROMIUM HEXACARBONYL ♦ HEXACARBONYL CHROMIUM

OSHA PEL: CL 0.1 $mg(CrO_3)/m^3$

ACGIH TLV: TWA 0.05 $mg(Cr)/m^3$; Confirmed Human Carcinogen

DFG TRK: Confirmed Animal Carcinogen with Unknown Relevance to Humans

NIOSH REL: (Chromium(VI)) TWA 0.001 $mg(Cr(VI))/m^3$

SAFETY PROFILE: Confirmed carcinogen. Poison by ingestion.

PROP: Colorless crystals from methylcyclohexane or by sublimation. Mp: 152–155°. Sltly sol in CCl_4; insol in H_2O, EtOH, and Et_2O.

HCC500 CAS: 118-74-1 C_6Cl_6 HR: 3

HEXACHLOROBENZENE

DOT: UN 2729

SYNS: PENTACHLOROPHENYL CHLORIDE ♦ PERCHLOROBENZENE ♦ PHENYL PERCHLORYL

ACGIH TLV: TWA 0.002 (skin); Animal Carcinogen

DFG MAK: Not Classifiable as a Human Carcinogen; BAT: 15 μg/dL in plasma/serum

DOT Classification: 6.1; Label: KEEP AWAY FROM FOOD

SAFETY PROFILE: Confirmed carcinogen. An experimental teratogen. Mildly toxic by inhalation. Combustible.

PROP: Needles from 2-propanaol. Mp: 226°, bp: 323–326°, flash p: 468°F, vap press: 1 mm @ 114.4°, vap d: 9.8, d: 2.44. Insol in water; sol in benzene; very sltly sol in hot alc; sol in hot ether and chloroform.

HCD250 CAS: 87-68-3 C_4Cl_6 HR: 3

HEXACHLOROBUTADIENE

DOT: UN 2279

SYNS: HEXACHLORO-1,3-BUTADIENE (MAK) ♦ PERCHLOROBUTADIENE

OSHA PEL: TWA 0.02 ppm

ACGIH TLV: TWA 0.02 ppm (skin); Animal Carcinogen

DFG MAK: Confirmed Animal Carcinogen with Unknown Relevance to Humans

DOT Classification: 6.1; Label: KEEP AWAY FROM FOOD

SAFETY PROFILE: Suspected carcinogen. Poison by ingestion. Moderately toxic by inhalation and skin contact. A skin and eye irritant. An experimental teratogen. Combustible.

PROP: A liquid. D: 1.682 @ 20°/4°, mp: –21°, bp: 211–215°, autoign temp: 1130°F, vap d: 8.99.

HCE500 CAS: 77-47-4 C_5Cl_6 HR: 3

HEXACHLOROCYCLOPENTADIENE

DOT: UN 2646

SYNS: HCCPD ♦ HEXACHLOROCYCLOPENTADIENE (ACGIH,DOT,OSHA) ♦ PCL ♦ PERCHLOROCYCLOPENTADIENE
OSHA PEL: TWA 0.01 ppm
ACGIH TLV: TWA 0.01 ppm; Not Classifiable as a Human Carcinogen
DOT Classification: 6.1; Label: Poison
SAFETY PROFILE: A deadly poison by inhalation. Moderately toxic by ingestion and skin contact. Experimental teratogenic effects. Corrosive. A severe skin and eye irritant.
PROP: Greenish-yellow to amber-colored liquid with a pungent odor. Fp: –2°, mp: 9.9°, bp: 234°, flash p: none (OC), d: 1.70 @ 25°/4°, vap d: 9.42.

HCI000 CAS: 67-72-1 C_2Cl_6 HR: 3
HEXACHLOROETHANE
SYNS: CARBON HEXACHLORIDE ♦ ETHANE HEXACHLORIDE ♦ PERCHLOROETHANE
OSHA PEL: TWA 1 ppm (skin)
ACGIH TLV: TWA 1 ppm; Suspected Human Carcinogen
DFG MAK: 1 ppm (9.8 mg/m^3)
NIOSH REL: (Hexachloroethane) Reduce to lowest level
SAFETY PROFILE: Confirmed carcinogen. Mildly toxic by ingestion.
PROP: Rhombic, triclinic, or cubic crystals from EtOH/Et_2O; colorless, camphor-like odor. Mp: 186.6° (subl), d: 2.091, vap press: 1 mm @ 32.7°, bp: 186.8° (triple point). Sol in alc, benzene, chloroform, ether, oils; insol in water.

HCK500 CAS: 1335-87-1 $C_{10}H_2Cl_6$ HR: 3
HEXACHLORONAPHTHALENE
OSHA PEL: TWA 0.2 mg/m^3 (skin)
ACGIH TLV: TWA 0.2 mg/m^3 (skin)
NIOSH REL: TWA 0.2 mg/m^3 (skin)
SAFETY PROFILE: A poison by ingestion, skin contact, and inhalation. Absorbed by skin.
PROP: White solid.

HCZ000 CAS: 684-16-2 C_3F_6O HR: 3
HEXAFLUOROACETONE
DOT: UN 2420
OSHA PEL: TWA 0.1 ppm (skin)
ACGIH TLV: TWA 0.1 ppm (skin)
DOT Classification: 2.3; Label: Poison Gas
SAFETY PROFILE: A poison by ingestion. Moderately toxic by inhalation. A poisonous irritant to the skin, eyes, and mucous membranes. An experimental teratogen.
PROP: A colorless, nonflammable gas. D: 1.65 @ 25°, fp: –129°, bp: –26°.

HDE000 CAS: 16940-81-1 F_6HP HR: 3
HEXAFLUOROPHOSPHORIC ACID
DOT: UN 1782
SYN: HYDROGEN HEXAFLUOROPHOSPHATE

OSHA PEL: TWA 2.5 mg(F)/m^3

ACGIH TLV: TWA 2.5 mg(F)/m^3; BEI: 3 mg/g creatinine of fluorides in urine prior to shift; 10 mg/g creatinine of fluorides in urine at end of shift.

NIOSH REL: (Inorganic Fluorides) TWA 2.5 mg(F)/m^3

DOT Classification: 8; Label: Corrosive

SAFETY PROFILE: A poison by all routes. A corrosive irritant to skin, eyes, and mucous membranes.

PROP: Clear oil. Mp: 31°, d: 1.65. Strong aq solns fume in air and gradually decomp.

HEK000 CAS: 680-31-9 $C_6H_{18}N_3OP$ HR: 3

HEXAMETHYLPHOSPHORAMIDE

SYNS: HEMPA ♦ HEXAMETHYLPHOSPHORIC ACID TRIAMIDE (MAK) ♦ HMPA ♦ HMPT ♦ HPT ♦ MEMPA ♦ PHOSPHORIC TRIS(DIMETHYLAMIDE) ♦ TRI(DIMETHYLAMINO)PHOSPHINE OXIDE

ACGIH TLV: Animal Carcinogen, Suspected Human Carcinogen

DFG MAK: Animal Carcinogen, Suspected Human Carcinogen

SAFETY PROFILE: Confirmed carcinogen. Moderately toxic by ingestion, skin contact.

PROP: Clear, colorless, mobile liquid; spicy odor. Mp: 7°, bp: 233°, d: 1.024 @ 25°/25°, vap d: 6.18. Misc in water.

HEN000 CAS: 110-54-3 C_6H_{14} HR: 3

n-HEXANE

DOT: UN 1208

OSHA PEL: TWA 50 ppm

ACGIH TLV: TWA 50 ppm (skin); BEI: 5 mg(2,5-hexanedione)/g creatinine in urine at end of shift; 40 ppm n-hexane in end-exhaled air during shift

DFG MAK: 50 ppm (180 mg/m^3)

NIOSH REL: TWA (Alkanes) 350 mg/m^3

DOT Classification: 3; Label: Flammable Liquid

SAFETY PROFILE: Slightly toxic by ingestion and inhalation. Experimental teratogenic effects. An eye irritant. Flammable liquid.

PROP: Colorless clear liquid; faint odor. Fp: –93.6°, bp: 69°, ULC: 90–95, lel: 1.2%, uel: 7.5%, flash p: –9.4°F, d: 0.655 @ 25°/4°, autoign temp: 437°F, vap press: 100 mm @ 15.8°, vap d: 2.97. Insol in water; misc in chloroform, ether, alc. Very volatile liquid.

HEO000 CAS: 124-09-4 $C_6H_{16}N_2$ HR: 3

1,6-HEXANEDIAMINE

DOT: UN 1783/UN 2280

SYNS: 1,6-DIAMINOHEXANE ♦ HEXAMETHYLENEDIAMINE, solid (UN 2280) (DOT) ♦ HMDA

ACGIH TLV: TWA 0.5 ppm

DOT Classification: 8; Label: Corrosive

SAFETY PROFILE: Moderately toxic by ingestion, inhalation, and skin contact. An experimental teratogen. A corrosive irritant to skin, eyes, and mucous membranes. Combustible.
PROP: Colorless leaflets, long needles by sublimation; odor of piperidine. Mp: 39–42°, bp: 205°. Absorbs water and CO_2 from air; very sol in water; sltly sol in alc, benzene.

HES000 CAS: 111-31-9 $C_6H_{14}S$ HR: 2
1-HEXANETHIOL
DOT: UN 1228/UN 3071 SYNS: HEXYL MERCAPTAN
NIOSH REL: (n-Alkane Mono Thiols) CL 0.5 ppm/15M
DOT Classification: 3; Label: Flammable Liquid, Poison (UN 1228); DOT Class: 6.1; Label: Poison, Flammable Liquid (UN 3071)
SAFETY PROFILE: Moderately toxic by inhalation and ingestion. A flammable liquid.

HEV000 CAS: 591-78-6 $C_6H_{12}O$ HR: 3
2-HEXANONE
SYNS: BUTYL METHYL KETONE ♦ MBK ♦ METHYL n-BUTYL KETONE (ACGIH)
OSHA PEL: TWA 5 ppm
ACGIH TLV: TWA 5 ppm; STEL 10 ppm
DFG MAK: 5 ppm (21 mg/m^3)
NIOSH REL: (Ketones) TWA 4 mg/m^3
DOT Classification: 3; Label: Flammable Liquid
SAFETY PROFILE: Moderately toxic by ingestion. Mildly toxic by inhalation and skin contact. Experimental teratogenic effects. A skin and eye irritant. Dangerous fire and explosion hazard.
PROP: Clear liquid; odor of nail-polish remover. Mp: –56.9°, bp: 127.2°, lel: 1.22%, uel: 8.0%, flash p: 95°F (OC), d: 0.830 @ 0°/4°, vap press: 10 mm @ 38.8°, vap d: 3.45, autoign temp: 991°F. Sltly sol in water; sol in alc and ether.

HFG500 CAS: 108-10-1 $C_6H_{12}O$ HR: 3
HEXONE
DOT: UN 1245
SYNS: ISOBUTYL METHYL KETONE ♦ ISOPROPYLACETONE ♦ METHYL ISOBUTYL KETONE (ACGIH, DOT) ♦ 4-METHYL-2-PENTANONE (FCC) ♦ MIBK ♦ MIK
OSHA PEL: TWA 50 ppm; STEL 75 ppm
ACGIH TLV: TWA 50 ppm; STEL 75 ppm; BEI: 2 mg/L of MIBK in urine at end of shift
DFG MAK: 20 ppm (83 mg/m^3)
NIOSH REL: (Ketones) TWA 200 mg/m^3 (Proposed: BEI 2 mg/L MIBK in urine, end of shift)
DOT Classification: 3; Label: Flammable Liquid
SAFETY PROFILE: Moderately toxic by ingestion. Mildly toxic by inhalation. Very irritating to the skin, eyes, and mucous membranes. An experimental teratogen. Flammable liquid. It can react violently with many substances.

PROP: Colorless mobile liquid; fruity, ethereal odor. Fp: –80.2°, bp: 116.8°, lel: 1.4%, uel: 7.5%, flash p: 62.6°F, d: 0.801, autoign temp: 858°F, vap press: 16 mm @ 20°. Misc with alc, ether; sol in water.

HFJ000 CAS: 108-84-9 $C_8H_{16}O_2$ HR: 3

sec-HEXYL ACETATE

DOT: UN 1233

SYNS: 1,3-DIMETHYLBUTYL ACETATE ♦ MAAC ♦ METHYL AMYL ACETATE (DOT) ♦ METHYLISOBUTYLCARBINOL ACETATE ♦ 4-METHYL-2-PENTANOL, ACETATE

OSHA PEL: TWA 50 ppm

ACGIH TLV: TWA 50 ppm

DFG MAK: 50 ppm (300 mg/m^3)

DOT Classification: 3; Label: Flammable Liquid

SAFETY PROFILE: Mildly toxic by ingestion, skin contact, and inhalation. A skin and human eye irritant. Flammable liquid.

PROP: Clear liquid; pleasant odor. Bp: 146.3°, fp: –63.8°, flash p: 113°F (COC), d: 0.8598 @ 20°/20°, vap press: 3.8 mm @ 20°, vap d: 4.97.

HFP875 CAS: 107-41-5 $C_6H_{14}O_2$ HR: 2

HEXYLENE GLYCOL

SYNS: 2,4-DIHYDROXY-2-METHYLPENTANE ♦ 1,2-HEXANEDIOL ♦ 2-METHYL-2,4-PENTANEDIOL

OSHA PEL: CL 25 ppm

ACGIH TLV: CL 25 ppm

DFG MAK: 10 ppm

SAFETY PROFILE: Moderately toxic by ingestion. Mildly toxic by skin contact. Combustible.

PROP: Mild odor, colorless liquid, water-sol. Bp: 197.1°, fp: –50°, flash p: 205°F (OC), d: 0.9234 @ 20°/20°, vap press: 0.05 mm @ 20°.

HGS000 CAS: 302-01-2 H_4N_2 HR: 3

HYDRAZINE

DOT: UN 2029

OSHA PEL: TWA 0.1 ppm (skin)

ACGIH TLV: TWA 0.1 ppm (skin); Suspected Human Carcinogen (Proposed: 0.01 ppm (skin); Suspected Human Carcinogen)

DFG TRK: Animal Carcinogen, Suspected Human Carcinogen

NIOSH REL: (Hydrazines) CL 0.04 mg/m^3/2H

DOT Classification: 3; Label: Flammable Liquid, Poison, Corrosive

SAFETY PROFILE: Confirmed carcinogen. A poison by ingestion and skin contact. Moderately toxic by inhalation. An experimental teratogen. Flammable liquid. It can react violently with many substances.

PROP: Colorless, oily, fuming liquid or white crystals. Mp: 254°, bp: 113.5°, flash p: 100°F (OC), d: 1.1011 @ 15° (liquid), autoign temp: can vary from 74°F in contact with iron rust, 270°F in contact with black iron, 313°F in contact with stainless steel, 518°F in contact with glass. Vap d: 1.1, lel: 4.7%, uel: 100%. Misc in H_2O, alcohols; sltly sol in org solvs.

HGU500 CAS: 10217-52-4 $H_4N_2•H_2O$ HR: 3

HYDRAZINE HYDRATE

DOT: UN 2030

SYNS: HYDRAZINE AQUEOUS SOLUTIONS, with not >64% hydrazine, by weight (DOT) ♦ HYDRAZINE HYDRATE, with not >64% hydrazine, by weight (DOT)

NIOSH REL: (Hydrazines) CL 0.04 mg/m^3/2H

DOT Classification: 8; Label: Corrosive, Poison

SAFETY PROFILE: A poison by ingestion, inhalation, skin contact. A corrosive irritant to the eyes, skin, and mucous membranes.

PROP: Colorless, fuming, refractive liquid. Mp: –51.7°, bp: 118.5° @ 740 mm, d: 1.03 @ 21°. Faint characteristic odor. A strong base, very corrosive; attacks glass, rubber, and cork. Very powerful reducing agent. Misc with water and alc; insol in chloroform and ether.

HGV000 CAS: 2644-70-4 $H_4N_2•ClH$ HR: 3

HYDRAZINE, HYDROCHLORIDE

SYNS: HYDRAZINE MONOCHLORIDE

NIOSH REL: (Hydrazines) CL 0.04 mg/m^3/2H

SAFETY PROFILE: A poison by ingestion.

HGW500 CAS: 10034-93-2 $H_4N_2•H_2O_4S$ HR: 3

HYDRAZINE SULFATE (1:1)

SYNS: HYDRAZINE HYDROGEN SULFATE ♦ HYDRAZONIUM SULFATE

NIOSH REL: (Hydrazines) CL 0.04 mg/m^3/2H

SAFETY PROFILE: Confirmed carcinogen. A poison by ingestion. An experimental teratogen. An eye irritant.

PROP: Colorless crystals. D: 1.378, mp: 254° (decomp). Sol in water; insol in alc; very sol in hot water.

HHG500 CAS: 7782-79-8 HN_3 HR: 3

HYDRAZOIC ACID

SYNS: DIAZOIMIDE ♦ HYDROGEN AZIDE ♦ HYDRONITRIC ACID ♦ TRIAZOIC ACID

ACGIH TLV: CL 0.1 ppm (vapor)

DFG MAK: 0.1 ppm (0.18 mg/m^3)

SAFETY PROFILE: Mildly toxic by inhalation. A severe irritant to skin, eyes, and mucous membranes. A dangerously sensitive explosive hazard when shocked or exposed to heat.

PROP: Colorless liquid; intolerable pungent odor. Mp: –80°, bp: 37°, d: 1.09 @ 25°/4°. Very sol in water.

HHJ000 CAS: 10035-10-6 BrH HR: 3

HYDROBROMIC ACID
DOT: UN 1048/UN 1788
SYNS: ANHYDROUS HYDROBROMIC ACID ♦ HYDROGEN BROMIDE (ACGIH,OSHA,MAK)
OSHA PEL: CL 3 ppm
ACGIH TLV: CL 3 ppm
DFG MAK: 2 ppm (6.7 mg/m^3)
DOT Classification: 8; Label: Corrosive (UN 1788); DOT Class: 2.3; Label: Poison Gas, Corrosive (UN 1048)
SAFETY PROFILE: A poison gas. A corrosive irritant to the eyes, skin, and mucous membranes.
PROP: Colorless with an acrid odor, or pale-yellow liquid. Mp: –87°, bp: –66.5°, d: 3.50 g/L @ 0°. Misc with water, alc. Keep protected from light.

HHL000 CAS: 7647-01-0 ClH HR: 3
HYDROCHLORIC ACID
DOT: UN 1050/UN 1789/UN 2186
SYNS: CHLOROHYDRIC ACID ♦ HYDROCHLORIDE ♦ HYDROGEN CHLORIDE, anhydrous (UN 1050) (DOT) ♦ HYDROGEN CHLORIDE, refrigerated liquid (UN 2186) (DOT) ♦ MURIATIC ACID
OSHA PEL: CL 5 ppm
ACGIH TLV: CL 5 ppm
DFG MAK: 5 ppm (7.6 mg/m^3)
DOT Classification: 8; Label: Corrosive (UN 1789); DOT Class: 2.3; Label: Poison Gas, Corrosive (UN 1050, UN 2186)
SAFETY PROFILE: A human poison. Mildly toxic to humans by inhalation. A corrosive irritant to the skin, eyes, and mucous membranes. An experimental teratogen. Nonflammable gas. It can react violently with many substances.
PROP: Colorless, corrosive, gas or fuming liquid; strongly corrosive with pungent odor. Dissolves in H_2O to give a strong, highly corrosive acid. Mp: –114.3°, bp: –84.8°, d: (gas) 1.639 g/L @ 0°, (liquid) 1.194 @ –26°, vap press: 4.0 atm @ 17.8°. Very sol in H_2O; sol in MeOH, EtOH, and Et_2O.

HHS000 CAS: 74-90-8 CHN HR: 3
HYDROCYANIC ACID
DOT: NA 1051/UN 1613/UN 1614
SYNS: HYDROGEN CYANIDE (ACGIH,OSHA) ♦ PRUSSIC ACID
OSHA PEL: STEL 4.7 ppm (skin)
ACGIH TLV: CL 4.7 ppm (skin)
DFG MAK: 10 ppm (11 mg/m^3)
NIOSH REL: (Cyanide) CL 5 mg(CN)/m^3/10M
DOT Classification: 6.1; Label: Poison (NA 1613, UN 1613, UN 1614); DOT Class: Forbidden (unstabilized); DOT Class: 6.1; Label: Poison, Flammable Liquid (UN 1051)
SAFETY PROFILE: A deadly human poison. A very dangerous fire hazard.

PROP: Very volatile liquid or colorless gas smelling of bitter almonds. Mp: –13°, bp: 25.7°, lel: 5.6%, uel: 40%, flash p: 0°F (CC), d: 0.715 @ 0°, autoign temp: 1000°F, vap press: 400 mm @ 9.8°, vap d: 0.932. Misc in water, alc, and ether.

HHU500 CAS: 7664-39-3 FH HR: 3

HYDROFLUORIC ACID

DOT: UN 1052/UN 1790

SYNS: HYDROGEN FLUORIDE, anhydrous (UN 1052) (DOT)

OSHA PEL: TWA 3 ppm; STEL 6 ppm (F)

ACGIH TLV: CL 3 ppm (F); BEI: 3 mg/g creatinine of fluorides in urine prior to shift; 10 mg/g creatinine of fluorides in urine at end of shift.

DFG MAK: 3 ppm (2.5 mg/m^3); BAT 7.0 mg/g creatinine in urine at end of shift

NIOSH REL: (HF) TWA 2.5 $mg(F)/m^3$; CL 5.0 $mg(F)/m^3$/15M

DOT Classification: 8; Label: Corrosive, Poison

SAFETY PROFILE: A poison by inhalation. A corrosive irritant to skin, eyes, and mucous membranes. Experimental teratogenic effects. It can react violently with many substances.

PROP: Clear, colorless, nonflammable, fuming, corrosive liquid or gas. One of the most acidic substances known, but aq solns are only weakly acid. Dissolves silica to give H_2SiF_6. Mp: –83.1°, bp: 19.54°, d: 0.901 g/L (gas), 0.699 @ 22° (liquid), vap press: 400 mm @ 2.5°. Very sol in H_2O, EtOH; sltly sol in Et_2O.

HHW800 CAS: 61788-32-7 HR: 3

HYDROGENATED TERPHENYLS

OSHA PEL: TWA 0.5 ppm

ACGIH TLV: TWA 0.5 ppm

NIOSH REL: (Hydrogenated Terphenyls) TWA 0.5 ppm

SAFETY PROFILE: Contact with hot coolant can cause severe damage to lungs, skin, and eyes from burns.

PROP: Complex mixtures of o-, m-, and p-terphenyls in various stages of hydrogenation. Five such stages exist for each of the three above isomers.

HHX000 CAS: 7647-01-0 ClH HR: 3

HYDROGEN CHLORIDE

OSHA PEL: CL 5 ppm

ACGIH TLV: CL 5 ppm

DFG MAK: 5 ppm (7 mg/m^3)

SAFETY PROFILE: A highly corrosive irritant to the eyes, skin, and mucous membranes. It can react violently with many substances.

PROP: Colorless, corrosive, nonflammable gas. Pungent odor, fumes in air. D: 1.639 @ –137.77°, bp: –154.37° @ 1.0 mm.

HIB050 CAS: 7722-84-1 H_2O_2 HR: 3

HYDROGEN PEROXIDE, 90%

DOT: UN 2014

SYNS: DIHYDROGEN DIOXIDE
OSHA PEL: TWA 1 ppm
ACGIH TLV: TWA 1 ppm; Animal Carcinogen
DFG MAK: 1 ppm (1.4 mg/m^3)
NIOSH REL: (Hydrogen peroxide) TWA 1.0 ppm
DOT Classification: 5.1; Label: Oxidizer, Corrosive
SAFETY PROFILE: Confirmed carcinogen. Moderately toxic by inhalation, ingestion, and skin contact. A corrosive irritant to skin, eyes, and mucous membranes. A dangerous fire hazard by chemical reaction. It can react violently with many substances.
PROP: Colorless, unstable, heavy liquid or, at low temp, a crystalline solid; bitter taste. D: 1.71 @ –20°, 1.46 @ 0°, vap press: 1 mm @ 15.3°, unstable, mp: –0.43°, bp: 152°. Misc with water; sol in ether; insol in pet ether. Decomposed by many org solvs.

HIC000 CAS: 7783-07-5 H_2Se HR: 3
HYDROGEN SELENIDE
DOT: UN 2202
SYNS: SELENIUM HYDRIDE
OSHA PEL: TWA 0.05 ppm (Se)
ACGIH TLV: TWA 0.05 ppm (Se)
DFG MAK: 0.05 ppm (0.17 mg/m^3)
DOT Classification: 2.3; Label: Poison Gas, Flammable Gas
SAFETY PROFILE: A deadly poison by inhalation. Very poisonous irritant to skin, eyes, and mucous membranes. Dangerous fire hazard.
PROP: Colorless gas. Mp: –64°, bp: –41.4°, d: 3.614 g/L (gas), 2.12 @ –42° (liquid), vap press: 10 atm @ 23.4°. Flammable. Disagreeable odor. Sol in carbonyl chloride and carbon disulfide.

HIC500 CAS: 7783-06-4 H_2S HR: 3
HYDROGEN SULFIDE
DOT: UN 1053
OSHA PEL: TWA 10 ppm; STEL 15 ppm
ACGIH TLV: TWA 10 ppm; STEL 15 ppm (Proposed: 5 ppm)
DFG MAK: 10 ppm (14 mg/m^3)
NIOSH REL: (Hydrogen Sulfide) CL 15 mg/m^3/10M
DOT Classification: 2.3; Label: Poison Gas, Flammable Gas
SAFETY PROFILE: A poison by inhalation. A severe irritant to eyes and mucous membranes. Very dangerous fire hazard. It can react violently with many substances.
PROP: Colorless, flammable, poisonous gas; offensive odor. Mp: –85.5°, bp: –60.4°, d: –60, (gas) 0.993, lel: 4%, uel: 46%, autoign temp: 500°F, d: 1.539 g/L @ 0°, vap press: 20 atm @ 25.5°, vap d: 1.189.

HIH000 CAS: 123-31-9 $C_6H_6O_2$ HR: 3
HYDROQUINONE
DOT: UN 2662

SYNS: p-BENZENEDIOL ♦ DIHYDROXYBENZENE (OSHA) ♦ p-DIOXYBENZENE ♦ HYDROQUINONE, liquid or solid (DOT) ♦ p-HYDROXYPHENOL

OSHA PEL: TWA 2 mg/m^3

ACGIH TLV: TWA 2 mg/m^3; Animal Carcinogen

DFG MAK: Animal Carcinogen, Suspected Human Carcinogen

NIOSH REL: (Hydroquinone) CL 2.0 mg/m^3/15M

DOT Classification: 6.1; Label: KEEP AWAY FROM FOOD

SAFETY PROFILE: Confirmed carcinogen. A poison by ingestion. A severe human skin irritant. Combustible.

???

HIM500 CAS: 107-16-4 C_2H_3NO HR: 3

HYDROXYACETONITRILE

SYNS: CYANOMETHANOL ♦ FORMALDEHYDE CYANOHYDRIN ♦ GLYCOLONITRILE

NIOSH REL: (Nitriles) CL 5 mg/m^3/15M

SAFETY PROFILE: A poison by ingestion, skin contact, and inhalation. An eye irritant. May undergo spontaneous and violent decomposition.

PROP: Bp: 183° (slt decomp).

PROP: Colorless, hexagonal prisms; needles from water. Mp: 172°, bp: 285° @ 730 mm, flash p: 329°F (CC), d: 1.358 @ 20°/4°, autoign temp: 960°F (CC), vap press: 1 mm @ 132.4°, vap d: 3.81. Very sol in alc and ether; sltly sol in benzene. Keep well closed and protected from light.

HNT600 CAS: 999-61-1 $C_6H_{10}O_3$ HR: 3

2-HYDROXYPROPYL ACRYLATE

SYNS: ACRYLIC ACID-2-HYDROXYPROPYL ESTER ♦ 1,2-PROPANEDIOL-1-ACRYLATE ♦ PROPYLENE GLYCOL MONOACRYLATE

OSHA PEL: TWA 0.5 ppm (skin)

ACGIH TLV: TWA 0.5 ppm (sensitizer)

SAFETY PROFILE: Poison by ingestion.

IAQ000 CAS: 96-45-7 $C_3H_6N_2S$ HR: 3

2-IMIDAZOLIDINETHIONE

SYNS: ETHYLENE THIOUREA ♦ ETU ♦ 2-MERCAPTOIMIDAZOLINE ♦ 2-THIOL-DIHYDROGLYOXALINE

DFG MAK: Confirmed Animal Carcinogen with Unknown Relevance to Humans

NIOSH REL: (ETU) Use encapsulated form; minimize exposure

SAFETY PROFILE: Confirmed carcinogen. Poison by ingestion. Experimental teratogenic effects. An eye irritant.

PROP: White crystals, needles, or prisms from EtOh or pentanol. Mp: 197–200°. Water solubility: 9 g/100 mL @ 30°. Often occurs as a main degradation product of the metal salts of ethylene bis-dithiocarbamic acid. Sol in H_2O; insol in Et_2O, $CHCl_3$, C_6H_6, and Me_2CO.

IBA000 CAS: 2465-27-2 $C_{17}H_{21}N_3{\cdot}ClH{\cdot}H_2O$ HR: 3

4,4'-(IMIDOCARBONYL)BIS(N,N-DIMETHYLAMINE) MONOHYDROCHLORIDE
SYNS: AURAMINE (MAK) ♦ AURAMINE HYDROCHLORIDE ♦ 1,1-BIS(p-DIMETHYLAMINOPHENYL)METHYLENIMINEHYDROCHLORIDE ♦ C.I. 41000
DFG MAK: Animal Carcinogen, Suspected Human Carcinogen
SAFETY PROFILE: Confirmed carcinogen. Poison by skin contact and ingestion.
PROP: Golden-yellow plates from H_2O. Mp: 267°. Sltly sol in cold H_2O.

IBB000 CAS: 492-80-8 $C_{17}H_{21}N_3$ HR: 3
4,4'-(IMIDOCARBONYL)BIS(N,N-DIMETHYLANILINE)
SYNS: AURAMINE (MAK) ♦ BIS(p-DIMETHYLAMINOPHENYL)METHYLENEIMINE ♦ C.I. 41000B ♦ TETRAMETHYLDIAMINODIPHENYLACETIMINE
DFG MAK: Animal Carcinogen, Suspected Human Carcinogen
SAFETY PROFILE: Confirmed human carcinogen.
PROP: Yellow needles or yellow plates from EtOH. Mp: 136°. Insol in water; sol in EtOH, Et_2O, and acids; sltly sol in H_2O.

IBX000 CAS: 95-13-6 C_9H_8 HR: 2
INDENE
SYN: INDONAPHTHENE
OSHA PEL: TWA 10 ppm
ACGIH TLV: TWA 10 ppm
NIOSH REL: (Indene) TWA 10 ppm
SAFETY PROFILE: Low toxicity by ingestion and inhalation. Irritating to skin, eyes, and mucous membranes.
PROP: Liquid from coal tars. D: 0.9968 @ 20°/4°, mp: –1.8°, bp: 181.6°. Water-insol, but misc in org solvs.

ICF000 CAS: 7440-74-6 In HR: 3
INDIUM
OSHA PEL: TWA 0.1 mg(In)/m^3
ACGIH TLV: TWA 0.1 mg(In)/m^3
SAFETY PROFILE: Teratogenic effects. Inhalation of indium compounds may cause damage to the respiratory system. Flammable in the form of dust when exposed to heat or flame.
PROP: Soft, silvery-white, malleable and ductile metal. Liquid wets glass; stable in dry air. Slowly oxidized in moist air. Reacts with halogens, S, Se, Te, As, P on heating. Dissolves in Hg. Not affected by alkalies. Has plastic properties at cryogenic temps. Mp: 156.61°, bp: 2080°, d: 7.31 @ 20°. Insol in H_2O in bulk form; sol in most acids.

IDM000 CAS: 7553-56-2 I_2 HR: 3
IODINE
SYNS: IODINE CRYSTALS ♦ IODINE SUBLIMED
OSHA PEL: CL 0.1 ppm
ACGIH TLV: CL 0.1 ppm

DFG MAK: 0.1 ppm (1.1 mg/m^3)
SAFETY PROFILE: A poison by ingestion. Moderately toxic by inhalation. It can react violently with many substances.
PROP: Rhombic, violet-black crystals with metallic luster; flakes with characteristic odor, sharp acrid taste. Sublimes slowly at room temp. Mp: 113.5°, bp: 185.24°, d: 4.93 (solid @ 25°), vap press: 1 mm @ 38.7°, vap press: (solid): 0.030 mm @ 0°. Sltly sol in H_2O. Sol in many org solvs.

IEP000 CAS: 75-47-8 CHI_3 HR: 3
IODOFORM
SYNS: TRIIODOMETHANE
OSHA PEL: TWA 0.6 ppm (skin)
ACGIH TLV: TWA 0.6 ppm
SAFETY PROFILE: A poison by ingestion. Moderately toxic by inhalation and skin contact.
PROP: Yellow powder or crystals; plates from Me_2CO; disagreeable odor. D: 4.1, mp: 120° (approx), bp: subl. Decomp at high temp, evolving iodine. Volatile with steam. Very sol in water, benzene, acetone; sltly sol in pet ether.

IGR499 HR: D
IRON COMPOUNDS
SAFETY PROFILE: Of varying toxicity. Exposure to iron oxides is potentially a serious risk in all industrial settings. Some iron compounds are suspected carcinogens. In general, ferrous compounds are more toxic than ferric compounds.

IHC450 CAS: 1309-37-1 Fe_2O_3 HR: 3
IRON OXIDE
SYNS: C.I. 77491 ♦ FERRIC OXIDE ♦ IRON(III) OXIDE ♦ JEWELER'S ROUGE
OSHA PEL: Dust and Fume: TWA 10 mg(Fe)/m^3; Rouge: TWA Total Dust: 10 mg/m^3; Respirable Fraction: 5 mg/m^3
ACGIH TLV: TWA 5 mg(Fe)/m^3 (vapor, dust); Not Classifiable as a Human Carcinogen; Rouge: 10 mg/m^3; Not Classifiable as a Human Carcinogen
DFG MAK: 1.5 mg/m^3 calculated as fine dust
NIOSH REL: (Iron Oxide, Dust and Fume) TWA 5 mg/m^3
SAFETY PROFILE: Dust or fumes are inhalation hazards. Dust can react violently with many substances.
PROP: Dark-red powder. Insol in H_2O.

IHG500 CAS: 13463-40-6 C_5FeO_5 HR: 3
IRON PENTACARBONYL
DOT: UN 1994
SYNS: IRON CARBONYL
OSHA PEL: TWA 0.1 ppm (Fe); STEL 0.2 ppm

ACGIH TLV: TWA 0.1 ppm (Fe); STEL 0.2 ppm
DFG MAK: 0.1 ppm (0.81 mg/m^3)
DOT Classification: 6.1; Label: Poison, Flammable Liquid
SAFETY PROFILE: A poison by inhalation, skin contact, and ingestion. A very dangerous fire and moderate explosion hazard.
PROP: Yellow to dark-red viscous liquid. Mp: –25°, Fp: –20° (to –19°), bp: 103.0°, flash p: 5°F, d: 1.453 @ 25°/4°, vap press: 40 mm @ 30.3°. Sol in hexane.

IHO850 CAS: 123-92-2 $C_7H_{14}O_2$ HR: 3
ISOAMYL ACETATE
SYNS: ACETIC ACID, ISOPENTYL ESTER ♦ BANANA OIL ♦ ISOAMYL ETHANOATE ♦ ISOPENTYL ALCOHOL ACETATE ♦ 3-METHYL-1-BUTYL ACETATE ♦ PEAR OIL
OSHA PEL: TWA 100 ppm
ACGIH TLV: TWA 100 ppm; (Proposed: TWA 50 ppm; STEL 100 ppm)
DFG MAK: 50 ppm
SAFETY PROFILE: Mildly toxic by ingestion and inhalation. Highly flammable liquid.
PROP: Colorless liquid; banana-like odor. Bp: 142.0°, ULC: 55–60, lel: 1% @ 212°F, uel: 7.5%, flash p: 77°F, d: 0.876, refr index: 1.400, autoign temp: 680°F, vap d: 4.49. Misc in alc, ether, ethyl acetate, fixed oils; sltly sol in water; insol in glycerin, propylene glycol.

IHP000 CAS: 123-51-3 $C_5H_{12}O$ HR: 3
ISOAMYL ALCOHOL
SYNS: ISOBUTYLCARBINOL ♦ ISOPENTYL ALCOHOL ♦ 3-METHYL BUTANOL
OSHA PEL: TWA 100 ppm; STEL 125 ppm
ACGIH TLV: TWA 100 ppm; STEL 125 ppm
DFG MAK: 100 ppm (370 mg/m^3)
SAFETY PROFILE: Moderately toxic by ingestion and skin contact. A skin and human eye irritant. Flammable liquid.
PROP: Clear liquid; pungent, repulsive taste. Bp: 132°, ULC: 35–40, lel: 1.2%, uel: 9.0% @ 212°F, flash p: 109°F (CC), d: 0.813, autoign temp: 662°F, vap d: 3.04, mp: –117.2°. Sol in water @ 14°; misc in alc and ether.

IHP010 CAS: 584-02-1 $C_5H_{12}O$ HR: 3
ISOAMYL ALCOHOL
SYNS: DIETHYL CARBINOL ♦ 3-PENTANOL
OSHA PEL: TWA 100 ppm; STEL 125 ppm
ACGIH TLV: TWA 100 ppm; STEL 125 ppm
DFG MAK: 100 ppm (360 mg/m^3)
SAFETY PROFILE: Moderately toxic by ingestion and skin contact. A severe eye and mild skin irritant. Dangerous fire and explosion hazard.
PROP: Liquid; acetone-like odor. Bp: 115.6°, d: 0.815 @ 25°/4°, flash p: 66°F, lel: 1.2%, uel: 9%. Sol in alc, ether; sltly sol in water.

IIJ000 CAS: 110-19-0 $C_6H_{12}O_2$ HR: 3

ISOBUTYL ACETATE

DOT: UN 1213

SYNS: ACETIC ACID-2-METHYLPROPYL ESTER ♦ 2-METHYLPROPYL ACETATE

OSHA PEL: TWA 150 ppm

ACGIH TLV: TWA 150 ppm

DFG MAK: 100 ppm (480 mg/m^3)

DOT Classification: 3; Label: Flammable Liquid

SAFETY PROFILE: Mildly toxic by ingestion and inhalation. A skin and eye irritant. Highly flammable liquid.

PROP: Colorless, neutral liquid; fruit-like odor. Mp: –98.9°, bp: 118°, flash p: 64°F (CC) (18°), d: 0.8685 @ 15°, refr index: 1.389, vap press: 10 mm @ 12.8°, autoign temp: 793°F, vap d: 4.0, lel: 2.4%, uel: 10.5%. Very sol in alc, fixed oils, propylene glycol; sltly sol in water.

IIL000 CAS: 78-83-1 $C_4H_{10}O$ HR: 3

ISOBUTYL ALCOHOL

DOT: UN 1212

SYNS: 1-HYDROXYMETHYLPROPANE ♦ ISOBUTANOL (DOT) ♦ ISOPROPYLCARBINOL ♦ 2-METHYL PROPANOL

OSHA PEL: TWA 50 ppm

ACGIH TLV: TWA 50 ppm

DFG MAK: 100 ppm (310 mg/m^3)

DOT Classification: 3; Label: Flammable Liquid

SAFETY PROFILE: Moderately toxic by ingestion and skin contact. Mildly toxic by inhalation. A severe skin and eye irritant. Flammable liquid.

PROP: Clear, colorless, refractive, mobile liquid; sweet odor. Flammable. Bp: 107.90°, flash p: 82°F, ULC: 40–45, lel: 1.2%, uel: 10.9% @ 212°F, fp: –108°, d: 0.800, autoign temp: 800°F, vap press: 10 mm @ 21.7°, vap d: 2.55. Sltly sol in water; misc with alc and ether.

IIM000 CAS: 78-81-9 $C_4H_{11}N$ HR: 3

ISOBUTYLAMINE

DOT: UN 1214

SYNS: 1-AMINO-2-METHYLPROPANE ♦ MONOISOBUTYLAMINE ♦ 1-PROPANAMINE, 2-METHYL-

DFG MAK: 5 ppm (15 mg/m^3)

DOT Classification: 3; Label: Flammable Liquid

SAFETY PROFILE: A poison by ingestion. A powerful irritant to skin, eyes, and mucous membranes. Skin contact can cause blistering. A very dangerous fire hazard.

PROP: Colorless liquid. Mp: –85.5°, bp: 68.6°, flash p: 15°F, d: 0.731 @ 20°/20°, vap press: 100 mm @ 18.8°, autoign temp: 712°F, vap d: 2.5. Misc with water, alc, and ether.

IJX000 CAS: 78-82-0 C_4H_7N HR: 3

ISOBUTYRONITRILE

DOT: UN 2284

SYNS: 2-CYANOPROPANE ♦ ISOPROPYL CYANIDE ♦ ISOPROPYL NITRILE ♦ 2-METHYLPROPANENITRILE
NIOSH REL: (Nitriles) TWA 22 mg/m^3
DOT Classification: 3; Label: Flammable Liquid, Poison
SAFETY PROFILE: A poison by ingestion and skin contact. Mildly toxic by inhalation. A skin irritant. A very dangerous fire hazard.
PROP: Colorless liquid, sltly sol in water, very sol in alc and ether. D: 0.773 @ 20°/20°, bp: 107°, mp: –75°, flash p: 46.4°F.

IKS600 CAS: 107-83-5 C_6H_{14} HR: 3
ISOHEXANE
SYNS: 1,2-DIMETHYLBUTANE ♦ 2-METHYLPENTANE
OSHA PEL: TWA 500 ppm; STEL 1000 ppm
ACGIH TLV: TWA 500 ppm; STEL 1000 ppm (hexane isomer)
DFG MAK: 200 ppm (720 mg/m^3)
NIOSH REL: (Alkanes) TWA 350 mg/m^3
SAFETY PROFILE: A human eye irritant. A very dangerous fire hazard.
PROP: Liquid or oil. Fp: –154°, bp: 60.3°, lel: 1.0%, uel: 7.0%, flash p: 20°F (CC), d: 0.669, vap d: 3.00, autoign temp: 583°F.

ILL000 CAS: 26952-21-6 $C_8H_{18}O$ HR: 2
ISOOCTYL ALCOHOL
SYN: ISOOCTANOL
OSHA PEL: TWA 50 ppm (skin)
ACGIH TLV: TWA 50 ppm (skin)
NIOSH REL: (Isooctyl Alcohol) TWA 50 ppm (skin)
SAFETY PROFILE: Moderately toxic by ingestion and skin contact. A skin and severe eye irritant.

IMF400 CAS: 78-59-1 $C_9H_{14}O$ HR: 3
ISOPHORONE
SYNS: ISOACETOPHORONE ♦ 1,1,3-TRIMETHYL-3-CYCLOHEXENE-5-ONE
OSHA PEL: TWA 4 ppm
ACGIH TLV: CL 5 ppm
DFG MAK: 2 ppm (11 mg/m^3)
NIOSH REL: TWA (Ketones) 23 mg/m^3
SAFETY PROFILE: Moderately toxic by ingestion. Mildly toxic by inhalation. A skin and severe eye irritant. Flammable liquid.
PROP: Practically water-white liquid. Bp: 215.2°, flash p: 184°F (OC), d: 0.9229, autoign temp: 864°F, vap press: 1 mm @ 38.0°, vap d: 4.77, lel: 0.8%, uel: 3.8%.

IMG000 CAS: 4098-71-9 $C_{12}H_{18}N_2O_2$ HR: 3
ISOPHORONE DIISOCYANATE
DOT: UN 2290/UN 2906

SYNS: IPDI ♦ 3-ISOCYANATOMETHYL-3,5,5-TRIMETHYLCYCLOHEXYLISOCYANATE ♦ ISOPHORONE DIAMINE DIISOCYANATE ♦ TRIISOCYANATOISOCYANURATE, solution, 70%, by weight (DOT)
OSHA PEL: TWA 0.005 ppm (skin)
ACGIH TLV: TWA 0.005 ppm (skin)
DFG MAK: 0.01 ppm (0.092 mg/m^3)
NIOSH REL: (Diisocyanates) 10H TWA 0.005 ppm; CL 0.02 ppm/10M
DOT Classification: 3; Label: Flammable Liquid; DOT Class: 6.1; Label: KEEP AWAY FROM FOOD (UN 2290)
SAFETY PROFILE: Poison by inhalation. Moderately toxic by skin contact. A flammable liquid.
PROP: D: 1.062 @ 20°/4°, bp: 217° @ 100 mm.

INA500 CAS: 109-59-1 $C_5H_{12}O_2$ HR: 2

2-ISOPROPOXYETHANOL

SYNS: ETHYLENE GLYCOL ISOPROPYL ETHER ♦ ETHYLENE GLYCOL, MONOISOPROPYL ETHER ♦ ISOPROPYL CELLOSOLVE ♦ MONOISOPROPYL ETHER of ETHYLENE GLYCOL
OSHA PEL: TWA 25 ppm
ACGIH TLV: TWA 25 ppm
DFG MAK: 5 ppm (22 mg/m^3)
SAFETY PROFILE: Moderately toxic by skin contact. Mildly toxic by inhalation and ingestion. A skin and eye irritant.
PROP: Bp: 144° @ 743 mm.

INE100 CAS: 108-21-4 $C_5H_{10}O_2$ HR: 3

ISOPROPYL ACETATE

DOT: UN 1220
SYNS: ACETIC ACID-1-METHYLETHYL ESTER (9CI) ♦ 2-ACETOXYPROPANE ♦ 2-PROPYL ACETATE
OSHA PEL: TWA 250 ppm; STEL 310 ppm
ACGIH TLV: TWA 250 ppm; STEL 310 ppm; (Proposed: TWA 100 ppm; STEL 200 ppm)
DFG MAK: 200 ppm (850 mg/m^3)
DOT Classification: 3; Label: Flammable Liquid
SAFETY PROFILE: Moderately toxic by ingestion. Mildly toxic by inhalation. Highly flammable liquid.
PROP: Colorless, aromatic liquid. Mp: –73°, bp: 88.4°, lel: 1.8%, uel: 7.8%, fp: –69.3°, flash p: 40°F, d: 0.874 @ 20°/20°, autoign temp: 860°F, vap press: 40 mm @ 17.0°. Sltly sol in water; misc in alc, ether, fixed oils.

INJ000 CAS: 67-63-0 C_3H_8O HR: 3

ISOPROPYL ALCOHOL

DOT: UN 1219
SYNS: DIMETHYLCARBINOL ♦ ISOPROPANOL (DOT) ♦ 2-PROPANOL
OSHA PEL: TWA 400 ppm; STEL 500 ppm

ACGIH TLV: TWA 400 ppm; STEL 500 ppm (Proposed: 200 ppm; STEL 400 ppm; Not Classifiable as a Human Carcinogen)
DFG MAK: 200 ppm (500 mg/m^3)
NIOSH REL: (Isopropyl Alcohol) TWA 400 ppm; CL 800 ppm/15M
DOT Classification: 3; Label: Flammable Liquid
SAFETY PROFILE: Mildly toxic by inhalation skin contact. Experimental teratogenic effects. An eye and skin irritant. A flammable liquid. It can react violently with many substances.
PROP: Clear, colorless liquid; slt odor, sltly bitter taste. Mp: –88.5 to –89.5°, bp: 82.5°, lel: 2.5%, uel: 12%, flash p: 53°F (CC), d: 0.7854 @ 20°/4°, refr index: 1.377 @ 20°, vap d: 2.07, ULC: 70, fp: –89.5°, autoign temp: 852°F. Misc with water, alc, ether, chloroform; insol in salt solns.

INK000 CAS: 75-31-0 C_3H_9N HR: 3
ISOPROPYLAMINE
DOT: UN 1221
SYNS: 2-AMINOPROPANE ♦ 1-METHYLETHYLAMINE ♦ 2-PROPANAMINE
OSHA PEL: TWA 5 ppm; STEL 10 ppm
ACGIH TLV: TWA 5 ppm; STEL 10 ppm
DFG MAK: 5 ppm (12 mg/m^3)
DOT Classification: 3; Label: Flammable Liquid; DOT Class: 3; Label: Flammable Liquid, Corrosive
SAFETY PROFILE: Poison by skin contact. Moderately toxic by ingestion. Mildly toxic by inhalation. A severe skin and eye irritant. Very dangerous fire hazard.
PROP: Colorless liquid; amino odor. Mp: –101.2°, flash p: –35°F (OC), d: 0.694 @ 15°/4°, autoign temp: 756°F, d: 2.03, bp: 33–34°, lel: 2.3%, uel: 10.4%. Misc with water, alc, and ether.

INX000 CAS: 768-52-5 $C_9H_{13}N$ HR: 2
N-ISOPROPYLANILINE
OSHA PEL: TWA 2 ppm (10 mg/m^3)(skin)
ACGIH TLV: TWA 2 ppm (skin)
NIOSH REL: (N-Isopropylaniline) TWA 2 ppm (skin)
SAFETY PROFILE: Moderately toxic by ingestion and inhalation.
PROP: Oil. Bp: 206–208°.

IOB000 CAS: 80-15-9 $C_9H_{12}O_2$ HR: 3
ISOPROPYLBENZENE HYDROPEROXIDE
SYNS: CUMENE HYDROPEROXIDE (DOT) ♦ CUMENYL HYDROPEROXIDE
DFG MAK: Moderate Skin Effects
SAFETY PROFILE: A poison by ingestion. Moderately toxic by skin contact and inhalation. A skin and eye irritant. Flammable liquid. It can react violently with many substances.
PROP: A liquid. Bp: 153°, flash p: 175°F, d: 1.05. The hydroperoxide of cumene.

IOZ750 CAS: 108-20-3 $C_6H_{14}O$ HR: 3

ISOPROPYL ETHER

DOT: UN 1159

SYNS: DIISOPROPYL ETHER ♦ DIISOPROPYL OXIDE ♦ 2-ISOPROPOXYPROPANE

OSHA PEL: TWA 500 ppm

ACGIH TLV: TWA 250 ppm; STEL 310 ppm

DFG MAK: 500 ppm (2100 mg/m^3)

DOT Classification: 3; Label: Flammable Liquid

SAFETY PROFILE: Mildly toxic by ingestion, inhalation, and skin contact. A skin irritant. A very dangerous fire hazard and severe explosion hazard.

PROP: Colorless liquid; ethereal odor. Mp: –60°, bp: 68.5°, lel: 1.4%, uel: 7.9%, flash p: –18°F (CC), d: 0.719 @ 25°, autoign temp: 830°F, vap press: 150 mm @ 25°, vap d: 3.52. Misc in water.

IPD000 CAS: 4016-14-2 $C_6H_{12}O_2$ HR: 2

ISOPROPYL GLYCIDYL ETHER

SYNS: 1,2-EPOXY-3-ISOPROPOXYPROPANE ♦ IGE (OSHA) ♦ 3-ISOPROPYLOXYPROPYLENE OXIDE

OSHA PEL: TWA 50 ppm; STEL 75 ppm

ACGIH TLV: TWA 50 ppm; STEL 75 ppm

DFG MAK: Confirmed Animal Carcinogen with Unknown Relevance to Humans

NIOSH REL: (Glycidyl Ethers) CL 240 mg/m^3/15M

SAFETY PROFILE: Moderately toxic by ingestion. Mildly toxic by inhalation and skin contact. A skin and eye irritant.

PROP: A liquid.

KEK000 CAS: 8008-20-6 HR: 3

KEROSENE

DOT: UN 1223

SYNS: STRAIGHT-RUN KEROSENE

NIOSH REL: (Kerosene) TWA 100 mg/m^3

DOT Classification: 3; Label: Flammable Liquid

SAFETY PROFILE: Moderately toxic by ingestion. A severe skin irritant. Combustible.

PROP: A pale-yellow to water-white, oily liquid. Bp: 175–325°, ULC: 40, flash p: 150–185°F, d: 0.80 to <1.0, lel: 0.7%, uel: 5.0%, autoign temp: 410°F, vap d: 4.5. Insol in water; misc with other pet solvents. A mixture of petroleum hydrocarbons, chiefly of the methane series having from 10–16 carbon atoms per molecule.

KEU000 CAS: 463-51-4 C_2H_2O HR: 3

KETENE

SYNS: CARBOMETHENE ♦ ETHENONE ♦ KETO-ETHYLENE

OSHA PEL: TWA 0.5 ppm; STEL 1.5 ppm

ACGIH TLV: TWA 0.5 ppm; STEL 1.5 ppm

DFG MAK: 0.5 ppm (0.87 mg/m^3)

SAFETY PROFILE: Poison by inhalation. Moderately toxic by ingestion.

PROP: Colorless gas with disagreeable taste and pungent odor. Decomp in water. Mp: –150°, bp: –56°, vap d: 1.45. Decomp in alc. Fairly sol in Me_2CO; sol in H_2O, ether, and acetone.

LBR000 CAS: 105-74-8 $C_{24}H_{46}O_4$ HR: 3

LAUROYL PEROXIDE

SYNS: BIS(1-OXODODECYL)PEROXIDE ♦ DILAUROYL PEROXIDE ♦ DODECANOYL PEROXIDE

DFG MAK: Mild skin effects

SAFETY PROFILE: It is a corrosive irritant to the eyes and mucous membranes and can cause burns. A dangerous fire hazard.

PROP: White, tasteless, coarse powder; faint odor. Mp: 53–55°.

LBX000 CAS: 112-55-0 $C_{12}H_{26}S$ HR: 3

LAURYL MERCAPTAN

DOT: UN 1228/UN 3071

SYNS: 1-DODECANETHIOL ♦ DODECYL MERCAPTAN ♦ 1-MERCAPTODODECANE

NIOSH REL: (n-Alkane Mono Thiols) CL 0.5 ppm/15M

DOT Classification: 3; Label: Flammable Liquid, Poison (UN 1228); DOT Class: 6.1; Label: Poison, Flammable Liquid (UN 3071)

SAFETY PROFILE: Inhalation hazard. Combustible when exposed to heat or flame.

PROP: Water-white to pale-yellow liquid. Mp: –7°, bp: 115–177°, flash p: 262°F (OC), d: 0.849 @ 15.5°/15.5°.

LCF000 CAS: 7439-92-1 Pb HR: 3

LEAD

SYNS: C.I. 77575

OSHA PEL: TWA 0.05 mg(Pb)/m^3

ACGIH TLV: TWA 0.15 mg(Pb)/m^3; BEI: 50 μg(lead)/L in blood; 150 μg(lead)/g creatinine in urine

DFG MAK: 0.1 mg/m^3; BAT: 70 μg(lead)/L in blood; 30 μg(lead)/L in blood of women less than 45 years old

NIOSH REL: TWA (Inorganic Lead) 0.10 mg(Pb)/m^3

SAFETY PROFILE: Poison by ingestion. An experimental teratogen. Very heavy intoxication can sometimes be detected by formation of a dark line on the gum margins, the so-called lead line. Flammable in the form of dust.

PROP: Bluish-gray, soft, weak, ductile metal which tarnishes in moist air. Otherwise stable to O_2 and H_2O at ordinary temp. Mp: 327.43°, bp: 1740°, d: 11.34 @ 20°/4°, vap press: 1 mm @ 973°. Dissolves in dil HNO_3, acetic acid, HCl (slowly). Sol in alkali solns. Attacked at room temp by F_2 and Cl_2.

LCM000 CAS: 13424-46-9 N_6Pb HR: 3

LEAD(II) AZIDE

DOT: UN 0129

OSHA PEL: TWA 0.05 mg(Pb)/m^3
ACGIH TLV: TWA 0.15 mg(Pb)/m^3
NIOSH REL: (Inorganic Lead) TWA 0.10 mg(Pb)/m^3
DOT Classification: Forbidden (dry)
SAFETY PROFILE: A deadly poison. An explosive sensitive to shock.
PROP: Colorless needles or white powder. Explodes @ 350° or when shocked. Sltly sol in cold water; very sol in acetic acid; insol in NH_4OH.

LCR000 CAS: 7758-97-6 CrO_4•Pb HR: 3
LEAD CHROMATE
SYNS: C.I. 77600
OSHA PEL: TWA 0.05 mg(Pb)/m^3; CL 0.1 mg(CrO_3)/m^3
ACGIH TLV: 0.05 mg(Cr)/m^3; Human Carcinogen
DFG MAK: Confirmed Animal Carcinogen with Unknown Relevance to Humans
NIOSH REL: (Chromium(VI)) TWA 0.001 mg(Cr(VI))/m^3; (Inorganic Lead) TWA 0.10 mg(Pb)/m^3
SAFETY PROFILE: Confirmed carcinogen. Mildly toxic by ingestion. It can react violently with many substances.
PROP: Yellow or orange-yellow powder. Stable orange-yellow monoclinic cryst; unstable yellow orthorhombic form, and orange-red tetragonal form, stable above 7°. Mp: 844°, bp: decomp, d: 6.3. One of the most insol salts. Insol in acetic acid; sol in solns of fixed alkali hydroxides, dil HNO_3.

LCT000 HR: 3
LEAD COMPOUNDS
SAFETY PROFILE: Lead poisoning is one of the commonest of occupational diseases. Some lead compounds are carcinogens of the lungs and kidneys. Lead is a cumulative poison. Increasing amounts build up in the body and eventually reach a point at which symptoms and disability occur.

LGM000 CAS: 68476-85-7 HR: 3
LIQUEFIED PETROLEUM GAS
DOT: UN 1075
SYNS: LPG ♦ L.P.G. (OSHA, ACGIH) ♦ PETROLEUM GAS, LIQUEFIED
OSHA PEL: TWA 1000 ppm
NIOSH REL: TWA 350 mg/m^3; CL 1800 mg/m^3/15M
ACGIH TLV: TWA 1000 ppm
DOT Classification: 2.1; Label: Flammable Gas
SAFETY PROFILE: A simple asphyxiant. A very dangerous fire hazard when exposed to heat or flame.

LHF000 CAS: 7789-24-4 FLi HR: 3
LITHIUM FLUORIDE

OSHA PEL: TWA 2.5 mg(F)/m^3

ACGIH TLV: TWA 2.5 mg(F)/m^3; BEI: 3 mg/g creatinine of fluorides in urine prior to shift; 10 mg/g creatinine of fluorides in urine at end of shift.

NIOSH REL: (Inorganic Fluorides) TWA 2.5 mg(F)/m^3

SAFETY PROFILE: Poison by ingestion.

PROP: Fine, white powder or cubic crystals. Mp: 848°, bp: 1676°, d: 2.635 @ 20°, vap press: 1 mm @ 1047°. Sol in acids. Sltly sol in H_2O; insol in EtOH.

LHH000 CAS: 7580-67-8 HLi HR: 3

LITHIUM HYDRIDE

DOT: UN 1414/UN 2805

OSHA PEL: TWA 0.025 mg/m^3

ACGIH TLV: TWA 0.025 mg/m^3

DFG MAK: 0.025 mg/m^3

DOT Classification: 4.3; Label: Dangerous When Wet

SAFETY PROFILE: Poison by inhalation. A severe eye, skin, and mucous membrane irritant. The powder ignites spontaneously in air.

PROP: White, translucent, moisture sensitive crystals. Mp: 688.7° (decomp), d: 0.76–0.77. Darkens rapidly on exposure to light. Reacts with H_2O to form LiOH and H_2, dissoc above mp to form Li metal and H_2.

MAH500 CAS: 1309-48-4 MgO HR: 2

MAGNESIUM OXIDE

SYNS: CALCINED MAGNESITE ♦ MAGNESIUM OXIDE FUME (ACGIH)

OSHA PEL: Fume: Total Dust: TWA 10 mg/m^3; Respirable Fraction: 5 mg/m^3

ACGIH TLV: TWA 10 mg/m^3 (fume)

DFG MAK: 1.5 mg/m^3 (fume)

SAFETY PROFILE: Inhalation of the fumes can be toxic.

PROP: White, bulky, very fine, odorless powder; or colorless cubic crystals, moisture sensitive. Mp: 2832°, bp: 3600°, d: 3.65–3.75. Very sltly sol in water; sol in dil acids; insol in alc.

MAM000 CAS: 108-31-6 $C_4H_2O_3$ HR: 3

MALEIC ANHYDRIDE

DOT: UN 2215

SYNS: cis-BUTENEDIOIC ANHYDRIDE ♦ 2,5-FURANDIONE ♦ MALEIC ACID ANHYDRIDE (MAK)

OSHA PEL: TWA 0.25 ppm

ACGIH TLV: TWA 0.25 ppm; (Proposed: TWA 0.1 ppm (skin, sensitizer); Not Classifiable as a Human Carcinogen)

DFG MAK: 0.1 ppm (0.41 mg/m^3)

DOT Classification: 8; Label: Corrosive

SAFETY PROFILE: Poison by ingestion and inhalation. Moderately toxic by skin contact. A corrosive irritant to eyes and skin. Combustible. It can react violently with many substances.

PROP: Fused black or white crystals. Orthorhombic needles from $CHCl_3$ or by subl. Mp: 52.8°, bp: 202°, flash p: 215°F (CC), d: 1.48 @ 20°/4°, autoign temp: 890°F, vap press: 1 mm @ 44.0°, vap d: 3.4, lel: 1.4%, uel: 7.1%. Sol in dioxane, water @ 30° forming maleic acid; very sltly sol in alc and ligroin.

MAO250 CAS: 109-77-3 $C_3H_2N_2$ HR: 3

MALONONITRILE

DOT: UN 2647

SYNS: CYANOACETONITRILE ♦ MALONIC DINITRILE ♦ METHYLENE CYANIDE ♦ PROPANEDINITRILE

NIOSH REL: (Nitriles) TWA 8 mg/m^3

DOT Classification: 6.1; Label: Poison

SAFETY PROFILE: Poison by ingestion and skin contact. A severe eye irritant. Combustible.

PROP: White powder or crystals. D: 1.049 @ 34°/4°, mp: 30.5°, bp: 220°, flash p: 266°F (TOC). Sol in H_2O, EtOH, Et_2O, and C_6H_6.

MAP750 CAS: 7439-96-5 Mn HR: 3

MANGANESE

SYNS: COLLOIDAL MANGANESE

OSHA PEL: Fume: TWA 1 mg/m^3; STEL 3 mg/m^3; Compounds: CL 5 mg/m^3

ACGIH TLV: Fume: 1 mg/m^3; STEL 3 mg/m^3; Dust and Compounds: TWA 5 mg/m^3 (Proposed: TWA 0.2 mg/m^3)

DFG MAK: 0.5 mg/m^3

SAFETY PROFILE: A skin and eye irritant. Flammable and moderately explosive in the form of dust or powder.

PROP: Reddish-gray or silvery, brittle, metallic element. Reacts with H_2O or steam to give H_2. Oxidizes superficially in air. Mp: 1244°, bp: 2060°, d: 7.20, vap press: 1 mm @ 1292°.

MAR500 HR: 3

MANGANESE COMPOUNDS

SAFETY PROFILE: Can cause central nervous system and pulmonary system damage by inhalation of fumes and dust.

MAV750 CAS: 12108-13-3 $C_9H_7MnO_3$ HR: 3

MANGANESE TRICARBONYL METHYLCYCLOPENTADIENYL

SYNS: METHYLCYCLOPENTADIENYL MANGANESE TRICARBONYL (OSHA) ♦ 2-METHYLCYCLOPENTADIENYL MANGANESE TRICARBONYL (ACGIH) ♦ MMT ♦ TRICARBONYL(METHYLCYCLOPENTADIENYL)MANGANESE

OSHA PEL: TWA 0.2 $mg(Mn)/m^3$ (skin)

ACGIH TLV: TWA 5 mg(Mn)/m^3
SAFETY PROFILE: Poison by ingestion, inhalation, and skin contact. A skin irritant.
PROP: Yellow liquid. D: 1.388 @ 20°/4°, mp: 1.5°, bp: 233°. Almost insol in H_2O; misc in nonpolar solvs.

MCS750 CAS: 1600-27-7 $C_4H_6O_4 \cdot Hg$ HR: 3
MERCURIC ACETATE
DOT: UN 1629
SYNS: BIS(ACETYLOXY)MERCURY ♦ DIACETOXYMERCURY ♦ MERCURIC DIACETATE ♦ MERCURYL ACETATE
OSHA PEL: CL 0.1 mg(Hg)/m^3 (skin)
ACGIH TLV: TWA 0.1 mg(Hg)/m^3 (skin); BEI: 35 µg/g creatinine total inorganic mercury in urine preshift; 15 µg/g creatinine total inorganic mercury in blood at end of shift at end of workweek.
DFG MAK: Confirmed Animal Carcinogen with Unknown Relevance to Humans
NIOSH REL: (Mercury, Aryl and Inorganic) CL 0.1 mg/m^3 (skin)
DOT Classification: 6.1; Label: Poison
SAFETY PROFILE: Poison by ingestion. Moderately toxic by skin contact. An experimental teratogen.
PROP: White crystals or powder; photosensitive; slt acetic odor. D: 3.280, mp: 178–180° (overheating causes decomp). Sol in H_2O and AcOH.

MCU250 CAS: 592-85-8 $C_2HgN_2S_2$ HR: 3
MERCURIC SULFOCYANATE
DOT: UN 1646
SYNS: MERCURIC THIOCYANATE ♦ MERCURY DITHIOCYANATE ♦ MERCURY(II) THIOCYANATE
OSHA PEL: CL 0.1 mg(Hg)/m^3 (skin)
ACGIH TLV: TWA 0.1 mg(Hg)/m^3 (skin); BEI: 35 µg/g creatinine total inorganic mercury in urine preshift; 15 µg/g creatinine total inorganic mercury in blood at end of shift at end of workweek.
DFG MAK: Confirmed Animal Carcinogen with Unknown Relevance to Humans
NIOSH REL: (Mercury, Aryl and Inorganic) CL 0.1 mg/m^3 (skin)
DOT Classification: 6.1; Label: Poison
SAFETY PROFILE: A poison by ingestion. Moderately toxic by skin contact.
PROP: White, odorless powder; sltly sol in cold water; more sol in boiling water (decomp); sol in dil HCl. Protect from light.

MCW250 CAS: 7439-97-6 Hg HR: 3
MERCURY
DOT: NA 2809
SYNS: MERCURY, METALLIC (DOT) ♦ QUICK SILVER
OSHA PEL: Vapor: TWA 0.05 mg/m^3 (skin)

ACGIH TLV: TWA 0.025 mg(Hg)/m^3 (skin); Not Classifiable as a Carcinogen (Proposed: BEI 35 μg/g creatinine total inorganic mercury in urine, preshift)
NIOSH REL: (Mercury, Aryl and Inorganic) CL 0.1 mg/m^3 (skin)
DOT Classification: 8; Label: Corrosive
SAFETY PROFILE: Poison by inhalation. An experimental teratogen.
PROP: Silvery, heavy, mobile liquid at room temp, freezing to a white solid. Solid: tin-white, ductile, malleable mass that can be cut with a knife. A liquid metallic element. Colorless vapor. When heated, reacts with O_2 (historically important reaction), S, halogens. Reacts with conc HNO_3, but not with dil non-oxidizing acids Mp: –38.89°, bp: 356.9°, d: 13.534 @ 25°, vap press: 2 X 10^{-3} mm @ 25°.

MCY000 CAS: 7789-47-1 Br_2Hg HR: 3

MERCURY(II) BROMIDE (1:2)

SYNS: MERCURIC BROMIDE ♦ MERCURIC BROMIDE, solid
OSHA PEL: CL 0.1 mg(Hg)/m^3 (skin)
ACGIH TLV: TWA 0.1 mg(Hg)/m^3 (skin); BEI: 35 μg/g creatinine total inorganic mercury in urine preshift; 15 μg/g creatinine total inorganic mercury in blood at end of shift at end of workweek.
NIOSH REL: (Mercury, Aryl and Inorganic) CL 0.1 mg/m^3 (skin)
SAFETY PROFILE: A poison by ingestion and skin contact.
PROP: White crystals or sublimable, colorless, crystalline powder or yellow liquid. Sensitive to light. Mp: 238°, bp: 318° (subl), d: 6.109 @ 25°, vap press: 1 mm @ 136.5°. Very sol in hot alc, methanol, HCl, HBr, alkali bromide solns; sltly sol in chloroform.

MCZ000 HR: 3

MERCURY COMPOUNDS, INORGANIC

SAFETY PROFILE: A number of mercury compounds can cause skin irritation and be absorbed through the skin. Acute toxicity: Soluble salts have violent corrosive effects on skin and mucous membranes, cause severe nausea, vomiting, abdominal pain, bloody diarrhea, kidney damage, and death usually within 10 days. Many mercury compounds are explosively unstable or undergo hazardous reactions.

MDA000 HR: 3

MERCURY COMPOUNDS, ORGANIC

DFG MAK: 0.01 mg/m^3
SAFETY PROFILE: Alkyl mercurials have very high toxicity; aryl compounds, particularly the phenyls, are much less toxic, and the organomercurials used in therapeutics are less toxic. The alkyls and aryls commonly cause skin burns and other forms of irritation, and both can be absorbed through the skin. Many mercury compounds are explosively unstable.

MDE750 CAS: 10415-75-5 $NO_3 \cdot Hg$ HR: 3

MERCURY(I) NITRATE (1:1)

DOT: UN 1627

SYNS: NITRIC ACID, MERCURY(I) SALT

OSHA PEL: CL 0.1 mg(Hg)/m^3 (skin)

ACGIH TLV: TWA 0.1 mg(Hg)/m^3 (skin); BEI: 35 µg/g creatinine total inorganic mercury in urine preshift; 15 µg/g creatinine total inorganic mercury in blood at end of shift at end of workweek.

NIOSH REL: (Mercury, Aryl and Inorganic) CL 0.1 mg/m^3 (skin)

DOT Classification: 6.1; Label: Poison

SAFETY PROFILE: Poison by ingestion. Moderately toxic by skin contact.

PROP: Crystals.

MDF750 CAS: 15829-53-5 Hg_2O HR: 3

MERCURY(I) OXIDE

OSHA PEL: CL 0.1 mg(Hg)/m^3 (skin)

ACGIH TLV: TWA 0.1 mg(Hg)/m^3 (skin); BEI: 35 µg/g creatinine total inorganic mercury in urine preshift; 15 µg/g creatinine total inorganic mercury in blood at end of shift at end of workweek.

NIOSH REL: (Mercury, Aryl and Inorganic) CL 0.1 mg/m^3 (skin)

SAFETY PROFILE: A poison. Flammable by chemical reaction; an oxidizer. It can react violently with many substances.

PROP: Black to grayish-black powder. Mp: decomp @ 100°, d: 9.8. Insol in water; sol in HNO_3. Protect from light.

MDG500 CAS: 7783-35-9 $O_4S{\cdot}Hg$ HR: 3

MERCURY(II) SULFATE (1:1)

SYNS: MERCURY BISULFATE ♦ MERCURY PERSULFATE

OSHA PEL: CL 0.1 mg(Hg)/m^3 (skin)

ACGIH TLV: TWA 0.1 mg(Hg)/m^3 (skin); BEI: 35 µg/g creatinine total inorganic mercury in urine preshift; 15 µg/g creatinine total inorganic mercury in blood at end of shift at end of workweek.

NIOSH REL: (Mercury, Aryl and Inorganic) CL 0.1 mg/m^3 (skin)

SAFETY PROFILE: Poison by ingestion. Moderately toxic by skin contact.

PROP: White, light-sensitive crystalline powder; odorless. Mp: decomp, d: 6.47. Sol in HCl, hot dilute H_2SO_4, concentrated solns of NaCl. Protect from light.

MDJ750 CAS: 141-79-7 $C_6H_{10}O$ HR: 3

MESITYL OXIDE

DOT: UN 1229

SYNS: ISOBUTENYL METHYL KETONE ♦ METHYL ISOBUTENYL KETONE ♦ 4-METHYL-3-PENTENE-2-ONE

OSHA PEL: TWA 15 ppm; STEL 25 ppm

ACGIH TLV: TWA 15 ppm; STEL 25 ppm

DFG MAK: 25 ppm (100 mg/m^3)

NIOSH REL: (Ketones) TWA 40 mg/m^3
DOT Classification: 3; Label: Flammable Liquid
SAFETY PROFILE: Moderately toxic by ingestion. Mildly toxic by inhalation and skin contact. May damage the eyes and lungs to a serious degree. Dangerous fire hazard.
PROP: Oily, colorless liquid; strong odor. Mp: –59°, bp: 130.0°, flash p: 87°F (CC), d: 0.8539 @ 20°/4°, autoign temp: 652°F, vap press: 10 mm @ 26.0°, vap d: 3.38. Solidifies @ 41.5°; somewhat sol in water @ 20°. Misc in alc and ether and with most organic liquids.

MDN250 CAS: 79-41-4 $C_4H_6O_2$ HR: 3
METHACRYLIC ACID
DOT: UN 2531
SYNS: 2-METHYLPROPENOIC ACID
OSHA PEL: TWA 20 ppm (skin)
ACGIH TLV: TWA 20 ppm
DOT Classification: 8; Label: Corrosive
SAFETY PROFILE: Moderately toxic by ingestion and skin contact. Corrosive to skin, eyes, and mucous membranes. Flammable liquid.
PROP: Corrosive liquid or colorless crystals; repulsive odor. Mp: 16°, bp: 163°, flash p: 171°F (COC), d: 1.014 @ 25° (glacial), vap press: 1 mm @ 25.5°. Sol in warm water; misc with alc, ether.

MDU600 CAS: 16752-77-5 $C_5H_{10}N_2O_2S$ HR: 3
METHOMYL
SYNS: METHYL-N-((METHYLCARBAMOYL)OXY)THIOACETIMIDATE
OSHA PEL: TWA 2.5 mg/m^3
ACGIH TLV: TWA 2.5 mg/m^3; Not Classifiable as a Human Carcinogen
SAFETY PROFILE: Poison by ingestion and inhalation. Mildly toxic by skin contact.
PROP: White, crystalline solid; sulfurous odor. Mp: 78–79°. Moderately sol in water; very sol in Me_2CO, EtOH, MeOH, and toluene.

MFC700 CAS: 150-76-5 $C_7H_8O_2$ HR: 3
4-METHOXYPHENOL
SYNS: HYDROQUINONE MONOMETHYL ETHER ♦ p-METHOXYPHENOL ♦ MME ♦ MONOMETHYL ETHER HYDROQUINONE
OSHA PEL: TWA 5 mg/m^3
ACGIH TLV: TWA 5 mg/m^3
SAFETY PROFILE: A skin irritant.
PROP: White, waxy solid, or leaflets from water. Mp: 52.5°, bp: 246°, d: 1.55 @ 20°/20°.

MFW100 CAS: 79-20-9 $C_3H_6O_2$ HR: 3
METHYL ACETATE
DOT: UN 1231

SYNS: ACETIC ACID METHYL ESTER ♦ METHYL ETHANOATE
OSHA PEL: TWA 200 ppm; STEL 250 ppm
ACGIH TLV: TWA 200 ppm; STEL 250 ppm
DFG MAK: 200 ppm (610 mg/m^3)
DOT Classification: 3; Label: Flammable Liquid
SAFETY PROFILE: Moderately toxic by inhalation. A moderate skin and eye irritant. Dangerous fire hazard.
PROP: Pleasant smelling, colorless, volatile liquid. Mp: –98.7°, lel: 3.1%, uel: 16%, bp: 57.8°, ULC: 85–90, flash p: 14°F, d: 0.92438, autoign temp: 935°F, vap press: 100 mm @ 9.4°, vap d: 2.55. Moderately sol in water; misc in alc, ether.

MFX590 CAS: 74-99-7 C_3H_4 HR: 3

METHYL ACETYLENE

SYNS: PROPINE ♦ PROPYNE (OSHA)
OSHA PEL: TWA 1000 ppm
ACGIH TLV: TWA 1000 ppm
DFG MAK: 1000 ppm (1700 mg/m^3)
SAFETY PROFILE: Dangerous fire hazard. Explosive in the form of vapor
PROP: Gas. Mp: –104°, lel: 1.7%, bp: –23.3°, vap press: 3876 mm @ 20°, d: 1.787 g/L @ 0°, vap d: 1.38. Mod sol in H_2O; sol in EtOH and Et_2O.

MFX600 CAS: 59355-75-8 HR: 3

METHYL ACETYLENE-PROPADIENE MIXTURE

DOT: UN 1060
SYNS: MAPP (OSHA) ♦ METHYLACETYLENE and PROPADIENE MIXTURES, stabilized (DOT)
OSHA PEL: TWA 1000 ppm; STEL 1250 ppm
ACGIH TLV: TWA 1000 ppm; STEL 1250 ppm
DFG MAK: 1000 ppm (1650 mg/m^3)
DOT Classification: 2.1; Label: Flammable Gas
SAFETY PROFILE: A flammable gas mixture.

MGA500 CAS: 96-33-3 $C_4H_6O_2$ HR: 3

METHYL ACRYLATE

DOT: UN 1919
SYNS: ACRYLIC ACID METHYL ESTER (MAK) ♦ METHYL PROPENATE ♦ METHYL PROPENOATE ♦ PROPENOIC ACID METHYL ESTER
OSHA PEL: TWA 10 ppm (skin)
ACGIH TLV: TWA 2 ppm (skin); Not Classifiable as a Human Carcinogen; (Proposed: TWA 2 ppm (skin, sensitizer); Not Classifiable as a Human Carcinogen)
DFG MAK: 5 ppm (18 mg/m^3)
DOT Classification: 3; Label: Flammable Liquid
SAFETY PROFILE: Poison by ingestion. Moderately toxic by skin contact. A skin and eye irritant. Dangerously flammable.

PROP: Colorless liquid; acrid odor. D: 0.9561 @ 20°/4°, mp: –76.5°, bp: 85° @ 608 mm, lel: 2.8%, uel: 25%, fp: –75°, flash p: 27°F (OC), vap press: 100 mm @ 28°, vap d: 2.97. Sol in alc and ether.

MGA750 CAS: 126-98-7 C_4H_5N HR: 3
METHYLACRYLONITRILE
SYNS: ISOPROPENE CYANIDE ♦ ISOPROPENYLNITRILE ♦ 2-METHYLPROPENENITRILE
OSHA PEL: TWA 1 ppm (skin)
ACGIH TLV: TWA 1 ppm (skin)
DOT Classification: 3; Label: Flammable Liquid, Poison
SAFETY PROFILE: Poison by ingestion, inhalation, and skin contact. An eye irritant. A dangerous fire hazard.
PROP: Colorless liquid. Mp: –36°, bp: 90.3°, d: 0.805, vap press: 40 mm @ 12.8°, flash p: 55°F. Insol in water.

MGA850 CAS: 109-87-5 $C_3H_8O_2$ HR: 3
METHYLAL
DOT: UN 1234
SYNS: DIMETHYL FORMAL ♦ FORMAL ♦ METHYLENE DIMETHYL ETHER
OSHA PEL: TWA 1000 ppm
ACGIH TLV: TWA 1000 ppm
DFG MAK: 1000 ppm (3200 mg/m^3)
DOT Classification: 3; Label: Flammable Liquid
SAFETY PROFILE: Moderately toxic by ingestion and inhalation. A very dangerous fire hazard.
PROP: Colorless volatile liquid; pungent odor. Mp: –104.8°, bp: 42.3°, d: 0.864 @ 20°/4°, vap press: 330 mm @ 20°, vap d: 2.63, autoign temp: 459°F, flash p: –0.4°F.

MGB150 CAS: 67-56-1 CH_4O HR: 3
METHYL ALCOHOL
DOT: UN 1230
SYNS: METHANOL ♦ WOOD ALCOHOL (DOT) ♦ WOOD NAPHTHA ♦ WOOD SPIRIT
OSHA PEL: TWA 200 ppm; STEL 250 ppm (skin)
ACGIH TLV: TWA 200 ppm; STEL 250 ppm (skin); BEI: 15 mg/L of methanol in urine at end of shift
DFG MAK: 200 ppm (270 mg/m^3); BAT: 30 mg/L in urine at end of shift
NIOSH REL: TWA 200 ppm; CL 800 ppm/15M
DOT Classification: 3; Label: Flammable Liquid, Poison
SAFETY PROFILE: A human poison by ingestion. Mildly toxic by inhalation. An experimental teratogen. An eye and skin irritant. A flammable liquid. It can react violently with many substances.
PROP: Clear, colorless, very mobile liquid; slt alcoholic odor when pure; crude material may have a repulsive pungent odor. Bp: 64.8°, lel: 6.0%, uel: 36.5%, ULC: 70, fp: –97.8°, d: 0.7915 @ 20°/4°, flash p: 54°F, autoign temp: 878°F, vap press: 100 mm @

21.2°, vap d: 1.11. Misc in water, ethanol, ether, benzene, ketones, and most other org solvs. Part misc in pet ether.

MGC250 CAS: 74-89-5 CH_5N HR: 3
METHYLAMINE
DOT: UN 1061/UN 1235
SYNS: METHYLAMINE (ACGIH,OSHA) ♦ MONOMETHYLAMINE
OSHA PEL: TWA 10 ppm
ACGIH TLV: TWA 5 ppm; STEL 15 ppm
DFG MAK: 10 ppm (13 mg/m^3)
DOT Classification: 2.3; Label: Poison Gas, Flammable Gas (UN 1061); DOT Class: 3; Label: Flammable Liquid, Corrosive (UN 1235)
SAFETY PROFILE: Poison. Moderately toxic by inhalation. A severe skin irritant. A flammable gas.
PROP: Colorless, flammable gas or liquid; powerful ammonia-like odor. Frequently encountered as strong aq soln. Bp: 6.3°, lel: 4.95%, uel: 20.75%, mp: –93.5°, flash p: 32°F (CC), d: 0.662 @ 20°/4°, autoign temp: 806°F, vap d: 1.07. Fuming liquid when liquefied: d: 0.699 @ –10.8°/4°. Sol in alc; misc with ether.

MGN500 CAS: 110-43-0 $C_7H_{14}O$ HR: 3
METHYL n-AMYL KETONE
DOT: UN 1110
SYNS: n-AMYL METHYL KETONE ♦ AMYL METHYL KETONE (DOT) ♦ 2-HEPTANONE ♦ METHYL PENTYL KETONE
OSHA PEL: TWA 100 ppm
ACGIH TLV: TWA 50 ppm
NIOSH REL: (Ketones) TWA 465 mg/m^3
DOT Classification: 3; Label: Flammable Liquid
SAFETY PROFILE: Moderately toxic by ingestion. Mildly toxic by inhalation and skin contact. A flammable liquid.
PROP: Colorless, mobile liquid; penetrating, fruity odor; or light yellow oil. Bp: 151.5°, flash p: 120°F (OC), autoign temp: 991°F, vap d: 3.94, d: 0.8197 @ 15°/4°. Very sltly sol in water; sol in alc and ether.

MGN750 CAS: 100-61-8 C_7H_9N HR: 3
METHYLANILINE
DOT: UN 2294
SYNS: N-METHYLAMINOBENZENE ♦ N-METHYLBENZENAMINE ♦ N-METHYLPHENYLAMINE ♦ MONOMETHYL ANILINE (OSHA) ♦ N-PHENYLMETHYLAMINE
OSHA PEL: TWA 0.5 ppm (skin)
ACGIH TLV: TWA 0.5 ppm (skin)
DFG MAK: 0.5 ppm (2.2 mg/m^3)
DOT Classification: 6.1; Label: KEEP AWAY FROM FOOD
SAFETY PROFILE: Poison by ingestion.

PROP: Colorless or sltly yellow liquid; becomes brown on exposure to air. Mp: –57°, d: 0.989 @ 20°/4°, bp: 194–197°. Sol in alc, ether; sltly sol in water.

MHR200 CAS: 74-83-9 CH_3Br HR: 3

METHYL BROMIDE

SYNS: BROMOMETHANE ♦ MONOBROMOMETHANE
OSHA PEL: TWA 5 ppm (skin)
ACGIH TLV: TWA 1 ppm (skin); Not Classifiable as a Human Carcinogen
DFG MAK: Confirmed Animal Carcinogen with Unknown Relevance to Humans
NIOSH REL: (Monohalomethanes) Reduce to lowest level
A human poison by inhalation. Corrosive to skin.
Mixtures of 10–15 percent with air may be ignited with difficulty. Moderately explosive when exposed to sparks or flame. Forms explosive mixtures with air within narrow limits at atmospheric pressure, with wider limits at higher pressure. The explosive sensitivity of mixtures with air may be increased by the presence of aluminum, magnesium, zinc, or their alloys. Incompatible with metals, dimethyl sulfoxide, ethylene oxide.
PROP: Colorless, transparent, volatile liquid or gas; burning taste, chloroform-like odor. Bp: 3.56°, lel: 13.5%, uel: 14.5%, fp: –93°, flash p: none, d: 1.732 @ 0°/0°, autoign temp: 998°F, vap d: 3.27, vap press: 1824 mm @ 25°. Sltly sol in water.

MIF765 CAS: 74-87-3 CH_3Cl HR: 3

METHYL CHLORIDE

DOT: UN 1063
SYNS: CHLOROMETHANE ♦ MONOCHLOROMETHANE
OSHA PEL: TWA 50 ppm; STEL 100 ppm
ACGIH TLV: TWA 50 ppm; STEL 100 ppm; Not Classifiable as a Human Carcinogen
DFG MAK: 50 ppm (100 mg/m^3); Suspected Carcinogen
NIOSH REL: (Monohalomethanes) TWA Reduce to lowest level
DOT Classification: 2.1; Label: Flammable Gas
SAFETY PROFILE: Suspected carcinogen. Very mildly toxic by inhalation. An experimental teratogen. A flammable gas. It can react violently with many substances.
PROP: Colorless gas; ethereal odor and sweet taste. D: 0.918 @ 20°/4°, mp: –97°, bp: –23.7°, flash p: <32°F, lel: 8.1%, uel: 17%, autoign temp: 1170°F, vap d: 1.78. Sltly sol in water; misc with chloroform, ether, glacial acetic acid; sol in alc.

MIH275 CAS: 71-55-6 $C_2H_3Cl_3$ HR: 3

METHYL CHLOROFORM

DOT: UN 2831
SYNS: METHYLTRICHLOROMETHANE ♦ 1,1,1-TCE ♦ 1,1,1-TRICHLOROETHANE
OSHA PEL: TWA 350 ppm; STEL 450 ppm
ACGIH TLV: TWA 350 ppm; STEL 450 ppm; BEI: 10 mg/L trichloroacetic acid in urine at end of workweek; Not Classifiable as a Human Carcinogen
DFG MAK: 200 ppm (1100 mg/m^3); BAT: 55 µg/dL in blood after several shifts
NIOSH REL: (1,1,1-Trichloroethane) CL 350 ppm/15M
DOT Classification: 6.1; Label: KEEP AWAY FROM FOOD

SAFETY PROFILE: Poison. Moderately toxic by ingestion, inhalation, and skin contact. An experimental teratogen.
PROP: Colorless, nonflammable liquid. Bp: 74.1°, fp: −32.5°, flash p: none, d: 1.3376 @ 20°/4°, vap press: 100 mm @ 20.0°. Insol in water; sol in acetone, benzene, carbon tetrachloride, methanol, ether.

MIQ075 CAS: 137-05-3 $C_9H_{13}NO_2$ HR: 2
METHYL 2-CYANOACRYLATE
SYNS: α-CYANOACRYLIC ACID METHYL ESTER ♦ 2-CYANOACRYLIC ACID, METHYL ESTER ♦ SUPER GLUE
OSHA PEL: TWA 2 ppm; STEL 4 ppm
ACGIH TLV: TWA 0.2 ppm
DFG MAK: 2 ppm (9.2 mg/m^3)
SAFETY PROFILE: Moderately toxic by ingestion and inhalation routes. A human eye irritant.
PROP: Thick, clear colorless liquid; sharp odor. Bp: 47–48° @ 2 mm.

MIQ740 CAS: 108-87-2 C_7H_{14} HR: 3
METHYLCYCLOHEXANE
DOT: UN 2296
SYNS: CYCLOHEXYLMETHANE ♦ HEXAHYDROTOLUENE ♦ TOLUENE HEXAHYDRIDE
OSHA PEL: TWA 400 ppm
ACGIH TLV: TWA 400 ppm
DFG MAK: 500 ppm (2000 mg/m^3)
DOT Classification: 3; Label: Flammable Liquid
SAFETY PROFILE: Moderately toxic by ingestion. Mildly toxic by inhalation and skin contact. Dangerous fire hazard.
PROP: Colorless liquid. Mp: −126.4°, lel: 1.2%, uel: 6.7%, bp: 100.3°, flash p: 25°F (CC), d: 0.7864 @ 0°/4°, 0.769 @ 20°/4°, vap press: 40 mm @ 22.0°, vap d: 3.39, autoign temp: 482°F.

MIQ745 CAS: 25639-42-3 $C_7H_{14}O$ HR: 3
METHYLCYCLOHEXANOL
DOT: UN 2617
SYNS: HEXAHYDROCRESOL ♦ HEXAHYDROMETHYLPHENOL
OSHA PEL: TWA 50 ppm
ACGIH TLV: TWA 50 ppm
DFG MAK: 50 ppm (235 mg/m^3)
DOT Classification: 3; Label: Flammable Liquid
SAFETY PROFILE: Moderately toxic by ingestion. Mildly toxic by skin contact and by inhalation. Combustible.
PROP: Colorless, viscous liquid; aromatic, menthol-like odor. Bp: 155–180°, flash p: 154°F (CC), autoign temp: 565°F, d: 0.924 @ 15.5°/15.5°, vap d: 3.93.

MIR500 CAS: 583-60-8 $C_7H_{12}O$ HR: 3

2-METHYLCYCLOHEXANONE

SYNS: 1-METHYLCYCLOHEXAN-2-ONE

OSHA PEL: TWA 50 ppm; STEL 75 ppm (skin)

ACGIH TLV: TWA 50 ppm (skin)

DFG MAK: 50 ppm (230 mg/m^3)

SAFETY PROFILE: Moderately toxic by ingestion and skin contact.

PROP: Liquid. D: 0.925 @ 20°/4°, mp: −14°, bp: 165.1°. Insol in water; sol in alc and ether.

MJM200 CAS: 101-14-4 $C_{13}H_{12}Cl_2N_2$ HR: 3

4,4'-METHYLENE BIS(2-CHLOROANILINE)

SYNS: DI(-4-AMINO-3-CHLOROPHENYL)METHANE ♦ MBOCA ♦ METHYLENE-4,4'-BIS(o-CHLOROANILINE) ♦ p,p'-METHYLENEBIS(o-CHLOROANILINE) ♦ METHYLENE-BIS-ORTHOCHLOROANILINE ♦ MOCA

OSHA PEL: TWA 0.02 ppm (skin)

ACGIH TLV: TWA 0.02 ppm (skin); Suspected Human Carcinogen (Proposed: 0.01 ppm; Suspected Human Carcinogen)

DFG MAK: Animal Carcinogen, Suspected Human Carcinogen

NIOSH REL: (MOCA): TWA 0.003 mg/m^3 (Skin)

SAFETY PROFILE: Confirmed carcinogen. Poison by ingestion. Flammable liquid.

PROP: Tan solid.

MJM600 CAS: 5124-30-1 $C_{15}H_{22}NO_2$ HR: 3

METHYLENE BIS(4-CYCLOHEXYLISOCYANATE)

SYNS: BIS(4-ISOCYANATOCYCLOHEXYL)METHANE ♦ DICYCLOHEXYLMETHANE-4,4'-DIISOCYANATE

OSHA PEL: CL 0.01

ACGIH TLV: TWA 0.005 ppm

NIOSH REL: (Dicyclohexylmethane 4,4'-diisocyanate) TWA CL 0.01 ppm

SAFETY PROFILE: Poison by inhalation. Mildly toxic by ingestion.

PROP: Colorless liquid.

MJN000 CAS: 101-61-1 $C_{17}H_{22}N_2$ HR: 3

4,4'-METHYLENE BIS(N,N'-DIMETHYLANILINE)

SYNS: p,p'-BIS(DIMETHYLAMINO)DIPHENYLMETHANE ♦ BIS(p-DIMETHYLAMINOPHENYL)METHANE ♦ p,p-DIMETHYLAMINODIPHENYLMETHANE ♦ MICHLER'S BASE ♦ 4,4'-TETRAMETHYLDIAMINODIPHENYLMETHANE

DFG MAK: Confirmed Animal Carcinogen, Suspected Human Carcinogen

DOT Classification: 3; Label: Flammable Liquid

SAFETY PROFILE: Confirmed carcinogen. Moderately toxic by ingestion. A flammable liquid.

PROP: Crystals, leaflets, or plates from EtOH or ligroin. Mp: 91°, bp: 390°. Sol in Me_2CO.

MJN750 CAS: 139-25-3 $C_{17}H_{14}N_2O_2$ HR: 3

5,5'-METHYLENEBIS(2-ISOCYANATO)TOLUENE

SYNS: 3,3'-DIMETHYLDIPHENYLMETHANE-4,4'-DIISOCYANATE
NIOSH REL: (Diisocyanates) TWA 0.005 ppm; CL 0.02 ppm/10M
SAFETY PROFILE: A sensitizer.

MJO250 CAS: 838-88-0 $C_{15}H_{18}N_2$ HR: 3

4,4'-METHYLENEBIS(2-METHYLANILINE)

SYNS: MBOT ♦ ME-MDA ♦ 4,4'-METHYLENE DI-o-TOLUIDINE
DFG MAK: Confirmed Animal Carcinogen, Suspected Human Carcinogen
SAFETY PROFILE: Confirmed carcinogen. Moderately toxic by ingestion. An eye irritant.
PROP: Pale-amber crystals from EtOH. Mp: 158–159°.

MJP400 CAS: 101-68-8 $C_{15}H_{10}N_2O_2$ HR: 3

METHYLENE BISPHENYL ISOCYANATE

DOT: UN 2489
SYNS: BIS(4-ISOCYANATOPHENYL)METHANE ♦ MDI ♦ METHYLENEBIS(p-PHENYLENE ISOCYANATE) ♦ METHYLENE DI(PHENYLENE ISOCYANATE) (DOT) ♦ 4,4'-METHYLENEDIPHENYL ISOCYANATE
OSHA PEL: CL 0.02 ppm
ACGIH TLV: 0.005 ppm
DFG MAK: 0.05 mg/m^3; Confirmed Animal Carcinogen with Unknown Relevance to Humans
NIOSH REL: (Diisocyanates) TWA 0.005 ppm; CL 0.02 ppm/10M
DOT Classification: 6.1; Label: KEEP AWAY FROM FOOD; DOT Class: 6.1; Label: Poison; DOT Class: 6.1; Label: Poison, Flammable Liquid; DOT Class: 3; Label: Flammable Liquid, Poison
SAFETY PROFILE: Poison by inhalation. Mildly toxic by ingestion. A skin and eye irritant. An allergic sensitizer. A flammable liquid.
PROP: Crystals or yellow fused solid. Mp: 37.2°, bp: 184° @ 3 mm, d: 1.19 @ 50°, vap press: 0.001 mm @ 40°.

MJP450 CAS: 75-09-2 CH_2Cl_2 HR: 3

METHYLENE CHLORIDE

DOT: UN 1593
SYNS: DICHLOROMETHANE (MAK, DOT) ♦ FREON 30 ♦ METHANE DICHLORIDE ♦ METHYLENE DICHLORIDE
OSHA PEL: 25 ppm
ACGIH TLV: TWA 50 ppm; Animal Carcinogen
DFG MAK: Confirmed Animal Carcinogen with Unknown Relevance to Humans; 100 ppm (350 mg/m^3); BAT: 5% CO-Hb in blood at end of shift;
NIOSH REL: (Methylene Chloride) Reduce to lowest feasible level
DOT Classification: 6.1; Label: KEEP AWAY FROM FOOD

SAFETY PROFILE: Confirmed carcinogen. Moderately toxic by ingestion. Mildly toxic by inhalation. An experimental teratogen. An eye and severe skin irritant. It is flammable but ignition is difficult. It can react violently with many substances.

PROP: Colorless, volatile liquid; odor of chloroform. Bp: 39.8°, lel: 15.5% in O_2, uel: 66.4% in O_2, fp: –96.7°, d: 1.326 @ 20°/4°, autoign temp: 1139°F, vap press: 380 mm @ 22°, vap d: 2.93, refr index: 1.424 @ 20 L. Sol in water; misc with alc, acetone, chloroform, ether, and carbon tetrachloride.

MJQ000 CAS: 101-77-9 $C_{13}H_{14}N_2$ HR: 3

4,4'-METHYLENEDIANILINE

DOT: UN 2651

SYNS: BIS(4-AMINOPHENYL)METHANE ♦ DAPM ♦ 4,4'-DIAMINODIPHENYLMETHANE (DOT) ♦ DI-(4-AMINOPHENYL)METHANE ♦ MDA ♦ 4,4'-METHYLENEBISANILINE

ACGIH TLV: TWA 0.1 ppm (skin); Animal Carcinogen

DFG MAK: Animal Carcinogen, Suspected Human Carcinogen

DOT Classification: 6.1; Label: KEEP AWAY FROM FOOD

SAFETY PROFILE: Confirmed carcinogen. A poison by ingestion. An eye irritant. Combustible.

PROP: Tan flakes, lumps, or pearly leaflets from benzene; faint amine-like odor. Mp: 93°, flash p: 440°F, bp: 232° @ 9 mm.

MKA400 CAS: 78-93-3 C_4H_8O HR: 3

METHYL ETHYL KETONE

DOT: UN 1193

SYNS: 2-BUTANONE (OSHA) ♦ ETHYL METHYL KETONE (DOT) ♦ MEK ♦ METHYL ACETONE (DOT)

OSHA PEL: TWA 200 ppm; STEL 300 ppm

ACGIH TLV: TWA 200 ppm; STEL 300 ppm; BEI: 2 mg(MEK)/L in urine at end of shift

DFG MAK: 200 ppm (600 mg/m^3)

NIOSH REL: (Ketones) TWA 590 mg/m^3

DOT Classification: 3; Label: Flammable Liquid

SAFETY PROFILE: Moderately toxic by ingestion, and skin contact. An experimental teratogen. Human eye irritation. Highly flammable liquid.

PROP: Colorless liquid; acetone-like odor. Fp: –85.9°, bp: 79.57°, lel: 1.8%, uel: 11.5%, flash p: 22°F (TOC), d: 0.80615 @ 20°/20°, vap press: 71.2 mm @ 20°, autoign temp: 960°F, vap d: 2.42, ULC: 85–90. Misc with alc, ether, fixed oils, and water.

MKA500 CAS: 1338-23-4 $C_8H_{16}O_4$ HR: 3

METHYL ETHYL KETONE PEROXIDE

SYNS: ETHYL METHYL KETONE PEROXIDE ♦ MEK PEROXIDE ♦ MEKP (OSHA) ♦ METHYL ETHYL KETONE HYDROPEROXIDE

OSHA PEL: CL 0.7 ppm

ACGIH TLV: CL 0.2 ppm

DFG MAK: Organic Peroxide, moderate skin irritant

DOT Classification: Forbidden
SAFETY PROFILE: Poison. Moderately toxic by ingestion and inhalation. A moderate skin and eye irritant. A shock-sensitive explosive.
PROP: Colorless liquid.

MKG750 CAS: 107-31-3 $C_2H_4O_2$ HR: 3
METHYL FORMATE
DOT: UN 1243
SYNS: METHYL METHANOATE
OSHA PEL: TWA 100 ppm; STEL 150 ppm
ACGIH TLV: TWA 100 ppm; STEL 150 ppm
DFG MAK: 50 ppm (120 mg/m^3)
DOT Classification: 3; Label: Flammable Liquid
SAFETY PROFILE: Moderately toxic by ingestion. Poison by inhalation. Flammable liquid.
PROP: Colorless liquid; agreeable odor. Mp: –99.8°, bp: 31.5°, lel: 5.9%, uel: 20%, flash p: –2.2°F, d: 0.98149 @ 15°/4°, 0.975 @ 20°/4°, autoign temp: 869°F, vap press: 400 mm @ 16°/0°, vap d: 2.07. Solidifies at about 100°. Moderately sol in water, methyl alcohol; misc in alc.

MKN000 CAS: 60-34-4 CH_6N_2 HR: 3
METHYL HYDRAZINE
DOT: UN 1244
SYNS: HYDRAZOMETHANE ♦ METHYLHYDRAZINE (DOT) ♦ MMH ♦ MONOMETHYL HYDRAZINE
OSHA PEL: CL 0.2 ppm (skin)
ACGIH TLV: CL 0.2 ppm; Suspected Human Carcinogen (Proposed: CL 0.01 ppm; Suspected Human Carcinogen)
NIOSH REL: CL 0.08 mg/m^3/2H
DOT Classification: 6.1; Label: Poison, Flammable Liquid, Corrosive
SAFETY PROFILE: Suspected carcinogen. Poison by inhalation, ingestion, and skin contact. An experimental teratogen. Corrosive to skin, eyes, and mucous membranes. May self-ignite in air.
PROP: Colorless, hydroscopic liquid; ammonia-like odor. D: 0.874 @ 25°, mp: –20.9°, bp: 87.8°, vap d: 1.6, flash p: 73.4°F, fp: –52.4°, autoign temp: 196°, lel: 2.5%, uel: 97 ± 2%. Sltly sol in water; sol in alc, hydrocarbons, and ether; misc with hydrazine. Strong reducing agent.

MKN250 CAS: 7339-53-9 $CH_6N_2{\cdot}ClH$ HR: 3
METHYLHYDRAZINE HYDROCHLORIDE
NIOSH REL: (Hydrazines) CL 0.08 mg/m^3/2H
SAFETY PROFILE: Poison by ingestion.

MKW200 CAS: 78-88-4 CH_3I HR: 3
METHYL IODIDE

DOT: UN 2644
SYNS: IODOMETHANE
OSHA PEL: TWA 2 ppm (skin)
ACGIH TLV: TWA 0.01 ppm; Animal Carcinogen
DFG MAK: Animal Carcinogen, Suspected Human Carcinogen
NIOSH REL: (Monohalomethanes) Reduce to lowest level
DOT Classification: 6.1; Label: Poison
SAFETY PROFILE: Confirmed carcinogen. A poison by ingestion. Moderately toxic by inhalation and skin contact. A human skin irritant.
PROP: Colorless liquid with pleasant odor, turns brown on exposure to light. Mp: –66.4°, bp: 42.5°, d: 2.279 @ 20°/4°, vap press: 400 mm @ 25.3°, vap d: 4.89. Sol in water @ 15°, misc in alc and ether.

MKW450 CAS: 110-12-3 $C_7H_{14}O$ HR: 3
METHYL ISOAMYL KETONE
DOT: UN 2302
SYNS: ISOAMYL METHYL KETONE ♦ ISOPENTYL METHYL KETONE ♦ 2-METHYL-5-HEXANONE ♦ 5-METHYLHEXAN-2-ONE (DOT) ♦ MIAK
OSHA PEL: TWA 50 ppm
ACGIH TLV: TWA 50 ppm
NIOSH REL: Ketones (Methyl Isoamyl Ketone) TWA 230 mg/m^3
DOT Classification: 3; Label: Flammable Liquid
SAFETY PROFILE: Moderately toxic by ingestions. Mildly toxic by inhalation and skin contact. A flammable liquid.
PROP: Colorless, stable liquid; pleasant odor. Bp: 144°, d: 0.8132 @ 20°/20°, fp: –73.9°, flash p: 110°F (OC). Sltly sol in water; misc with most org solvs.

MKW600 CAS: 108-11-2 $C_6H_{14}O$ HR: 3
METHYL ISOBUTYL CARBINOL
DOT: UN 2053
SYNS: ISOBUTYL METHYL CARBINOL ♦ MAOH ♦ METHYL AMYL ALCOHOL ♦ 4-METHYL-2-PENTANOL (MAK) ♦ MIBC ♦ MIC
OSHA PEL: TWA 25 ppm; STEL 40 ppm (skin)
ACGIH TLV: TWA 25 ppm; STEL 40 ppm (skin)
DFG MAK: 25 ppm (110 mg/m^3)
DOT Classification: 3; Label: Flammable Liquid
SAFETY PROFILE: Moderately toxic by ingestion and skin contact. Mildly toxic by inhalation. A skin and severe eye irritant. Flammable liquid.
PROP: Clear liquid. Bp: 131.8°, fp: <–90° (sets to a glass), flash p: 106°F, d: 0.8079 @ 20°/20°, vap press: 2.8 mm @ 20°, vap d: 3.53, lel: 1.0%, uel: 5.5%.

MKX250 CAS: 624-83-9 C_2H_3NO HR: 3
METHYL ISOCYANATE
DOT: UN 2480
SYNS: ISOCYANIC ACID, METHYL ESTER ♦ MIC
OSHA PEL: TWA 0.02 ppm (skin)

ACGIH TLV: TWA 0.02 ppm (skin)
DFG MAK: 0.01 ppm (0.025 mg/m^3)
DOT Classification: 6.1; Label: Poison, Flammable Liquid; DOT Class: 3; Label: Flammable Liquid, Poison
SAFETY PROFILE: Poison by inhalation, ingestion, and skin contact. An experimental teratogen. A severe eye, skin, and mucous membrane irritant and a sensitizer. A flammable liquid.
PROP: Liquid; sharp, unpleasant odor. D: 0.9599 @ 20°/20°, bp: 43–45°, flash p: <5°F.

MLA750 CAS: 563-80-4 $C_5H_{10}O$ HR: 3
METHYL ISOPROPYL KETONE
DOT: UN 2397
SYNS: ISOPROPYL METHYL KETONE ♦ 3-METHYL BUTAN-2-ONE (DOT) ♦ MIPK
OSHA PEL: TWA 200 ppm
ACGIH TLV: TWA 200 ppm
DOT Classification: 3; Label: Flammable Liquid
SAFETY PROFILE: Poison by ingestion. Mildly toxic by inhalation and skin contact. A skin and eye irritant. Flammable.
PROP: Colorless liquid; acetone-like odor. D: 0.805 @ 16°/4°, bp: 93–94°.

MLC750 CAS: 75-86-5 C_4H_7NO HR: 3
2-METHYLLACTONITRILE
DOT: UN 1541
SYNS: ACETONE CYANOHYDRIN (ACGIH,DOT) ♦ 2-HYDROXY-2-METHYLPROPIONITRILE
ACGIH TLV: CL 4.7 ppm (skin)
NIOSH REL: (Nitriles) CL 4 mg/m^3/15M
DOT Classification: 6.1; Label: Poison
SAFETY PROFILE: Poison by ingestion, skin contact, and inhalation. Combustible.
PROP: Mp: –20°, bp: 82° @ 23 mm, d: 0.932 @ 19°, autoign temp: 1270°F, flash p: 165°F, vap d: 2.93. Very sol in H_2O; spar sol in pet ether.

MLE650 CAS: 74-93-1 CH_4S HR: 3
METHYL MERCAPTAN
DOT: UN 1064
SYNS: METHANETHIOL ♦ THIOMETHANOL
OSHA PEL: TWA 0.5 ppm
DFG MAK: 0.5 ppm (1 mg/m^3)
NIOSH REL: (n-Alkane Monothiols) CL 0.5 ppm/15M
DOT Classification: 2.3; Label: Poison Gas, Flammable Gas
SAFETY PROFILE: Poison by inhalation. Very dangerous fire hazard.
PROP: Gas; odor of rotten cabbage. Mp: –123.1°, vap d: 1.66, lel: 3.9%, uel: 21.8%, bp: 5.95°, d: 0.8665 @ 20°/4°, solidifies @ –123°, flash p: –0.4°F. Sol in water.

MLF550 CAS: 22967-92-6 CH_3Hg HR: 3

METHYLMERCURY

OSHA PEL: TWA 0.01 mg(Hg)/m^3; STEL 0.03 mg/m^3 (skin)

ACGIH TLV: TWA 0.01 mg(Hg)/m^3; BEI: 35 µg/g creatinine total inorganic mercury in urine preshift; 15 µg/g creatinine total inorganic mercury in blood at end of shift at end of workweek.

DFG MAK: Confirmed Animal Carcinogen with Unknown Relevance to Humans

NIOSH REL: (Mercury, Organo) TWA 0.01 mg/m^3; STEL 0.03 mg/m^3 (skin)

SAFETY PROFILE: A poison. An experimental teratogen.

MLH750 CAS: 80-62-6 $C_5H_8O_2$ HR: 3

METHYL METHACRYLATE

DOT: NA 1247

SYNS: METHACRYLIC ACID, METHYL ESTER (MAK) ♦ METHYL METHACRYLATE MONOMER, INHIBITED (DOT) ♦ METHYL-2-METHYL-2-PROPENOATE ♦ MME

OSHA PEL: TWA 100 ppm

ACGIH TLV: TWA 100 ppm; Not Classifiable as a Human Carcinogen; (Proposed: TWA 50 ppm; STEL 100 ppm (sensitizer); Not Classifiable as a Human Carcinogen)

DFG MAK: 50 ppm (210 mg/m^3)

SAFETY PROFILE: Moderately toxic by inhalation. Mildly toxic by ingestion. Experimental teratogenic effects. A skin and eye irritant. A very dangerous fire hazard. It can react violently with many substances.

PROP: Colorless liquid; sharp, fruity odor. Mp: –50°, bp: 101.0°, flash p: 50°F (OC), d: 0.936 @ 20°/4°, vap press: 40 mm @ 25.5°, vap d: 3.45, lel: 2.1%, uel: 12.5%. Very sltly sol in water. Sol in Me_2CO.

MMU250 CAS: 614-00-6 $C_7H_8N_2O$ HR: 3

N-METHYL-N-NITROSOANILINE

SYNS: N-METHYL-N-NITROSOBENZENAMINE ♦ MNA ♦ N-NITROSOMETHYLPHENYLAMINE (MAK) ♦ NMA ♦ PHENYLMETHYLNITROSAMINE

DFG MAK: Animal Carcinogen, Suspected Human Carcinogen

SAFETY PROFILE: Confirmed carcinogen. Poison by ingestion. An experimental teratogen.

PROP: A liquid. D: 1.124 @ 20°/4°, mp: 13°, bp: 128–128.4° @ 19 mm.

MPF200 CAS: 872-50-4 C_5H_9NO HR: 3

N-METHYLPYRROLIDONE

SYNS: N-METHYL-2-PYRROLIDINONE ♦ 1-METHYL-5-PYRROLIDINONE ♦ 1-METHYL-2-PYRROLIDONE ♦ NMP

DFG MAK: 19 ppm (80 mg/m^3)

SAFETY PROFILE: Moderately toxic by ingestion. Mildly toxic by skin contact. An experimental teratogen. Combustible.

PROP: Colorless liquid; mild odor. Fp: –24°, mp: –17°, bp: 202°, flash p: 204°F (OC), d: 1.027 @ 25°/4°, vap d: 3.4.

MPI750 CAS: 681-84-5 $C_4H_{12}O_4Si$ HR: 3

METHYL SILICATE

DOT: UN 2606

SYNS: METHYL ORTHOSILICATE (DOT) ♦ TETRAMETHOXYSILANE ♦ TETRAMETHYL SILICATE

OSHA PEL: TWA 1 ppm

ACGIH TLV: TWA 1 ppm

DOT Classification: 3; Label: Flammable Liquid, Poison

SAFETY PROFILE: Moderately toxic by inhalation. Mildly toxic by skin contact. A severe eye irritant. A flammable liquid.

PROP: Clear liquid. Vap d: 5.25, d: 1.03 @ 22°/4°, mp: 4–5°, bp: 120–121°.

MPK250 CAS: 98-83-9 C_9H_{10} HR: 3

α-METHYL STYRENE

DOT: UN 2303

SYNS: ISOPROPENYLBENZENE ♦ 2-PHENYLPROPENE ♦ 2-PHENYLPROPYLENE

OSHA PEL: TWA 50 ppm; STEL 100 ppm

ACGIH TLV: TWA 50 ppm; STEL 100 ppm

DFG MAK: 100 ppm (490 mg/m^3)

DOT Classification: 3; Label: Flammable Liquid

SAFETY PROFILE: Mildly toxic by inhalation. A skin and eye irritant. Flammable.

PROP: Colorless liquid. D: 0.913 @ 17°/4°, mp: −24.0°, bp: 167–170°. Insol in water; misc in alc and ether.

MQS250 CAS: 12001-26-2 HR: 2

MICA

OSHA PEL: TWA Respirable Fraction: 3 mg/m^3

ACGIH TLV: TWA Respirable Fraction: 3 mg/m^3

NIOSH REL: (Silicates <1% Crystalline Silica) TWA 3 mg/m^3

SAFETY PROFILE: The dust is injurious to lungs.

PROP: Containing less than 1% crystalline silica (FEREAC 39,23540,74).

MQS500 CAS: 90-94-8 $C_{17}H_{20}N_2O$ HR: 3

MICHLER'S KETONE

SYNS: 4,4′-BIS(DIMETHYLAMINO)BENZOPHENONE ♦ BIS(4-(DIMETHYLAMINO)PHENYL)METHANONE ♦ TETRAMETHYLDIAMINOBENZOPHENONE

DFG MAK: Confirmed Animal Carcinogen with Unknown Relevance to Humans

DOT Classification: 3; Label: Flammable Liquid

SAFETY PROFILE: Confirmed human carcinogen. A poison by ingestion. A flammable liquid.

PROP: Leaves from ethanol. Mp: 179°, bp: >360° decomp. Insol in water; very sol in benzene; sol in alc; very sltly sol in ether.

MQV750 CAS: 8012-95-1 HR: 2

MINERAL OIL
SYNS: OIL MIST, MINERAL (OSHA, ACGIH) ♦ PARAFFIN OIL ♦ PETROLATUM, liquid ♦ WHITE MINERAL OIL
OSHA PEL: Oil Mist: TWA 5 mg/m^3
ACGIH TLV: (oil mist) TWA 5 mg/m^3; STEL 10 mg/m^3 (Proposed: TWA 5 mg/m^3)
SAFETY PROFILE: A human teratogen by inhalation. Inhalation of vapor or particulates can cause aspiration pneumonia. A skin and eye irritant. Highly purified food grades are of low toxicity. Slightly combustible liquid.
PROP: Colorless, oily liquid; practically tasteless and odorless. D: 0.83–0.86 (light), 0.875–0.905 (heavy), flash p: 444°F (OC), ULC: 10–20. Insol in water and alc; sol in benzene, chloroform, and ether. A mixture of liquid hydrocarbons from petroleum.

MRC250 CAS: 7439-98-7 Mo HR: 3
MOLYBDENUM
SYN: MOLYBDATE
OSHA PEL: Soluble Compounds: TWA 5 mg(Mo)/m^3; Insoluble Compounds: TWA Total Dust: 10 mg/m^3; Respirable Fraction: 5 mg/m^3
ACGIH TLV: Soluble Compounds: TWA 5 mg(Mo)/m^3; Insoluble Compounds: TWA 10 mg(Mo)/m^3; (Proposed: Soluble Compounds: TWA 0.5 mg(Mo)/m^3 Confirmed Animal Carcinogen with Unknown Relevance to Humans; Insoluble Compounds: TWA 10 mg(Mo)/m^3)
DFG MAK: (Insoluble Compounds) 4 mg/m^3; (Soluble Compounds) 5 mg/m^3
SAFETY PROFILE: An experimental teratogen. Flammable or explosive in the form of dust.
PROP: Lustrous, cubic, silver-white metallic crystals or gray-black powder. Fairly soft when pure. Less reactive than Cr to acids. Combines with O_2 on heating to give MoO_3. Mp: 2626°, bp: 5560°, d: 10.2, vap press: 1 mm @ 3102°.

MRC750 HR: 3
MOLYBDENUM COMPOUNDS
SAFETY PROFILE: Molybdenum and its compounds are highly toxic. Inhalation of molybdenum dust from alloys or carbides can cause "hard-metal lung disease. Hexavalent molybdenum compounds are readily absorbed through the gastrointestinal tract.

MRE000 CAS: 1313-27-5 MoO_3 HR: 3
MOLYBDENUM TRIOXIDE
SYNS: MOLYBDENUM(VI) OXIDE ♦ MOLYBDIC ANHYDRIDE ♦ MOLYBDIC TRIOXIDE
OSHA PEL: TWA 5 mg(Mo)/m^3
ACGIH TLV: Soluble Compounds: TWA 5 mg(Mo)/m^3; (Proposed: TWA Soluble Compounds: TWA 0.5 mg(Mo)/m^3 Confirmed Animal Carcinogen with Unknown Relevance to Humans)
SAFETY PROFILE: Poison by ingestion. A powerful irritant.

PROP: White or yellow to sltly bluish powder, granules, or solid. Photosensitive. An acidic oxide. Mp: 795°; bp: 1155°; d: 4.696 @ 26°/4°. Sol in 1000 parts water, in concentrated mineral acids, solutions of alkali hydroxides. Sol in ammonia or potassium bitartrate, solidifying to a yellowish-white mass.

MRG000 CAS: 55398-86-2 $C_{12}H_9ClO$ HR: 2
MONOCHLORODIPHENYL OXIDE
SYNS: MONOCHLOROPHENYLETHER
OSHA PEL: TWA 500 μg/m^3
SAFETY PROFILE: Moderately toxic by ingestion.

MRP750 CAS: 110-91-8 C_4H_9NO HR: 3
MORPHOLINE
DOT: UN 2054
SYNS: DIETHYLENE OXIMIDE ♦ TETRAHYDRO-p-ISOXAZINE
OSHA PEL: TWA 20 ppm (skin); STEL 30 ppm (skin)
ACGIH TLV: TWA 20 ppm (skin); Not Classifiable as a Human Carcinogen
DFG MAK: 10 ppm (36 mg/m^3)
DOT Classification: 3; Label: Flammable Liquid
SAFETY PROFILE: Moderately toxic by ingestion, inhalation, and skin contact. A corrosive irritant to skin, eyes, and mucous membranes. A flammable liquid.
PROP: Colorless, hygroscopic oil; amine odor. Fp: –7.5°, bp: 128.9°, flash p: 100°F (OC), autoign temp: 590°F, vap press: 10 mm @ 23°, vap d: 3.00, mp: –4.9°, d: 1.007 @ 20°/4°. Volatile with steam; misc with water evolving some heat; misc with acetone, benzene, ether, castor oil, methanol, ethanol, ethylene, glycol, linseed oil, turpentine, pine oil. Immiscible with concentrated NaOH solns.

NAH600 CAS: 8030-30-6 HR: 3
NAPHTHA
DOT: UN 1255/UN 1256/UN 1270/UN 2553
SYNS: HI-FLASH NAPHTHA ♦ NAPHTHA COAL TAR (OSHA) ♦ PETROLEUM BENZIN ♦ PETROLEUM DISTILLATES (NAPHTHA) ♦ SUPER VMP
OSHA PEL: TWA 100 ppm
ACGIH TLV: TWA 300 ppm
NIOSH REL: (Refined Petroleum Solvents) 10H TWA 350 mg/m^3; CL 1800 mg/m^3/15M
DOT Classification: 3; Label: Flammable Liquid
SAFETY PROFILE: Carcinogenic effects reported by skin contact. Mildly toxic by inhalation. A flammable liquid.
PROP: Dark straw-colored to colorless liquid. Bp: 149–216°, flash p: 107°F (CC), d: 0.862–0.892, autoign temp: 531°F. Sol in benzene, toluene, xylene, etc. Made from American coal oil and consists chiefly of pentane, hexane, and heptane (XPHPAW 255,43,40).

NAJ500 CAS: 91-20-3 $C_{10}H_8$ HR: 3

NAPHTHALENE
DOT: UN 1334/UN 2304
SYNS: MOTH BALLS (DOT) ♦ NAPHTHALINE
OSHA PEL: TWA 10 ppm; STEL 15 ppm
ACGIH TLV: TWA 10 ppm; STEL 15 ppm; Not Classifiable as a Human Carcinogen
DFG MAK: Confirmed Animal Carcinogen with Unknown Relevance to Humans
DOT Classification: 4.1; Label: Flammable Solid
SAFETY PROFILE: Human poison by ingestion. An experimental teratogen. An eye and skin irritant. Flammable.
PROP: Aromatic odor; white, crystalline, volatile flakes. Plates from EtOH with characteristic odor. Mp: 80.1°, bp: 217.9°, flash p: 174°F (OC), d: 1.162, lel: 0.9%, uel: 5.9%, vap press: 1 mm @ 52.6°, vap d: 4.42, autoign temp: 1053°F (567°C). Sol in alc, benzene; insol in water; very sol in ether, CCl_4, CS_2, hydronaphthalenes, and in fixed and volatile oils.

NAM500 CAS: 3173-72-6 $C_{12}H_6N_2O_2$ HR: 2
1,5-NAPHTHALENE DIISOCYANATE
SYNS: 1,5-DIISOCYANATONAPHTHALENE
DFG MAK: 0.01 ppm (0.087 mg/m^3)
SAFETY PROFILE: A powerful allergen. An irritant.
PROP: White to light-yellow crystals.

NBE500 CAS: 91-59-8 $C_{10}H_9N$ HR: 3
β-NAPHTHYLAMINE
DOT: UN 1650
SYNS: 2-AMINONAPHTHALENE ♦ C.I. 37270 ♦ 2-NAPHTHALAMINE
ACGIH TLV: Confirmed Human Carcinogen
DFG MAK: Human Carcinogen
NIOSH REL: (β-Naphthylamine) TWA use 29 CFR 1910.1009
DOT Classification: 6.1; Label: Poison
SAFETY PROFILE: Confirmed human carcinogen. Moderately toxic by ingestion. Combustible.
PROP: White to faint pink, lustrous leaflets from water; faint aromatic odor. Mp: 113°, d: 1.061 @ 98°/4°, vap press: 1 mm @ 108.0°, bp: 294°. Sol in hot water, alc, and ether.

NBE700 CAS: 134-32-7 $C_{10}H_9N$ HR: 3
1-NAPHTHYLAMINE
DOT: UN 2077
SYNS: 1-AMINONAPHTHALENE
OSHA PEL: Cancer Suspect Agent
NIOSH REL: (α-Naphthylamine) TWA use 29 CFR 1910.1004
DOT Classification: 6.1; Label: KEEP AWAY FROM FOOD
SAFETY PROFILE: Confirmed carcinogen. Moderately toxic by ingestion. Combustible.

PROP: White crystals, reddening on exposure to air; unpleasant odor. Needles from EtOH (aq) or Et_2O. Mp: 50°, bp: 300.8°, flash p: 315°F, d: 1.131, vap press: 1 mm @ 104.3°, vap d: 4.93. Sublimes, volatile with steam. Sol in 590 parts water; very sol in alc, ether. Keep well closed and away from light.

NCH000 CAS: 463-82-1 C_5H_{12} HR: 3

NEOPENTANE

DOT: UN 2044

SYNS: 2,2-DIMETHYLPROPANE ♦ tert-PENTANE

ACGIH TLV: TWA 600 ppm

DFG MAK: 1000 ppm (3000 mg/m^3)

NIOSH REL: TWA 120 ppm; CL 610 ppm/15M

DOT Classification: 2.1; Label: Flammable Gas

SAFETY PROFILE: An inhalation hazard. Both the gas and the liquid are flammable.

PROP: Liquid or gas at room temp. Solidifies @ –19.8°, bp: 9.5°, d: 0.613° @ 0°/0° (liquid), flash p: <19.4°F, lel: 1.4%, uel: 7.5%. Insol in water.

NCI500 CAS: 126-99-8 C_4H_5Cl HR: 3

NEOPRENE

DOT: UN 1991

SYNS: CHLOROBUTADIENE ♦ 2-CHLORO-1,3-BUTADIENE ♦ β-CHLOROPRENE (OSHA, MAK)

OSHA PEL: TWA 10 ppm (skin)

ACGIH TLV: TWA 10 ppm (skin)

DFG MAK: Animal Carcinogen, Suspected Human Carcinogen

NIOSH REL: CL (Chloroprene) 1 ppm/15M

DOT Classification: 3; Label: Flammable Liquid, Poison (UN 1991); DOT Class: Forbidden

SAFETY PROFILE: Suspected carcinogen. Poison by ingestion. Moderately toxic by inhalation. An experimental teratogen. A skin and eye irritant. A very dangerous fire hazard.

PROP: Colorless liquid. An oil-resistant synthetic rubber made by the polymerization of chloroprene. D: 0.958 @ 20°/20°, bp: 59.4°, flash p: –4°F, lel: 4.0%, uel: 20%, vap d: 3.0, brittle point: –35°, softens @ approx 80°. Sltly sol in water; misc in alc and ether.

NCW500 CAS: 7440-02-0 Ni HR: 3

NICKEL

SYNS: C.I. 77775 ♦ NICKEL SPONGE

OSHA PEL: TWA Soluble Compounds: 0.1 mg(Ni)/m^3; Insoluble Compounds: 1 mg(Ni)/m^3

ACGIH TLV: TWA 0.1 mg(Ni)/m^3; Not Suspected as a Human Carcinogen)

DFG TRK: Human Carcinogen

NIOSH REL: (Inorganic Nickel) TWA 0.015 mg(Ni)/m^3

SAFETY PROFILE: Confirmed carcinogen. Poison by ingestion. An experimental teratogen. Hypersensitivity to nickel is common and can cause allergic contact dermatitis, pulmonary asthma, and conjunctivitis. It can react violently with many substances.
PROP: A silvery-white, hard, malleable, and ductile metal. Crystallizes as metallic cubes. D: 8.90 @ 25°, vap press: 1 mm @ 1810°, mp: 1455°, bp: 2920°. Stable in air at room temp.

NCY500 CAS: 3333-67-3 $CNiO_3$ HR: 3
NICKEL(II) CARBONATE (1:1)
SYNS: CARBONIC ACID, NICKEL SALT (1:1) ♦ C.I. 77779
OSHA PEL: TWA 1 mg(Ni)/m^3
ACGIH TLV: TWA 1 mg(Ni)/m^3 (Proposed: TWA 0.05 mg(Ni)/m^3; Human Carcinogen)
DFG TRK: 0.5 mg/m^3; Human Carcinogen
NIOSH REL: (Inorganic Nickel) TWA 0.015 mg(Ni)/m^3
SAFETY PROFILE: Confirmed carcinogen.
PROP: Rhombic, light-green crystals or solid. Mp: decomp.

NCZ000 CAS: 13463-39-3 C_4NiO_4 HR: 3
NICKEL CARBONYL
DOT: UN 1259
SYNS: NICKEL TETRACARBONYL ♦ TETRACARBONYL NICKEL
OSHA PEL: TWA 0.001 ppm (Ni)
ACGIH TLV: TWA 0.05 mg(Ni)/m^3
DFG TRK: Animal Carcinogen, Suspected Human Carcinogen
NIOSH REL: (Nickel Carbonyl) TWA 0.001 ppm
DOT Classification: 6.1; Label: Poison, Flammable Liquid
SAFETY PROFILE: Confirmed carcinogen. An experimental teratogen. A human poison by inhalation. An experimental teratogen. A very dangerous fire hazard.
PROP: Colorless, volatile liquid or needles; oxidizes in air. Mp: –19.3°, bp: 43°, lel: 2% @ 20°, d: 1.3185 @ 17°, vap press: 400 mm @ 25.8°, flash p: <–4°. Oxidizes in air. Sol in alc, benzene, chloroform, acetone, and carbon tetrachloride.

NDB000 HR: 3
NICKEL COMPOUNDS
OSHA PEL: TWA Soluble: Compounds: 0.1 mg(Ni)/m^3; Insoluble Compounds: 1 mg(Ni)/m^3
ACGIH TLV: (insoluble) TWA 0.05 mg(Ni)/m^3; Human Carcinogen; (soluble) TWA 0.1 mg/m^3,Not Classifiable as a Carcinogen
DFG MAK: Human Carcinogen
NIOSH REL: (Inorganic Nickel) TWA 0.015 mg(Ni)/m^3
SAFETY PROFILE: Nickel and many of its compounds are poisons and carcinogens. All airborne nickel contaminating dusts are regarded as carcinogenic by inhalation.

NDF500 CAS: 1313-99-1 NiO HR: 3

NICKEL MONOXIDE

SYNS: C.I. 77777 ♦ NICKEL OXIDE (MAK)

OSHA PEL: TWA 1 mg(Ni)/m^3

ACGIH TLV: TWA 1 mg(Ni)/m^3 (Proposed: TWA 0.05 mg(Ni)/m^3; Human Carcinogen)

DFG TRK: 0.5 mg/m^3; Human Carcinogen

NIOSH REL: (Inorganic Nickel) TWA 0.015 mg(Ni)/m^3

SAFETY PROFILE: Confirmed carcinogen.

PROP: Cubic, green-black crystals; yellow when hot. Mp: 1984°, d: 7.45. Insol in water; sol in acids, NH_3 (aq).

NDN000 CAS: 54-11-5 $C_{10}H_{14}N_2$ HR: 3

NICOTINE

DOT: UN 1654

SYNS: (S)-3-(1-METHYL-2-PYRROLIDINYL)PYRIDINE (9CI)

OSHA PEL: TWA 0.5 mg/m^3 (skin)

ACGIH TLV: TWA 0.5 mg/m^3 (skin)

DFG MAK: 0.07 ppm; 0.47 mg/m^3

DOT Classification: 6.1; Label: Poison

SAFETY PROFILE: A poison by ingestion and skin contact. Human teratogenic effects. Can be absorbed by intact skin. An experimental teratogen. Combustible.

PROP: An alkaloid from tobacco. In its pure state, a colorless and almost odorless oil; sharp burning taste. Mp: <–80°, bp: 247.3° (partial decomp), lel: 0.75%, uel: 4.0%, d: 1.0092 @ 20°, autoign temp: 471°F, vap press: 1 mm @ 61.8°, vap d: 5.61. Volatile with steam; misc with water below 60°; very sol in alc, chloroform ether, pet ether, and kerosene oils.

NED500 CAS: 7697-37-2 HNO_3 HR: 3

NITRIC ACID

DOT: UN 2031

SYNS:

OSHA PEL: TWA 2 ppm; STEL 4 ppm

ACGIH TLV: TWA 2 ppm; STEL 4 ppm

DFG MAK: 2 ppm (5.2 mg/m^3)

NIOSH REL: (Nitric Acid) TWA 2 ppm

DOT Classification: 8; Label: Corrosive, Oxidizer, Poison

SAFETY PROFILE: Human poison by ingestion. An experimental teratogen. Corrosive to eyes, skin, mucous membranes, and teeth. Flammable by chemical reaction with reducing agents. It is a powerful oxidizing agent. It can react violently with many substances.

PROP: Transparent, colorless or yellowish, fuming, suffocating, caustic and corrosive liquid. Pure acid; decomp especially in light. Mp: –42°, bp: 83°, d: 1.50269 @ 25°/4°.

NEG100 CAS: 10102-43-9 NO HR: 3

NITRIC OXIDE
DOT: UN 1660
OSHA PEL: TWA 25 ppm
ACGIH TLV: TWA 25 ppm
NIOSH REL: (Oxides of Nitrogen) TWA 25 ppm
DOT Classification: 2.3; Label: Poison Gas
SAFETY PROFILE: A poison gas. A severe eye, skin, and mucous membrane irritant. It can react violently with many substances.

PROP: Colorless non-flammable gas or blue liquid and solid. With O_2 gives brown NO_2. Mp: –161°, bp: –151.18°, d: 1.3402 g/L; liquid, 1.269 @ –150°; gas, 1.04 g/L. Sltly sol in water.

NEJ500 CAS: 602-87-9 $C_{12}H_9NO_2$ HR: 3
5-NITROACENAPHTHENE
SYNS: 1,2-DIHYDRO-5-NITRO-ACENAPHTHYLENE ♦ 5-NITRONAPHTHALENE ETHYLENE
DFG MAK: Animal Carcinogen, Suspected Human Carcinogen
SAFETY PROFILE: Confirmed carcinogen.
PROP: A solid. Mp: 101.5–102.5°. Sol in Et_2O, EtOH, ligroin, and hot H_2O.

NEO500 CAS: 100-01-6 $C_6H_6N_2O_2$ HR: 3
p-NITROANILINE
DOT: UN 1661
SYNS: 1-AMINO-4-NITROBENZENE ♦ C.I. 37035 ♦ 4-NITROBENZENAMINE ♦ PNA
OSHA PEL: TWA 3 mg/m^3 (skin)
ACGIH TLV: TWA 3 mg/m^3 (skin); Not Classifiable as a Human Carcinogen
DFG MAK: 1 ppm (5.7 mg/m^3)
DOT Classification: 6.1; Label: Poison
SAFETY PROFILE: Poison by ingestion. Combustible.
PROP: Bright-yellow powder or pale-yellow needles from water. Mp: 148.5°, bp: 332°, flash p: 390°F (CC), d: 1.424, vap press: 1 mm @ 142.4°. Sol in water, alc, ether, benzene, methanol.

NEX000 CAS: 98-95-3 $C_6H_5NO_2$ HR: 3
NITROBENZENE
DOT: UN 1662
SYNS: NITROBENZOL (DOT) ♦ OIL of MIRBANE (DOT)
OSHA PEL: TWA 1 ppm (skin)
ACGIH TLV: TWA 1 ppm (skin); Animal Carcinogen; BEI: 5 mg/g creatinine of total p-nitrophenol in urine at end of shift at end of workweek
DFG MAK: Confirmed Animal Carcinogen with Unknown Relevance to Humans; BAT: 100 µg/L of aniline in blood after several shifts
DOT Classification: 6.1; Label: Poison

SAFETY PROFILE: Confirmed carcinogen. Moderately toxic by ingestion and skin contact. An experimental teratogen. An eye and skin irritant. It is absorbed rapidly through the skin. Combustible. It can react violently with many substances.
PROP: Bright-yellow crystals or pale-yellow to colorless, oily liquid; odor of volatile almond oil. Mp: 6°, bp: 210–211°, ULC: 20–30%, lel: 1.8% @ 200°F, flash p: 190°F (CC), d: 1.205 @ 15°/4°, autoign temp: 900°F, vap press: 1 mm @ 44.4°, vap d: 4.25. Volatile with steam; sol in about 500 parts water; very sol in alc, benzene, ether, oils.

NFQ000 CAS: 92-93-3 $C_{12}H_9NO_2$ HR: 3
p-NITROBIPHENYL
SYNS: 4-PHENYL-NITROBENZENE ♦ PNB
OSHA: Cancer Suspect Agent
ACGIH TLV: Confirmed Human Carcinogen
DFG MAK: Confirmed Animal Carcinogen, Suspected Human Carcinogen
NIOSH REL: (4-Nitrobiphenyl) TWA use 29 CFR 1910.1003
SAFETY PROFILE: Confirmed carcinogen. Moderately toxic by ingestion.
PROP: Yellow needles from alc. Mp: 113–114°, bp: 223–224°. Insol in water; sltly sol in cold alc; very sol in ether.

NFS525 CAS: 100-00-5 $C_6H_4ClNO_2$ HR: 3
p-NITROCHLOROBENZENE
DOT: UN 1578
SYNS: 4-CHLORONITROBENZENE ♦ 4-CHLORO-1-NITROBENZENE ♦ PNCB
OSHA PEL: TWA 1 mg/m^3 (skin)
ACGIH TLV: TWA 0.1 ppm (skin); Animal Carcinogen
DFG MAK: Confirmed Animal Carcinogen with Unknown Relevance to Humans
NIOSH REL: (p-Nitrochlorobenzene) lowest feasible conc. (skin)
DOT Classification: 6.1; Label: Poison
SAFETY PROFILE: Confirmed carcinogen A poison by ingestion.A fFlammable liquid.
PROP: D: 1.520, mp: 83°, bp: 242°, flash p: 110°. Insol in water; sltly sol in alc; very sol in CS_2 and ether.

NFY500 CAS: 79-24-3 $C_2H_5NO_2$ HR: 3
NITROETHANE
DOT: UN 2842
OSHA PEL: TWA 100 ppm
ACGIH TLV: TWA 100 ppm
DFG MAK: 100 ppm (310 mg/m^3)
DOT Classification: 3; Label: Flammable Liquid
SAFETY PROFILE: Moderately toxic by ingestion. Mildly toxic by inhalation. An eye and mucous membrane irritant. A flammable liquid.
PROP: Oily, colorless liquid with pleasant but slightly irritating odor. Mp: –90°, bp: 114.0°, fp: –50°, d: 1.046 @ 25°/25°, autoign temp: 778°F, flash p: 106°F, decomp @ 335–382°, lel: 4.0%, vap press: 15.6 mm @ 20°, vap d: 2.58. Misc in MeOH, EtOH, Et_2O, $CHCl_3$; sltly sol in hot water; insol in cold H_2O.

NGR500 CAS: 10102-44-0 NO_2 HR: 3

NITROGEN DIOXIDE

SYNS: NITROGEN PEROXIDE

OSHA PEL: STEL 1 ppm

ACGIH TLV: TWA 3 ppm; STEL 5 ppm; Not Classifiable as a Human Carcinogen

DFG MAK: 5 ppm (9 mg/m^3)

NIOSH REL: CL (Oxides of Nitrogen) 1 ppm/15M

SAFETY PROFILE: Moderately toxic to humans by inhalation. An experimental teratogen.

PROP: Brown gas or colorless solid to yellow liquid; irritating odor. Reacts with H_2O giving HNO_3 + NO. Mp: –9.3° (yellow liquid), bp: 21° (red-brown gas with decomp), d: 1.491 @ 0°, vap press: 400 mm @ 80°. Liquid below 21.15°. Sol in concentrated sulfuric acid, nitric acid. Corrosive to steel when wet.

NGU000 CAS: 10024-97-2 N_2O HR: 2

NITROGEN OXIDE

DOT: UN 1070/UN 2201

SYNS: DINITROGEN MONOXIDE ♦ LAUGHING GAS

OSHA PEL (Shipyard): Simple asphyxiant-inert gas and vapor

ACGIH TLV: 50 ppm; Not Classifiable as a Human Carcinogen

DFG MAK: 100 ppm (180 mg/m^3)

NIOSH REL: (Waste Anesthetic Gases and Vapors) TWA 25 ppm

DOT Classification: 2.2; Label: Nonflammable Gas

SAFETY PROFILE: Moderately toxic by inhalation. An experimental teratogen. Does not burn but is flammable by chemical reaction.

PROP: Colorless nonflammable gas, liquid, or cubic crystals; slt sweet odor. Mp: –90.8°, bp: –88.49°, d: 1.977 g/L (liquid 1.226 @ –89°). Sltly sol in H_2O; sol in Et_2O; freely sol in $CHCl_3$ and EtOH.

NGW000 CAS: 7783-54-2 F_3N HR: 3

NITROGEN TRIFLUORIDE

DOT: UN 2451

SYN: NITROGEN FLUORIDE

OSHA PEL: TWA 10 ppm

ACGIH TLV: TWA 10 ppm; BEI: 3 mg/g creatinine of fluorides in urine prior to shift; 10 mg/g creatinine of fluorides in urine at end of shift.

NIOSH REL: (Inorganic Fluorides) TWA 2.5 mg(F)/m^3

DOT Classification: 2.2; Label: Nonflammable Gas, Oxidizer

SAFETY PROFILE: A poison. Mildly toxic by inhalation. Severe explosion hazard by chemical reaction with reducing agents.

PROP: Colorless, odorous gas; odor of mold. Mp: –208.5°, bp: –129°, d (liquid): 1.537 @ –129°; d: (liquid @ bp) 1.885. Sltly sol in H_2O.

NGY000 CAS: 55-63-0 $C_3H_5N_3O_9$ HR: 3

NITROGLYCERIN

DOT: UN 0143/UN 0144/UN 1204/UN 3064

SYNS: BLASTING GELATIN (DOT) ♦ GLYCERYL TRINITRATE ♦ GTN ♦ NITROGLYCERINE ♦ TRINITROGLYCEROL

OSHA PEL: STEL 0.1 mg/m^3 (skin)

ACGIH TLV: TWA 0.05 ppm (skin)

DFG MAK: 0.05 ppm (0.47 mg/m^3) (skin)

NIOSH REL: CL (Nitroglycerin or EGDN) 0.1 mg/m^3/20M

DOT Classification: EXPLOSIVE 1.1D; Label: EXPLOSIVE 1.1D, Poison (UN 0143); DOT Class: Forbidden (not desensitized); DOT Class: 3; Label: Flammable Liquid (UN 3064, UN 1204); DOT Class: EXPLOSIVE 1.1D; Label: EXPLOSIVE 1.1D (UN 0144)

SAFETY PROFILE: Poison by ingestion. An experimental teratogen. A skin irritant. Toxic effects may occur by ingestion, inhalation of dust, or absorption through intact skin. A very dangerous fire hazard and severe explosion hazard when shocked.

PROP: Colorless to yellow liquid; sweet burning taste; sensitive to shock. Mp: 13°, bp: explodes @ 218°, d: 1.599 @ 15°/15°, vap press: 1 mm @ 127°, vap d: 7.84, autoign temp: 518°F, decomp @ 50–60°, fp: 13°. Volatile @ 100°. Misc with ether, acetone, glacial acetic acid, ethyl acetate, benzene, nitrobenzene, pyridine, chloroform, ethylene bromide, dichloroethylene; sltly sol in pet ether, glycerin. M(trinisc in most org solvs; prac insol in H_2O.

NHI500 CAS: 7046-61-9 $C_6H_8N_4O_4$ HR: 3

NITROIMINODIETHYLENEDIISOCYANIC ACID

SYN: 3-NITRO-3-AZAPENTANE-1,5-DIISOCYANATE

NIOSH REL: (Diisocyanates) TWA 0.005 ppm; CL 0.02 ppm/10M

SAFETY PROFILE: A sensitizer.

NHM500 CAS: 75-52-5 CH_3NO_2 HR: 3

NITROMETHANE

DOT: UN 1261

SYNS: NITROCARBOL

OSHA PEL: TWA 100 ppm

ACGIH TLV: TWA 20 ppm (Proposed: TWA 20 ppm; Confirmed Animal Carcinogen with Unknown Relevance to Humans)

DFG MAK: 100 ppm (250 mg/m^3)

DOT Classification: 3; Label: Flammable Liquid

SAFETY PROFILE: Poison by ingestion. Mildly toxic by inhalation. A very dangerous fire hazard. It can react violently with many substances.

PROP: A poisonous oily liquid; moderate to strong disagreeable odor. Bp: 101°, lel: 7.3%, fp: –29°, flash p: 95°F (CC), d: 1.1322 @ 25°/4°, autoign temp: 785°F, vap press: 27.8 mm @ 20°, vap d: 2.11. Sol in EtOH, Et_2O, DMF; sltly sol in H_2O.

NHQ000 CAS: 86-57-7 $C_{10}H_7NO_2$ HR: 3

1-NITRONAPHTHALENE

SYNS: α-NITRONAPHTHALENE
DFG MAK: Confirmed Animal Carcinogen with Unknown Relevance to Humans
SAFETY PROFILE: A skin, eye, and mucous membrane irritant. Flammable solid and combustible liquid.
PROP: Yellow crystals. Bp: 304°, flash p: 327°F (CC), d: 1.331 @ 4°/4°, vap d: 5.96, mp: 59–61°. Insol in water; sol in CS_2, alc, chloroform, and ether.

NHQ500 CAS: 581-89-5 $C_{10}H_7NO_2$ HR: 3

2-NITRONAPHTHALENE

SYN: β-NITRONAPHTHALENE
DFG TRK: Animal Carcinogen, Suspected Human Carcinogen
NIOSH REL: (2-Nitronaphthalene) lowest feasible conc
SAFETY PROFILE: Confirmed carcinogen. Moderately toxic by ingestion. A skin and lung irritant. Combustible.
PROP: Colorless solution when dissolved in ethanol. Mp: 79°, bp: 165° @ 15 mm. Insol in water; very sol in alc and ether.

NIX500 CAS: 108-03-2 $C_3H_7NO_2$ HR: 3

1-NITROPROPANE

SYN: 1-NP
OSHA PEL: TWA 25 ppm
ACGIH TLV: TWA 25 ppm; Not Classifiable as a Human Carcinogen
DFG MAK: 25 ppm (92 mg/m^3)
SAFETY PROFILE: Poison by ingestion. Mildly toxic by inhalation. A human eye irritant. Very dangerous fire hazard.
PROP: Colorless oily liquid; irritant to mucous membranes. Fp: –108°, bp: 132°, flash p: 93°F (TCC), d: 1.003 @ 20°/20°, autoign temp: 789°F, vap press: 7.5 mm @ 20°, vap d: 3.06, lel: 2.2%. Sltly sol in water; misc with alc, ether, and many org solvs.

NIY000 CAS: 79-46-9 $C_3H_7NO_2$ HR: 3

2-NITROPROPANE

DOT: UN 2608
SYNS: ISONITROPROPANE ♦ 2-NP
OSHA PEL: TWA 10 ppm
ACGIH TLV: TWA 10 ppm; Animal Carcinogen
DFG TRK: Animal Carcinogen, Suspected Human Carcinogen
NIOSH REL: (2-Nitropropane) TWA reduce to lowest feasible level
DOT Classification: 3; Label: Flammable Liquid
SAFETY PROFILE: Confirmed carcinogen. An experimental teratogen. Moderately toxic by ingestion and inhalation. An experimental teratogen. Very dangerous fire hazard.
PROP: Colorless liquid. Bp: 120°, fp: –93°, flash p: 82°F (TCC), d: 0.992 @ 20°/20°, autoign temp: 802°F, vap press: 10 mm @ 15.8°, vap d: 3.06, lel: 2.6%. Misc with org solvs; sol in water, alc, and ether.

NJA000 CAS: 5522-43-0 $C_{16}H_9NO_2$ HR: 3

3-NITROPYRENE

SYN: 1-NITROPYRENE

DFG MAK: Confirmed Animal Carcinogen with Unknown Relevance to Humans

SAFETY PROFILE: Confirmed carcinogen.

PROP: Yellow needles from MeCN. Mp: 151–152°.

NJW500 CAS: 55-18-5 $C_4H_{10}N_2O$ HR: 3

N-NITROSODIETHYLAMINE

SYNS: DENA ♦ DIETHYLNITROSAMINE ♦ N-ETHYL-N-NITROSO-ETHANAMINE ♦ NDEA

DFG MAK: Animal Carcinogen, Suspected Human Carcinogen

SAFETY PROFILE: Confirmed carcinogen. Poison by ingestion. An experimental teratogen.

PROP: Yellow oil. D: 0.9422 @ 20°/4°, bp: 176.9°. Sol in water, alc, and ether.

NKA600 CAS: 62-75-9 $C_2H_6N_2O$ HR: 3

N-NITROSODIMETHYLAMINE

SYNS: N,N-DIMETHYLNITROSAMINE ♦ DMN ♦ DMNA ♦ N-METHYL-N-NITROSOMETHANAMINE ♦ NDMA

OSHA PEL: Cancer Suspect Agent

ACGIH TLV: Animal Carcinogen

DFG MAK: Animal Carcinogen, Suspected Human Carcinogen

SAFETY PROFILE: Confirmed carcinogen. An experimental teratogen. A poison by ingestion and inhalation.

PROP: Yellow liquid; sol in water, alc, and ether. Bp: 152°, d: 1.005 @ 20°/4°.

NKB700 CAS: 621-64-7 $C_6H_{14}N_2O$ HR: 3

N-NITROSODI-N-PROPYLAMINE

SYNS: DI-n-PROPYLNITROSAMINE ♦ DPN ♦ DPNA ♦ NDPA ♦ N-NITROSO-N-DIPROPYLAMINE ♦ N-NITROSO-N-PROPYL-1-PROPANAMINE

DFG MAK: Animal Carcinogen, Suspected Human Carcinogen

SAFETY PROFILE: Confirmed carcinogen. Moderately toxic by ingestion. An experimental teratogen.

NKD000 CAS: 612-64-6 $C_8H_{10}N_2O$ HR: 3

N-NITROSO-N-ETHYL ANILINE

SYNS: N-ETHYL-N-NITROSOBENZENAMINE ♦ NEA ♦ N-NITROSOETHYLPHENYLAMINE (MAK)

DFG MAK: Animal Carcinogen, Suspected Human Carcinogen

SAFETY PROFILE: Confirmed carcinogen. Poison by ingestion. An experimental teratogen.

PROP: Yellow oil. D: 1.087 @ 20°/4°, bp: 119–120° @ 15 mm. Insol in water.

NKM000 CAS: 1116-54-7 $C_4H_{10}N_2O_3$ HR: 3

NITROSOIMINO DIETHANOL

SYNS: BIS(β-HYDROXYETHYL)NITROSAMINE ♦ 2,2'-IMINODI-N-NITROSOETHANOL ♦ NDELA ♦ N-NITROSODIETHANOLAMINE (MAK)
DFG MAK: Animal Carcinogen, Suspected Human Carcinogen
SAFETY PROFILE: Confirmed carcinogen. Mildly toxic by ingestion.

NKZ000 CAS: 59-89-2 $C_4H_8N_2O_2$ HR: 3
4-NITROSOMORPHOLINE
SYNS: NITROSOMORPHOLINE ♦ N-NITROSOMORPHOLINE (MAK) ♦ NMOR
DFG MAK: Animal Carcinogen, Suspected Human Carcinogen
SAFETY PROFILE: Confirmed carcinogen. Poison by ingestion. Moderately toxic by inhalation.

NLJ500 CAS: 100-75-4 $C_5H_{10}N_2O$ HR: 3
N-NITROSOPIPERIDINE
SYNS: 1-NITROSOPIPERIDINE ♦ NPIP
DFG MAK: Animal Carcinogen, Suspected Human Carcinogen
SAFETY PROFILE: Confirmed carcinogen. Poison by ingestion. An experimental teratogen.
PROP: Light-yellow oil. D: 1.063 @ 18.5°/4°, bp: 217–218°. Sol in water; very sol in acid solns.

NLP500 CAS: 930-55-2 $C_4H_8N_2O$ HR: 3
N-NITROSOPYRROLIDINE
SYNS: 1-NITROSOPYRROLIDINE ♦ NPYR ♦ TETRAHYDRO-N-NITROSOPYRROLE
DFG MAK: Animal Carcinogen, Suspected Human Carcinogen
SAFETY PROFILE: Confirmed carcinogen. Poison by ingestion.

NMO500 CAS: 99-08-1 $C_7H_7NO_2$ HR: 3
m-NITROTOLUENE
DOT: UN 1664
SYNS: 3-METHYLNITROBENZENE ♦ MNT ♦ 3-NITROTOLUOL
OSHA PEL: TWA 2 ppm (skin)
ACGIH TLV: TWA 2 ppm (skin)
DFG MAK: 5 ppm (28 mg/m^3)
DOT Classification: 6.1; Label: Poison
SAFETY PROFILE: Poison by ingestion. Moderately toxic by inhalation. Combustible.
PROP: Crystals or liquid. Mp: 16°, flash p: 233°F (CC), d: 1.1630 @ 15°/4°, vap press: 1 mm @ 50.2°, vap d: 4.72, bp: 231.9°. Misc with alc, ether; sol in benzene; sol in water @ 30°.

NMO525 CAS: 88-72-2 $C_7H_7NO_2$ HR: 3
o-NITROTOLUENE
DOT: UN 1664
SYNS: 2-METHYLNITROBENZENE ♦ 2-NITROTOLUENE ♦ ONT
OSHA PEL: TWA 2 ppm (skin)

ACGIH TLV: TWA 2 ppm (skin)
DFG MAK: Animal Carcinogen, Suspected Human Carcinogen
DOT Classification: 6.1; Label: Poison
SAFETY PROFILE: Confirmed carcinogen. Moderately toxic by ingestion. Mucous membrane effects by inhalation. Combustible.
PROP: Yellowish oily liquid. Crystals in two forms; transparent needles or snow white crystals. Mp: –10°, bp: 222.3°, flash p: 223°F (CC), d: 1.1622 @ 19°/15°, vap press: 1 mm @ 50°, vap d: 4.72. Insol in water; sol in SO_2 and pet ether; misc in alc, benzene, and ether. Sltly sol in NH_3.

NMO550 CAS: 99-99-0 $C_7H_7NO_2$ HR: 3
p-NITROTOLUENE
DOT: UN 1664
SYNS: p-METHYL NITROBENZENE ♦ 4-NITROTOLUENE ♦ PNT
OSHA PEL: TWA 2 ppm (skin)
ACGIH TLV: TWA 2 ppm (skin)
DFG MAK: 5 ppm (28 mg/m^3)
DOT Classification: 6.1; Label: Poison
SAFETY PROFILE: Moderately toxic by ingestion and inhalation. Mildly toxic by skin contact. Combustible.
PROP: Yellowish crystals from alc. Bp: 238.3°, flash p: 223°F (CC), d: 1.286, vap press: 1 mm @ 53.7°, vap d: 4.72. Mp: 53–54°. Insol in water; sol in alc, benzene, ether, chloroform, and acetone.

NMP500 CAS: 99-55-8 $C_7H_8N_2O_2$ HR: 3
5-NITRO-o-TOLUIDINE
SYNS: 1-AMINO-2-METHYL-5-NITROBENZENE ♦ C.I. 37105
DFG MAK: Animal Carcinogen, Suspected Human Carcinogen
SAFETY PROFILE: Confirmed carcinogen. Moderately toxic by ingestion.
PROP: Yellow prisms from EtOH. Mp: 107°.

NMX000 CAS: 111-84-2 C_9H_{20} HR: 3
NONANE
OSHA PEL: TWA 200 ppm
ACGIH TLV: TWA 200 ppm
DOT Classification: 3; Label: Flammable Liquid
SAFETY PROFILE: Mildly toxic by inhalation. A very dangerous fire hazard.
PROP: Colorless liquid. Mp: –53.7°, fp: –51°, bp: 150.7°, lel: 0.8%, uel: 2.9%, flash p: 88°F (CC), d: 0.718 @ 20°/4°, autoign temp: 374°F, vap press: 10 mm @ 38.0°, vap d: 4.41. Insol in water; sol in abs alc and ether.

OAP000 CAS: 2234-13-1 $C_{10}Cl_8$ HR: 3
OCTACHLORONAPHTHALENE
OSHA PEL: TWA 0.1 mg/m^3 (skin); STEL 0.3 mg/m^3

ACGIH TLV: TWA 0.1 mg/m^3 (skin); STEL 0.3 mg/m^3
NIOSH REL: TWA 0.1 mg/m^3; STEL 0.3 mg/m^3 (skin)
SAFETY PROFILE: Poison by inhalation, ingestion, and skin contact.
PROP: Crystals from cyclohexane. Mp: 197.5–198°, bp: 246–250° @ 0.5 mm. Sol in C_6H_6.

OBM000 CAS: 382-21-8 C_4F_8 HR: 3
OCTAFLUORO-sec-BUTENE
SYNS: OCTAFLUOROISOBUTENE ♦ PERFLUOROISOBUTYLENE (ACGIH) ♦ PFIB
ACGIH TLV: CL 0.01 ppm
SAFETY PROFILE: A deadly poison by inhalation. A skin, eye, and mucous membrane irritant.
PROP: A gas at room temp. D: 1.592 @ 0°, bp: 5–6° @ 740 mm.

OCU000 CAS: 111-65-9 C_8H_{18} HR: 3
OCTANE
DOT: UN 1262
OSHA PEL: TWA 300 ppm; STEL 375 ppm
ACGIH TLV: TWA 300 ppm
DFG MAK: 500 ppm (2400 mg/m^3)
NIOSH REL: (Alkanes) TWA 350 mg/m^3
DOT Classification: 3; Label: Flammable Liquid
SAFETY PROFILE: May act as a simple asphyxiant. A very dangerous fire hazard and severe explosion hazard.
PROP: Clear liquid. Bp: 125.8°, lel: 1.0%, uel: 4.7%, fp: –56.5°, flash p: 56°F, d: 0.7036 @ 20°/4°, autoign temp: 428°F, vap press: 10 mm @ 19.2°, vap d: 3.86. Insol in water; sltly sol in alc, ether; misc with benzene.

ODI000 CAS: 106-68-3 $C_8H_{16}O$ HR: 3
3-OCTANONE
DOT: UN 2271
SYNS: AMYL ETHYL KETONE ♦ EAK ♦ ETHYL AMYL KETONE ♦ 5-METHYL-3-HEPTANONE (OSHA)
OSHA PEL: TWA 25 ppm
DOT Classification: 3; Label: Flammable Liquid
SAFETY PROFILE: Moderately irritating to skin, eyes, and mucous membranes by inhalation. A flammable liquid.
PROP: Liquid; fruity odor. Bp: 157–162°, d: 0.822 @ 20°/20°, flash p: 138°F.

OKK000 CAS: 20816-12-0 O_4Os HR: 3
OSMIUM TETROXIDE
DOT: UN 2471
SYNS: OSMIC ACID ♦ OSMIUM(VIII) OXIDE
OSHA PEL: TWA 0.0002 ppm; STEL 0.0006 ppm (Os)
ACGIH TLV: TWA 0.0002 ppm; STEL 0.0006 ppm (Os)

DFG MAK: 0.0002 ppm (0.002 mg/m^3)
DOT Classification: 6.1; Label: Poison
SAFETY PROFILE: Poison by ingestion and inhalation.
PROP: (A) Yellow, monoclinic, colorless crystals; (B) yellow mass; pungent, chlorine-like odor. Mp (A): 39.5°, mp: (B): 41°, bp: 130° (subl), d: 4.906 @ 22°, vap press (A): 10 mm @ 26.0°, vap press (B): 10 mm @ 31.3°. Sol in CCl_4, C_6H_6, EtOH, Et_2O; spar sol in dil H_2SO_4 and H_2O.

OLA000 CAS: 144-62-7 $C_2H_2O_4$ HR: 3
OXALIC ACID
SYNS: ETHANEDIOIC ACID ♦ ETHANEDIONIC ACID
OSHA PEL: TWA 1 mg/m^3; STEL 2 mg/m^3
ACGIH TLV: TWA 1 mg/m^3; STEL 2 mg/m^3
SAFETY PROFILE: Moderately toxic by ingestion. A skin and severe eye irritant.
PROP: Orthorhombic colorless crystals from water. Mp: 101.5° (anhyd) 189°, d: 1.65 @ 18.5°/4°. Very sol in H_2O; mod sol in EtOH; spar sol in Et_2O.

OPM000 CAS: 101-80-4 $C_{12}H_{12}N_2O$ HR: 3
4,4′-OXYDIANILINE
SYNS: p-AMINOPHENYL ETHER ♦ BIS(4-AMINOPHENYL)ETHER ♦ 4,4-DIAMINODIPHENYL ETHER ♦ 4,4′-OXYBISBENZENAMINE ♦ 4,4′-OXYDIPHENYLAMINE
DFG MAK: Animal Carcinogen, Suspected Human Carcinogen
SAFETY PROFILE: Confirmed carcinogen. Moderately toxic by ingestion.
PROP: Colorless crystals. Mp: 187°, bp: >300°.

ORA000 CAS: 7783-41-7 F_2O HR: 3
OXYGEN DIFLUORIDE
DOT: UN 2190
SYNS: FLUORINE MONOXIDE ♦ FLUORINE OXIDE ♦ OXYGEN FLUORIDE
OSHA PEL: CL 0.05 ppm
ACGIH TLV: CL 0.05 ppm
DOT Classification: 2.3; Label: Poison Gas, Oxidizer
SAFETY PROFILE: Poison by inhalation. A corrosive skin, eye, and mucous membrane irritant. It can react violently with many substances.
PROP: Colorless gas or yellowish-brown liquid. Stable to 2°. Stable in dry glass vessels. Reacts with Hg, but is less reactive than Cl_2O. D: (liquid) 1.90 @ –224°, mp: –223.8°, bp: –144.8°. Sltly sol in water.

ORW000 CAS: 10028-15-6 O_3 HR: 3
OZONE
SYNS: TRIATOMIC OXYGEN
OSHA PEL: TWA 0.1 ppm; STEL 0.3 ppm
ACGIH TLV: TWA 0.05 ppm (heavy work), 0.08 ppm (moderate work), 0.10 (light work); all workloads < 2 hours 20 ppm; Not Classifiable as a Human Carcinogen

DFG MAK: Confirmed Animal Carcinogen with Unknown Relevance to Humans
SAFETY PROFILE: A human poison by inhalation. Experimental teratogenic effects. A skin, eye, upper respiratory system, and mucous membrane irritant. It can react violently with many substances.
PROP: Blue or violet-black solid or unstable colorless gas or dark-blue liquid; characteristic odor at low concentration. Mp: −193°, bp: −111.9°, d: (gas) 2.144 g/L, 1.71 @ −183°, d: (liquid) 1.614 g/mL @ −195.4°. Sltly sol in water.

PAH750 CAS: 8002-74-2 HR: 2
PARAFFIN
SYNS: PARAFFIN WAX ♦ PARAFFIN WAX FUME (ACGIH)
OSHA PEL: Fume: TWA 2 mg/m^3 (fume)
ACGIH TLV: Fume: TWA 2 mg/m^3 (fume)
NIOSH REL: (Paraffin Wax Fume) TWA 2 mg/m^3
SAFETY PROFILE: A skin and eye irritant. Many paraffin waxes contain carcinogens. Fumes cause lung damage.
PROP: Colorless or white, translucent wax; odorless. D: approx 0.90, mp: 50–57°. Insol in water, alc; sol in benzene, chloroform, ether, carbon disulfide, oils; misc with fats.

PAT750 CAS: 19624-22-7 B_5H_9 HR: 3
PENTABORANE(9)
DOT: UN 1380
SYN: PENTABORANE (ACGIH,DOT,OSHA)
OSHA PEL: TWA 0.005 ppm; STEL 0.015 ppm
ACGIH TLV: TWA 0.005 ppm; STEL 0.013 ppm
DFG MAK: 0.005 ppm (0.013 mg/m^3)
DOT Classification: 4.2; Label: Spontaneously Combustible, Poison
SAFETY PROFILE: Poison by inhalation. Dangerous fire hazard by chemical reaction; spontaneously flammable in air. Dangerous explosion hazard.
PROP: Colorless gas or liquid; bad odor. Mp: –46.6°, bp: 60°, d: 0.61 @ 0°, vap d: 2.2, vap press: 66 mm @ 0°, lel: 0.42%. Sol in THF, diglyme, Et_2O, and hexane.

PAW250 CAS: 42279-29-8 $C_{12}H_5Cl_5O$ HR: 3
PENTACHLORO DIPHENYL OXIDE
SYNS: PHENYL ETHER PENTACHLORO
OSHA PEL: TWA 0.5 mg/m^3
SAFETY PROFILE: A poison by ingestion.

PAW500 CAS: 76-01-7 C_2HCl_5 HR: 3
PENTACHLOROETHANE
DOT: UN 1669
SYNS: ETHANE PENTACHLORIDE ♦ PENTALIN
DFG MAK: 5 ppm (42 mg/m^3)
DOT Classification: 6.1; Label: Poison

SAFETY PROFILE: Poison by inhalation. Moderately toxic by ingestion. An irritant. A flammable liquid.
PROP: Colorless liquid; chloroform-like odor. Mp: –29°, bp: 161–162°, d: 1.6728 @ 25°/4°. Insol in water; misc in alc and ether.

PAW750 CAS: 1321-64-8 $C_{10}H_3Cl_5$ HR: 3
PENTACHLORONAPHTHALENE
OSHA PEL: TWA 0.5 mg/m^3 (skin)
ACGIH TLV: TWA 0.5 mg/m^3
DFG MAK: 0.5 mg/m^3
NIOSH REL: (Pentachloronaphthalen) TWA 0.5 mg/m^3 (skin)
SAFETY PROFILE: Poison by ingestion, inhalation, and skin contact. An irritant.
PROP: White solid.

PAX250 CAS: 87-86-5 C_6HCl_5O HR: 3
PENTACHLOROPHENOL
SYNS: PCP ♦ PENTA ♦ 2,3,4,5,6-PENTACHLOROPHENOL
OSHA PEL: TWA 0.5 mg/m^3 (skin)
ACGIH TLV: TWA 0.5 mg/m^3 (skin); BEI: 2 mg/g creatinine in urine prior to last shift of workweek; Animal Carcinogen
DFG MAK: Confirmed Animal Carcinogen, Suspected Human Carcinogen; BAT: 1000 μg/L in plasma/serum
SAFETY PROFILE: Confirmed human carcinogen. Poison by ingestion and skin contact. An experimental teratogen. A skin irritant.
PROP: Dark-colored flakes, monoclinic prisms, and sublimed needle crystals from benzene; characteristic odor. Mp: 174°, mp: 191° (anhydrous), bp: 310° (decomp), d: 1.978, vap press: 40 mm @ 211.2°. Sol in ether, benzene; very sol in alc; insol in water; sltly sol in cold pet ether.

PBB750 CAS: 115-77-5 $C_5H_{12}O_4$ HR: 3
PENTAERYTHRITOL
SYNS: 2,2-BIS(HYDROXYMETHYL)-1,3-PROPANEDIOL ♦ PENTAERYTHRITE ♦ TETRAKIS(HYDROXYMETHYL)METHANE
OSHA PEL: TWA Total Dust: 10 mg/m^3; Respirable Fraction: 5 mg/m^3
ACGIH TLV: TWA (nuisance particulate) 10 mg/m^3 of total dust (when toxic impurities are not present, e.g., quartz <1%)
SAFETY PROFILE: Mildly toxic by ingestion. A nuisance dust. Flammable.
PROP: Ditetragolan crystals from HCl (aq). Mp: 262°, d: 1.38 @ 25°/4°.

PBK250 CAS: 109-66-0 C_5H_{12} HR: 3
n-PENTANE
DOT: UN 1265
SYNS: AMYL HYDRIDE (DOT)
OSHA PEL: TWA 600 ppm; STEL 750 ppm

ACGIH TLV: TWA 600 ppm
DFG MAK: 1000 ppm (3000 mg/m^3)
NIOSH REL: (Alkanes) TWA 350 mg/m^3
DOT Classification: 3; Label: Flammable Liquid
SAFETY PROFILE: Moderately toxic by inhalation. The liquid can cause blisters on contact. Flammable liquid.
PROP: Colorless liquid. Bp: 36.1°, flash p: <–40°F, fp: –129.8°, d: 0.626 @ 20°/4°, autoign temp: 588°F, vap press: 400 mm @ 18.5°, vap d: 2.48, lel: 1.5%, uel: 7.8%. Sol in water; misc in alc, ether, org solv.

PBM000 CAS: 110-66-7 $C_5H_{12}S$ HR: 3

1-PENTANETHIOL

DOT: UN 1228/UN 3071
SYNS: AMYL MERCAPTAN (DOT) ♦ PENTYL MERCAPTAN
NIOSH REL: (Thiols (n-Alkane Mono)) CL 0.5 ppm/15M
DOT Classification: 3; Label: Flammable Liquid, Poison (UN 1228); DOT Class: 6.1; Label: Poison, Flammable Liquid (UN 3071)
SAFETY PROFILE: Moderately toxic by inhalation. A weak sensitizer and allergen. May cause contact dermatitis. A flammable liquid.
PROP: Water-white to yellow liquid. D: 0.857 @ 20°, bp: 123.64°, flash p: 65°F, vap press: 13.8 mm @ 25°, vap d: 3.59. Insol in water; misc in alc and ether.

PBN250 CAS: 107-87-9 $C_5H_{10}O$ HR: 3

2-PENTANONE

DOT: UN 1249
SYNS: ETHYL ACETONE ♦ METHYL PROPYL KETONE (ACGIH, DOT) ♦ MPK
OSHA PEL: TWA 200 ppm; STEL 250 ppm
ACGIH TLV: TWA 200 ppm; STEL 250 ppm
DFG MAK: 200 ppm (710 mg/m^3)
NIOSH REL: TWA 530 mg/m^3
DOT Classification: 3; Label: Flammable Liquid
SAFETY PROFILE: Moderately toxic by ingestion. Mildly toxic by skin contact and inhalation. A skin irritant. A highly flammable liquid.
PROP: Water-white liquid; fruity, ethereal odor. D: 0.801–0.806, vap d: 3.0, bp: 102°, flash p: 45°F, autoign temp: 941°F, lel: 1.5%, uel: 8.2%. Sltly sol in water; misc with alc, ether.

PCF275 CAS: 127-18-4 C_2Cl_4 HR: 3

PERCHLOROETHYLENE

DOT: UN 1897
SYNS: CARBON DICHLORIDE ♦ ETHYLENE TETRACHLORIDE ♦ PERCHLORETHYLENE ♦ PERK ♦ TETRACHLOROETHYLENE (DOT)
OSHA PEL: TWA 25 ppm

ACGIH TLV: TWA 50 ppm; STEL 200 ppm (Proposed: TWA 25 ppm; Animal Carcinogen); BEI: 3.5 mg/L trichloroacetic acid in urine at end of shift at end of workweek
DFG MAK: Confirmed Animal Carcinogen with Unknown Relevance to Humans; BAT: blood 100 µg/dL
NIOSH REL: (Tetrachloroethylene) Minimize workplace exposure
DOT Classification: 6.1; Label: KEEP AWAY FROM FOOD
SAFETY PROFILE: Confirmed carcinogen. Moderately toxic by ingestion and inhalation. An experimental teratogen. An eye and severe skin irritant. Can cause dermatitis.
PROP: Colorless liquid; chloroform-like odor. Mp: –23.35°, fp: –22.35°, bp: 121.20°, d: 1.6311 @ 15°/4°, vap press: 15.8 mm @ 22°, vap d: 5.83.

PCF300 CAS: 594-42-3 CCl_4S HR: 3
PERCHLOROMETHYL MERCAPTAN
DOT: UN 1670
SYNS: PCM ♦ TRICHLOROMETHANE SULFENYL CHLORIDE
OSHA PEL: TWA 0.1 ppm
ACGIH TLV: TWA 0.1 ppm
DOT Classification: 6.1; Label: Poison
SAFETY PROFILE: Poison by ingestion and inhalation. A severe skin, eye, and mucous membrane irritant.
PROP: Yellow, oily liquid. Bp: slt decomp @ 149°, d: 1.700 @ 20°, vap d: 6.414.

PCF750 CAS: 7616-94-6 $ClFO_3$ HR: 2
PERCHLORYL FLUORIDE
DOT: UN 3083
SYNS: CHLORINE FLUORIDE OXIDE ♦ CHLORINE OXYFLUORIDE
OSHA PEL: TWA 3 ppm; STEL 6 ppm
ACGIH TLV: TWA 3 ppm; STEL 6 ppm
DOT Classification: 2.3; Label: Poison Gas, Oxidizer
SAFETY PROFILE: A poison gas. Can be absorbed through the skin. While nonflammable, it supports combustion. It can react violently with many substances.
PROP: Colorless, noncorrosive gas stable to 5°; stable to water; characteristic sweet odor. Mp: –146°, bp: –46.8°, d: (liquid) 1.434. d: (gas) 0.637.

PCL500 CAS: 79-21-0 $C_2H_4O_3$ HR: 3
PEROXYACETIC ACID
DOT: UN 3149
SYNS: ACETIC PEROXIDE ♦ MONOPERACETIC ACID
DFG MAK: Very strong skin effects; Confirmed Animal Carcinogen with Unknown Relevance to Humans
DOT Classification: 5.1; Label: Oxidizer, Corrosive (UN3149)
SAFETY PROFILE: Poison by ingestion. Moderately toxic by inhalation and skin contact. A corrosive eye, skin, and mucous membrane irritant. A flammable liquid. It can react violently with many substances.

PROP: Not over 40% peracetic acid and not over 6% hydrogen peroxide (FEREAC 41,15972,76). Colorless liquid; strong odor. Fp: 0.1°, bp: 105°, explodes @ 110°, flash p: 105°F (OC), d: 1.15 @ 20°. Sol in H_2O, EtOH, Et_2O.

PCS250 CAS: 8002-05-9 HR: 3
PETROLEUM DISTILLATE
DOT: UN 1268
OSHA PEL: TWA 400 ppm
DOT Classification: 3; Label: Flammable Liquid
SAFETY PROFILE: Mildly toxic by inhalation and ingestion. Moderate skin and eye irritation. A flammable liquid.

PCT250 CAS: 8032-32-4 HR: 3
PETROLEUM SPIRITS
DOT: UN 1271
SYNS: BENZINE (LIGHT PETROLEUM DISTILLATE) ♦ LIGROIN ♦ PETROLEUM SPIRIT (DOT) ♦ VM and P NAPHTHA ♦ VM & P NAPHTHA ♦ VM&P NAPHTHA ♦ VM & P NAPHTHA (ACGIH,OSHA)
OSHA PEL: TWA 300 ppm; STEL 400 PPM
ACGIH TLV: TWA 300 ppm; Animal Carcinogen
NIOSH REL: (VM & P Naphtha) TWA 350 mg/m^3; CL 1800 mg/m^3/15M
DOT Classification: 3; Label: Flammable Liquid
SAFETY PROFILE: Confirmed carcinogen. Mildly toxic by inhalation. An eye irritant. A flammable liquid.
PROP: Volatile, clear, colorless and non-fluorescent liquid. Mp: <–73°, bp: 40–80°, ULC: 95–100, lel: 1.1%, uel: 5.9%, flash p: <0°F, d: 0.635–0.660, autoign temp: 550°F, vap d: 2.50.

PCW250 CAS: 85-01-8 $C_{14}H_{10}$ HR: 3
PHENANTHRENE
SYNS: COAL TAR PITCH VOLATILES: PHENANTHRENE ♦ PHENANTRIN
OSHA PEL: TWA 0.2 mg/m^3
SAFETY PROFILE: Moderately toxic by ingestion. A human skin photosensitizer. Combustible.
PROP: Solid or monoclinic crystals; plates from EtOH. Mp: 100°, bp: 339°, d: 1.179 @ 25°, vap press: 1 mm @ 118.3°, vap d: 6.14. Insol in water; sol in CS_2, benzene, and hot alc; very sol in ether.

PDN750 CAS: 108-95-2 C_6H_6O HR: 3
PHENOL
DOT: UN 1671/UN 2312/NA 2821
SYNS: CARBOLIC ACID ♦ OXYBENZENE
OSHA PEL: TWA 5 ppm (skin)
ACGIH TLV: TWA 5 ppm (skin); BEI: 250 mg(total phenol)/g creatinine in urine at end of shift; Not Classifiable as a Human Carcinogen

DFG MAK: Confirmed Animal Carcinogen with Unknown Relevance to Humans; BAT: 300 mg/L at end of shift

NIOSH REL: (Phenol) TWA 20 mg/m^3; CL 60 mg/m^3/15M

DOT Classification: 6.1; Label: Poison

SAFETY PROFILE: Human poison by ingestion. Moderately toxic by skin contact. A severe eye and skin irritant. An experimental teratogen. Combustible.

PROP: Deliquescent needles or white, crystalline mass that turns pink or red if not perfectly pure; burning taste, distinctive odor. Mp: 43°, fp: 41°, bp: 90.2° @ 25 mm, flash p: 175°F (CC), d: 1.072, autoign temp: 1319°F, vap press: 1 mm @ 40.1°, vap d: 3.24. Sol in water; misc in alc and ether.

PDP250 CAS: 92-84-2 $C_{12}H_9NS$ HR: 3

PHENOTHIAZINE

SYNS: DIBENZO-1,4-THIAZINE ♦ PHENTHIAZINE ♦ THIODIPHENYLAMINE

OSHA PEL: TWA 5 mg/m^3 (skin)

ACGIH TLV: TWA 5 mg/m^3 (skin)

NIOSH REL: (Phenothiazine) TWA 5 mg/m^3 (skin)

SAFETY PROFILE: Moderately toxic to humans by ingestion. Can cause skin irritation and photosensitization.

PROP: Yellow plates, rhombic leaflets, or diamond-shaped plates from toluene or butanol. Mp: 182°, sublimes at 130° at 1 mm, bp: 371°. Freely sol in benzene; sol in ether, hot acetic acid; sltly sol in alc and in mineral oils; practically insol in pet ether, chloroform, water.

PEY250 CAS: 95-54-5 $C_6H_8N_2$ HR: 3

o-PHENYLENEDIAMINE

DOT: UN 1673

SYNS: 2-AMINOANILINE ♦ 1,2-BENZENEDIAMINE ♦ C.I. 76010 ♦ o-DIAMINOBENZENE ♦ 1,2-PHENYLENEDIAMINE (DOT)

ACGIH TLV: TWA 0.1 mg/m^3; Animal Carcinogen

DFG MAK: Confirmed Animal Carcinogen with Unknown Relevance to Humans

DOT Classification: 6.1; Label: KEEP AWAY FROM FOOD

SAFETY PROFILE: Confirmed carcinogen. Poison by ingestion. Mildly toxic by skin contact.

PROP: Tan crystals or leaflets from water. Mp: 104°, bp: 257°. Sltly sol in water; very sol in alc, chloroform, ether.

PEY500 CAS: 106-50-3 $C_6H_8N_2$ HR: 3

p-PHENYLENEDIAMINE

DOT: UN 1673

SYNS: 4-AMINOANILINE ♦ C.I. 76060 ♦ p-DIAMINOBENZENE ♦ PHENYLENEDIAMINE, PARA, solid (DOT) ♦ PPD

OSHA PEL: TWA 0.1 mg/m^3 (skin)

ACGIH TLV: TWA 0.1 mg/m^3 (skin); Not Classifiable as a Human Carcinogen

DFG MAK: 0.1 mg/m^3 as total dust; Confirmed Animal Carcinogen with Unknown Relevance to Humans
DOT Classification: 6.1; Label: KEEP AWAY FROM FOOD
SAFETY PROFILE: Suspected carcinogen. Poison by ingestion. Mildly toxic by skin contact. A human skin irritant. Combustible.
PROP: White to sltly red crystals or leaflets from Et_2O. Mp: 146°, flash p: 312°F, vap d: 3.72, bp: 267°. Sol in alc, chloroform, ether.

PFA850 CAS: 101-84-8 $C_{12}H_{10}O$ HR: 2
PHENYL ETHER
SYNS: BIPHENYL OXIDE ♦ DIPHENYL ETHER ♦ DIPHENYL OXIDE
OSHA PEL: Vapor: TWA 1 ppm
ACGIH TLV: TWA 1 ppm; STEL 2 ppm (vapor)
DFG MAK: 1 ppm (7.1 mg/m^3)
NIOSH REL: (Phenyl Ether, vapor) TWA 1 ppm
SAFETY PROFILE: Moderately toxic by ingestion. A skin and eye irritant. Combustible.
PROP: Colorless low-melting crystals; geranium odor. Mp: 37–39°, bp: 259°, flash p: 239°F, d: 1.0728 @ 20°, vap d: 5.86, autoign temp: 1148°F, lel: 0.8%, uel: 1.5%.

PFA860 CAS: 8004-13-5 $C_{12}H_{10} \cdot C_{12}H_{10}O$ HR: 3
PHENYL ETHER-BIPHENYL MIXTURE
SYNS: DIPHENYL mixed with DIPHENYL OXIDE
OSHA PEL: Vapor: TWA 1 ppm
NIOSH REL: (Phenyl Ether-Biphenyl Mixture, vapor) TWA 1 ppm
SAFETY PROFILE: Poison by inhalation. Moderately toxic by ingestion. A mild skin and eye irritant.
PROP: Eutectic mixture 73.5% phenylether and 26.5% biphenyl by weight (MELAAD 48,247,57).

PFF360 CAS: 122-60-1 $C_9H_{10}O_2$ HR: 3
PHENYL GLYCIDYL ETHER
SYNS: 1,2-EPOXY-3-PHENOXYPROPANE ♦ PGE ♦ PHENOL GLYCIDYL ETHER (MAK) ♦ 3-PHENOXY-1,2-EPOXYPROPANE ♦ PHENYL-2,3-EPOXYPROPYL ETHER
OSHA PEL: TWA 1 ppm
ACGIH TLV: TWA 0.1 ppm (skin); Animal Carcinogen; (Proposed: TWA 0.1 ppm (skin, sensitizer) Confirmed Animal Carcinogen with Unknown Relevance to Humans)
DFG MAK: Confirmed Animal Carcinogen, Suspected Human Carcinogen
NIOSH REL: (Phenyl Glycidyl Ether) CL 1 ppm/15M
SAFETY PROFILE: Confirmed carcinogen. Moderately toxic by ingestion and skin contact. A severe eye and skin irritant.

PFI000 CAS: 100-63-0 $C_6H_8N_2$ HR: 3
PHENYLHYDRAZINE
DOT: UN 2572

SYNS: HYDRAZINOBENEZENE
OSHA PEL: TWA 5 ppm (skin); STEL 10 ppm
ACGIH TLV: TWA 0.1 ppm (skin); Animal Carcinogen
DFG MAK: Confirmed Animal Carcinogen with Unknown Relevance to Humans
NIOSH REL: CL 0.6 mg/m^3/2H
DOT Classification: 6.1; Label: Poison
SAFETY PROFILE: Confirmed carcinogen. Poison by ingestion. Flammable.
PROP: Yellow, monoclinic crystals, oil, or plates (usually an oil with aniline-like odor). Mp: 19.6°, bp: 243.5° (decomp), flash p: 192°F (CC), d: 1.0978 @ 20°/4°, vap press: 1 mm @ 71.8°, vap d: 3.7. Sltly sol in hot water; misc in alc, chloroform, ether, benzene.

PFI250 CAS: 59-88-1 $C_6H_8N_2 \cdot ClH$ HR: 3
PHENYLHYDRAZINE HYDROCHLORIDE
SYNS: PHENYLHYDRAZINE MONOHYDROCHLORIDE
NIOSH REL: (Hydrazines) CL 0.6 mg/m^3/2H
SAFETY PROFILE: Poison by ingestion.
PROP: Leaflets from water; crystals from alc. Mp: 250–254° (decomp). Very sol in water; sol in alc; insol in ether.

PFL850 CAS: 108-98-5 C_6H_6S HR: 3
PHENYL MERCAPTAN
DOT: UN 2337
SYNS: THIOPHENOL (DOT)
OSHA PEL: TWA 0.5 ppm
ACGIH TLV: TWA 0.5 ppm
NIOSH REL: CL 0.5 mg/m^3/15M
DOT Classification: 6.1; Label: Poison, Flammable Liquid
SAFETY PROFILE: Poison by ingestion, inhalation, and skin contact. A severe eye irritant. Can cause severe dermatitis.
PROP: A liquid; repulsive odor. Bp: 169.5°, d: 1.973 @ 25°/4°.

PFT500 CAS: 135-88-6 $C_{16}H_{13}N$ HR: 3
N-PHENYL-β-NAPHTHYLAMINE
SYNS: 2-ANILINONAPHTHALENE ♦ 2-NAPHTHYLPHENYLAMINE ♦ PBNA ♦ PHENYL-2-NAPHTHYLAMINE
ACGIH TLV: Not Classifiable as a Human Carcinogen
DFG MAK: Confirmed Animal Carcinogen with Unknown Relevance to Humans
SAFETY PROFILE: Suspected carcinogen. Moderately toxic by ingestion.
PROP: Rhombic crystals and needles from MeOH. Mp: 107–108°, bp: 395.5°. Insol in water; sol in hot benzene; very sol in hot alc, ether.

PFV250 CAS: 638-21-1 C_6H_7P HR: 3
PHENYLPHOSPHINE
OSHA PEL: CL 0.05 ppm
ACGIH TLV: CL 0.05 ppm

NIOSH REL: (Phenylphosphine) CL 0.05 ppm
SAFETY PROFILE: Poison by inhalation. Ignites spontaneously in air.
PROP: Foul smelling liquid. Bp: 160–161°, d: 1.001 @ 15 mm. Insol in water; sol in alkali; very sol in alc and ether.

PGX000 CAS: 75-44-5 CCl_2O HR: 3
PHOSGENE
DOT: UN 1076
SYNS: CARBON OXYCHLORIDE ♦ CHLOROFORMYL CHLORIDE
OSHA PEL: TWA 0.1 ppm
ACGIH TLV: TWA 0.1 ppm
DFG MAK: 0.02 ppm (0.082 mg/m^3)
NIOSH REL: (Phosgene) TWA 0.1 ppm; CL 0.2 ppm/15M
DOT Classification: 2.3; Label: Poison Gas, Corrosive
SAFETY PROFILE: A human poison by inhalation. A severe eye, skin, and mucous membrane irritant.
PROP: Colorless, poison gas or volatile liquid; odor of new-mown hay or green corn. Mp: –128°, bp: 7.6°, d: 1.419 @ 0°/4°, vap press: 1180 mm @ 20°, vap d: 3.4. Very sltly sol in water; very sol in benzene and acetic acid; decomp sltly in water.

PGY000 CAS: 7803-51-2 H_3P HR: 3
PHOSPHINE
DOT: UN 2199
SYNS: HYDROGEN PHOSPHIDE ♦ PHOSPHORUS TRIHYDRIDE
OSHA PEL: TWA 0.3 ppm; STEL 1 ppm
ACGIH TLV: TWA 0.3 ppm; STEL 1 ppm
DFG MAK: 0.1 ppm (0.14 mg/m^3)
DOT Classification: 2.3; Label: Poison Gas, Flammable Gas
SAFETY PROFILE: A poison by inhalation.Very dangerous fire hazard by spontaneous chemical reaction. It can react violently with many substances.
PROP: Colorless gas; foul odor of decaying fish. Extremely weak base. Mp: –132.5°, bp: –87.5°, d: 1.529 g/L @ 0°, autoign temp: 212°F, lel: 1%. Sltly sol in water (giving neutral soln).

PHB250 CAS: 7664-38-2 H_3O_4P HR: 3
PHOSPHORIC ACID
DOT: UN 1805
SYNS: ORTHOPHOSPHORIC ACID
OSHA PEL: TWA 1 mg/m^3; STEL 3 mg/m^3
ACGIH TLV: TWA 1 mg/m^3; STEL 3 mg/m^3
DOT Classification: 8; Label: Corrosive
SAFETY PROFILE: Human poison by ingestion. Moderately toxic by skin contact. A corrosive irritant to eyes, skin, and mucous membranes, and a systemic irritant by inhalation.

PROP: Colorless syrupy liquid or rhombic crystals. Mp: 42.35°, loses 1/2H_2O @ 213°, fp: 42.4°, d: 1.864 @ 25°, vap press: 0.0285 mm @ 20°. Misc with water and many org solvs.

PHO500 CAS: 7723-14-0 P HR: 3

PHOSPHORUS (red)

DOT: UN 1338

SYN: PHOSPHORUS, amorphous (DOT)

DFG MAK: 0.1 mg/m^3

DOT Classification: 4.1; Label: Flammable Solid

SAFETY PROFILE: Generally less reactive than white phosphorus. Dangerous fire hazard. It can react violently with many substances.

PROP: Reddish-brown powder. Bp: 280° (with ignition), mp: 590° @ 43 atm, d: 2.34, autoign temp: 500°F in air, vap d: 4.77.

PHP010 CAS: 7723-14-0 P_4 HR: 3

PHOSPHORUS (yellow)

DOT: UN 1381/UN 2447

SYNS: PHOSPHOROUS (WHITE) ♦ WHITE PHOSPHORUS ♦ YELLOW PHOSPHORUS

OSHA PEL: TWA 0.1 mg/m^3

ACGIH TLV: TWA 0.1 mg/m^3

DOT Classification: 4.2; Label: Spontaneously Combustible, Poison

SAFETY PROFILE: Human poison by ingestion. More reactive than red phosphorus. Dangerous fire hazard when exposed to heat, flame, or by chemical reaction with oxidizers. Ignites spontaneously in air. It can react violently with many substances.

PROP: Cubic, colorless crystals from ligroin or conc HCl; yellow leaflets from water; colorless to yellow, wax-like solid. Mp: 44.1°, bp: 280°, flash p: spontaneously flammable in air, d: 1.82, autoign temp: 86°F, vap press: 1 mm @ 76.6°, vap d: 4.42. Mod sol in water.

PHQ800 CAS: 10025-87-3 Cl_3OP HR: 3

PHOSPHORUS OXYCHLORIDE

DOT: UN 1810

SYNS: PHOSPHORUS OXYTRICHLORIDE ♦ PHOSPHORYL CHLORIDE

OSHA PEL: TWA 0.1 ppm

ACGIH TLV: TWA 0.1 ppm

DFG MAK: 0.2 ppm (1 mg/m^3)

DOT Classification: 8; Label: Corrosive, Poison

SAFETY PROFILE: Poison by inhalation and ingestion. A corrosive eye, skin, and mucous membrane irritant. Potentially explosive reaction with water. It can react violently with many substances.

PROP: Colorless to sltly yellow, mobile, fuming liquid. Very reactive to nucleophiles. Hydrolyzed rapidly by H_2O at room temp. Mp: 1.2°, bp: 105.1°, d: 1.685 @ 15.5°, vap press: 40 mm @ 27.3°, vap d: 5.3.

PHR500 CAS: 10026-13-8 Cl_5P HR: 3

PHOSPHORUS PENTACHLORIDE

DOT: UN 1806

SYNS: PHOSPHORIC CHLORIDE ♦ PHOSPHORUS PERCHLORIDE

OSHA PEL: TWA 1 mg/m^3

ACGIH TLV: TWA 0.85 mg/m^3

DFG MAK: 1 mg/m^3

DOT Classification: 8; Label: Corrosive

SAFETY PROFILE: Poison by inhalation. Moderately toxic by ingestion. A severe eye, skin, and mucous membrane irritant. Corrosive to body tissues. Flammable by chemical reaction. It can react violently with many substances.

PROP: Yellowish-white, fuming, crystalline mass; pungent odor. Mp: (under press) 148° decomp, bp: subl @ 160°, d: 4.65 g/L @ 296°, vap press: 1 mm @ 55.5°.

PHS000 CAS: 1314-80-3 P_2S_5 HR: 3

PHOSPHORUS PENTASULFIDE

DOT: UN 1340

SYNS: PHOSPHORIC SULFIDE ♦ PHOSPHORUS PERSULFIDE ♦ THIOPHOSPHORIC ANHYDRIDE

OSHA PEL: TWA 1 mg/m^3; STEL 3 mg/m^3

ACGIH TLV: TWA 1 mg/m^3; STEL 3 mg/m^3

DFG MAK: 1 mg/m^3

DOT Classification: 4.3; Label: Dangerous When Wet

SAFETY PROFILE: A poison by ingestion. A severe eye and skin irritant. Readily liberates toxic hydrogen sulfide and phosphorus pentoxide and evolves heat on contact with moisture. Dangerous fire hazard in the form of dust. It can react violently with many substances.

PROP: Gray to yellow-green, crystalline, deliquescent mass. Bp: 514°, d: 2.09, autoign temp: 287°F, mp: 286–290°.

PHS250 CAS: 1314-56-3 O_5P_2 HR: 3

PHOSPHORUS PENTOXIDE

DOT: NA 1807

SYNS: DIPHOSPHORUS PENTOXIDE ♦ PHOSPHORUS PENTAOXIDE

DFG MAK: 1 mg/m^3

DOT Classification: 8; Label: Corrosive

SAFETY PROFILE: Poison by inhalation. A corrosive irritant to the eyes, skin, and mucous membranes. It can react violently with many substances.

PROP: Deliq crystals. D: 2.30, mp: 340°, subl @ 360°.

PHT275 CAS: 7719-12-2 Cl_3P HR: 3

PHOSPHORUS TRICHLORIDE

DOT: UN 1809

SYNS: PHOSPHORUS CHLORIDE

OSHA PEL: TWA 0.2 ppm; STEL 0.5 ppm
ACGIH TLV: TWA 0.2 ppm; STEL 0.5 ppm
DFG MAK: 0.5 ppm (2.8 mg/m^3)
DOT Classification: 8; Label: Corrosive, Poison
SAFETY PROFILE: Poison by ingestion and inhalation. A corrosive irritant to skin, eyes, and mucous membranes. It can react violently with many substances.
PROP: Clear, colorless, fuming liquid. Mp: –111.8°, fp: –93.6°, bp: 76°, d: 1.574 @ 21°, vap press: 100 mm @ 21°, vap d: 4.75. Decomp by water and alc; sol in benzene, chloroform, and ether.

PHW750 CAS: 85-44-9 $C_8H_4O_3$ HR: 3

PHTHALIC ANHYDRIDE

DOT: UN 2214
SYNS: 1,2-BENZENEDICARBOXYLIC ACID ANHYDRIDE ♦ 1,3-DIOXOPHTHALAN ♦ PHTHALIC ACID ANHYDRIDE
OSHA PEL: TWA 1 ppm
ACGIH TLV: TWA 1 ppm; Not Classifiable as a Human Carcinogen; (Proposed: TWA 1 ppm (skin, sensitizer) Not Classifiable as a Human Carcinogen)
DFG MAK: 1 mg/m^3 as total dust
DOT Classification: 8; Label: Corrosive
SAFETY PROFILE: Poison by ingestion. Experimental teratogenic effects. A corrosive eye, skin, and mucous membrane irritant. Combustible.
PROP: White, crystalline needles from alc. Mp: 131.2°, lel: 1.7%, uel: 10.4%, bp: 284°, flash p: 305°F (CC), d: 1.527 @ 4°, autoign temp: 1058°F, vap press: 1 mm @ 96.5°, vap d: 5.10. Very sltly sol in water; sol in alc; sltly sol in ether.

PHX550 CAS: 626-17-5 $C_8H_4N_2$ HR: 3

m-PHTHALODINITRILE

SYNS: 1,3-BENZENEDICARBONITRILE ♦ 1,3-DICYANOBENZENE ♦ IPN ♦ ISOPHTHALONITRILE ♦ m-PDN
OSHA PEL: TWA 5 mg/m^3
ACGIH TLV: TWA 5 mg/m^3
SAFETY PROFILE: Poison by ingestion. An eye irritant.
PROP: Colorless crystals. Vap d: 4.42, mp: 161.5–162°, bp: sublimes. Insol in water; sol in benzene, acetone.

PID000 CAS: 88-89-1 $C_6H_3N_3O_7$ HR: 3

PICRIC ACID

DOT: UN 0154/UN 1344
SYNS: C.I. 10305 ♦ 2-HYDROXY-1,3,5-TRINITROBENZENE ♦ TRINITROPHENOL (UN 0154) (DOT) ♦ 2,4,6-TRINITROPHENYL (OSHA)
OSHA PEL: TWA 0.1 mg/m^3 (skin)
ACGIH TLV: TWA 0.1 mg/m^3
DFG MAK: 0.1 mg/m^3

DOT Classification: 4.1; Label: Flammable Solid (NA 1344, UN 1344)
SAFETY PROFILE: Poison by ingestion. An irritant and an allergen. A flammable solid. It can react violently with many substances.
PROP: Colorless crystals from ligroin or conc HCl; yellow leaflets from water; or yellow liquid; very bitter. Mp: 121.8°, bp: explodes >300°, flash p: 302°F, d: 1.763, autoign temp: 572°F, vap d: 7.90. Mod sol in water.

PIK000 CAS: 142-64-3 $C_4H_{10}N_2 \cdot 2ClH$ HR: 2
PIPERAZINE DIHYDROCHLORIDE
SYNS: PIPERAZINE HYDROCHLORIDE
OSHA PEL: TWA 5 mg/m^3
ACGIH TLV: TWA 5 mg/m^3
NIOSH REL: (Piperazine) TWA 5.0 mg/m^3
SAFETY PROFILE: Mildly toxic by ingestion.

PJD500 CAS: 7440-06-4 Pt HR: 2
PLATINUM
SYNS: C.I. 77795 ♦ PLATINUM BLACK ♦ PLATINUM SPONGE
OSHA PEL: TWA (metal) 1 mg/m^3; (soluble salts as Pt) 0.002 mg/m^3
ACGIH TLV: TWA (metal) 1 mg/m^3; (soluble salts as Pt) 0.002 mg/m^3
DFG MAK: 0.002 mg/m^3
NIOSH REL: (Platinum (as Pt), metal) TWA 1 mg/m^3; (Platinum (as Pt), soluble salts): TWA 0.002 mg/m^3
SAFETY PROFILE: Finely divided platinum is a powerful catalyst and can be dangerous to handle.
PROP: Silvery-white, malleable, ductile metal. Unaffected by air or H_2O. Platinum-black is velvety-black, finely divided, and this is attacked by O_2 at 5°. At 1° HBr, HI, Br_2, $FeCl_3$, NaCN (+ O_2) are sltly corrosive. No reaction with SO_2. Does not form an amalgam with Hg. Mp: 1772°, bp: 3827°, d: 21.45 @ 20°. Sol in aq regia, HCl in air, fused alkali.

PJE250 CAS: 13454-96-1 Cl_4Pt HR: 3
PLATINUM(IV) CHLORIDE
SYN: PLATINUM TETRACHLORIDE
OSHA PEL: TWA 0.002 $mg(Pt)/m^3$
ACGIH TLV: TWA 0.002 $mg(Pt)/m^3$
SAFETY PROFILE: Poison by ingestion. A severe skin irritant.
PROP: Hygrosopic red-brown crystals. Sol in H_2O and Me_2CO.

PJL750 CAS: 1336-36-3 HR: 3
POLYCHLORINATED BIPHENYLS
DOT: UN 2315

SYNS: AROCLOR 1016 ♦ AROCLOR 1221 ♦ AROCLOR 1232 ♦ AROCLOR 1242 ♦ AROCLOR 1248 ♦ AROCLOR 1254 ♦ AROCLOR 1260 ♦ AROCLOR 1262 ♦ AROCLOR 1268 ♦ AROCLOR 2565 ♦ AROCLOR 4465 ♦ AROCLOR 5442 ♦ PCB ♦ PCBs ♦ POLYCHLOROBIPHENYL

DFG MAK: Suspected Carcinogen

NIOSH REL: TWA (Polychlorinated Biphenyls) 0.001 mg/m^3

DOT Classification: 9; Label: CLASS 9

SAFETY PROFILE: Confirmed carcinogen. Moderately toxic by ingestion. Some are poisons by other routes. Combustible.

PROP: Bp: 340–375°, flash p: 383°F (COC), d: 1.44 @ 30°. A series of technical mixtures consisting of many isomers and compounds that vary from mobile oily liquids to white crystalline solids and hard noncrystalline resins. Technical products vary in composition, in the degree of chlorination, and possibly according to batch (IARC** 7,262,74).

PKQ059 CAS: 9002-86-2 $(C_2H_3Cl)_n$ HR: 2

POLYVINYL CHLORIDE

SYNS: POLY(CHLOROETHYLENE) ♦ PVC (MAK) ♦ VINYL CHLORIDE HOMOPOLYMER ♦ VINYL CHLORIDE POLYMER

DFG MAK: 1.5 mg/m^3 (dust)

SAFETY PROFILE: Chronic inhalation of dusts can cause pulmonary damage and abnormal liver function. "Meat wrapper's asthma$$CLOSE QUOTES has resulted from the cutting of PVC films with a hot knife. Can cause allergic dermatitis.

PROP: Polymers with molecular weights ranging from 60,000 to 150,000 (CNREA8 15,333,55). White powder, d: 1.406.

PKS750 CAS: 65997-15-1 HR: 1

PORTLAND CEMENT

SYNS: CEMENT, PORTLAND

OSHA PEL: TWA Total Dust: 10 mg/m^3; Respirable Fraction: 5 mg/m^3

ACGIH TLV: TWA (nuisance particulate) 10 mg/m^3 of total dust (when toxic impurities are not present, e.g., quartz <1%)

DFG MAK: 5 mg/m^3

NIOSH REL: (Portland Cement, respirable fraction) TWA 5 mg/m^3; (Portland Cement, total dust): TWA 10 mg/m^3

SAFETY PROFILE: A nuisance dust. A skin irritant.

PROP: Fine gray powder composed of compounds of lime, aluminum, silica, and iron oxide as $(4CaO \cdot Al_2O_3 \cdot Fe_2)_3$, $(3CaOAl_2O_3)$, $(3CaO \cdot SiO_2)$, and $(2CaOSiO_2)$. Small amounts of magnesia, sodium, potassium, chromium, and sulfur are also present in combined form. Containing less than 1% crystalline silica (FEREAC 39,23540,74).

PKU250 CAS: 7789-29-9 FK•FH HR: 3

POTASSIUM ACID FLUORIDE

DOT: NA 1811

SYNS: HYDROGEN POTASSIUM FLUORIDE ♦ POTASSIUM BIFLUORIDE, solid or solution (DOT) ♦ POTASSIUM FLUORIDE ♦ POTASSIUM HYDROGEN FLUORIDE
OSHA PEL: TWA 2.5 mg(F)/m^3
ACGIH TLV: TWA 2.5 mg(F)/m^3; BEI: 3 mg/g creatinine of fluorides in urine prior to shift; 10 mg/g creatinine of fluorides in urine at end of shift.
DOT Classification: 8; Label: Corrosive, Poison
NIOSH REL: TWA 2.5 mg(F)/m^3
SAFETY PROFILE: A poison by all routes. Corrosive to the eyes, skin, and mucous membranes. A very reactive, dangerous material.
PROP: Colorless, deliquescent, cubic crystals. Undergoes tetragonal to cubic transition at 1°. Mp: 238.8° (decomp). Very sol in H_2O; insol in EtOH.

PLC500 CAS: 151-50-8 CN•K HR: 3
POTASSIUM CYANIDE
DOT: UN 1680
SYNS: CYANIDES (OSHA) ♦ HYDROCYANIC ACID, POTASSIUM SALT
OSHA PEL: TWA 5 mg(CN)/m^3
ACGIH TLV: CL 5 mg(CN)/m^3 (skin)
DFG MAK: 5 mg(CN)/m^3
NIOSH REL: CL (Cyanide) 5 mg(CN)/m^3/10M
DOT Classification: 6.1; Label: Poison
SAFETY PROFILE: A deadly human poison by ingestion. Experimental teratogenic and reproductive effects. Reacts with acids or acid fumes to liberate deadly HCN.
PROP: Colorless water soln. Deliquescent colorless cubic crystals. Undergoes cubic to orthorhombic transition. Slt odor of bitter almonds. Mp: 622°. Very sol in H_2O; sparingly sol in EtOH.

PLF500 CAS: 7789-23-3 FK HR: 3
POTASSIUM FLUORIDE
DOT: UN 1812
OSHA PEL: TWA 2.5 mg(F)/m^3
ACGIH TLV: TWA 2.5 mg(F)/m^3; BEI: 3 mg/g creatinine of fluorides in urine prior to shift; 10 mg/g creatinine of fluorides in urine at end of shift.
NIOSH REL: TWA (Inorganic Fluorides) 2.5 mg(F)/m^3
DOT Classification: 6.1; Label: KEEP AWAY FROM FOOD
SAFETY PROFILE: Poison by ingestion. Experimental teratogenic effects. A corrosive irritant to the eyes, skin, and mucous membranes. It can react violently with many substances.
PROP: White, crystalline, deliq powder; sharp saline taste; or deliq colorless crystals. Bp: 1500°, d: 2.48, vap press: 1 mm @ 885°, mp: 859.9°. Very sol in boiling water; insol in alc.

PLH000 CAS: 16924-00-8 $F_7Ta \cdot 2K$ HR: 3

POTASSIUM HEPTAFLUOROTANTALATE
SYNS: POTASSIUM FLUOTANTALATE ♦ TANTALUM POTASSIUM FLUORIDE
OSHA PEL: TWA 2.5 mg(F)/m^3
ACGIH TLV: TWA 2.5 mg(F)/m^3; BEI: 3 mg/g creatinine of fluorides in urine prior to shift; 10 mg/g creatinine of fluorides in urine at end of shift.
NIOSH REL: TWA 2.5 mg(F)/m^3
SAFETY PROFILE: Poison by ingestion.
PROP: Colorless crystals. Mp: 730°.

PLH750 CAS: 16871-90-2 $F_6Si{\bullet}2K$ HR: 3
POTASSIUM HEXAFLUOROSILICATE
DOT: UN 2655
SYNS: POTASSIUM FLUOSILICATE ♦ POTASSIUM SILICOFLUORIDE (DOT)
OSHA PEL: TWA 2.5 mg(F)/m^3
ACGIH TLV: TWA 2.5 mg(F)/m^3; BEI: 3 mg/g creatinine of fluorides in urine prior to shift; 10 mg/g creatinine of fluorides in urine at end of shift.
NIOSH REL: TWA (Inorganic Fluorides) 2.5 mg(F)/m^3
DOT Classification: 6.1; Label: KEEP AWAY FROM FOOD
SAFETY PROFILE: A poison by ingestion. A strong irritant.
PROP: White, fine powder or colorless, cubic crystals. Moisture-sensitive. D: 2.27, mp: decomp. Sltly sol in cold water; practically insol in alc.

PLJ500 CAS: 1310-58-3 HKO HR: 3
POTASSIUM HYDROXIDE
DOT: UN 1813/UN 1814
SYNS: CAUSTIC POTASH ♦ LYE ♦ POTASSIUM HYDRATE (DOT)
OSHA PEL: CL 2 mg/m^3
ACGIH TLV: CL 2 mg/m^3
DOT Classification: 8; Label: Corrosive
SAFETY PROFILE: Poison by ingestion. Very corrosive to the eyes, skin, and mucous membranes. It can react violently with many substances.
PROP: White or colorless, orthorhombic, deliquescent pieces, lumps, or sticks having crystalline fracture. Mp: 406°, bp: 1324°, d: 2.044. Very sol in water, alc; sol in EtOH; insol in Et_2O.

PLP000 CAS: 7722-64-7 $MnO_4{\bullet}K$ HR: 3
POTASSIUM PERMANGANATE
DOT: UN 1490
SYNS: C.I. 77755 ♦ PERMANGANATE of POTASH (DOT)
OSHA PEL: CL 5 mg(Mn)/m^3
ACGIH TLV: TWA 5 mg(Mn)/m^3
DOT Classification: 5.1; Label: Oxidizer

SAFETY PROFILE: A human poison by ingestion. A strong irritant. Flammable by chemical reaction. A powerful oxidizer and dangerous explosion hazard. It can react violently with many substances.PROP: Air-stable, dark-purple crystals with a blue metallic sheen; sweetish astringent taste. Aq solns slowly deposit MnO_2. Mp: decomp @ <240°, d: 2.703. Sol in H_2O; mod sol in MeOH, AcOH, Me_2CO, and Py.

PLS250 CAS: 506-61-6 $C_2AgN_2 \cdot K$ HR: 3

POTASSIUM SILVER CYANIDE

SYNS: SILVER POTASSIUM CYANIDE

DOT Classification: 6.1; Label: Poison, KEEP AWAY FROM FOOD

OSHA PEL: TWA 5 mg(CN)/m^3

ACGIH TLV: CL 5 mg(CN)/m^3 (skin)

DFG MAK: 5 mg/m^3

NIOSH REL: (Cyanide) CL 5 mg(CN)/m^3/10M

SAFETY PROFILE: Poison by ingestion. A severe skin and eye irritant.

PROP: White crystals, light-sensitive. Sol in water, acids.

PMJ750 CAS: 74-98-6 C_3H_8 HR: 3

PROPANE

DOT: UN 1978

OSHA PEL: TWA 1000 ppm

ACGIH TLV: TWA 2500 ppm

DFG MAK: 1000 ppm (1800 mg/m^3)

DOT Classification: 2.1; Label: Flammable Gas

SAFETY PROFILE: An asphyxiant. Flammable gas.

PROP: Colorless gas. Bp: –44.5°, flash p: –156°F, lel: 2.3%, uel: 9.5%, autoign temp: 842°F, d: 0.5852 @ –44.5°/4°, vap d: 1.56. Sol in water, alc, ether.

PML500 CAS: 107-03-9 C_3H_8S HR: 3

1-PROPANETHIOL

DOT: UN 1228/UN 3071

SYNS: 1-MERCAPTOPROPANE ♦ PROPYL MERCAPTAN ♦ N-PROPYLTHIOL

NIOSH REL: (Thiols (n-Alkane Mono)) CL 0.5 ppm/15M

DOT Classification: 3; Label: Flammable Liquid, Poison (UN 1228); DOT Class: 6.1; Label: Poison, Flammable Liquid (UN 3071)

SAFETY PROFILE: A poison. Mildly toxic by inhalation. A severe eye irritant. A flammable.

PROP: Flash p: –4°F.

PMN450 CAS: 107-19-7 C_3H_4O HR: 3

PROPARGYL ALCOHOL

SYNS: 1-PROPYNE-3-OL ♦ 3-PROPYNOL ♦ 2-PROPYNYL ALCOHOL

OSHA PEL: TWA 1 ppm (skin)

ACGIH TLV: TWA 1 ppm (skin)

DFG MAK: 2 ppm (4.7 mg/m^3)
SAFETY PROFILE: Poison by ingestion and skin contact. Moderately toxic by inhalation. A skin and mucous membrane irritant. Flammable liquid.
PROP: Moderately volatile liquid; geranium odor. D: 0.9715 @ 20°/4°, mp: -48—52°, bp: 114–115°, flash p: 33°C (97°F) (OC), vap press: 11.6 mm @ 20°, vap d: 1.93.

PMT100 CAS: 57-57-8 $C_3H_4O_2$ HR: 3
β-PROPIOLACTONE
SYNS: BPL ♦ 3-HYDROXYPROPIONIC ACID LACTONE ♦ 1,3-PROPIOLACTONE
OSHA: Carcinogen
ACGIH TLV: TWA 0.5 ppm; Animal Carcinogen
DFG MAK: Animal Carcinogen, Suspected Human Carcinogen
SAFETY PROFILE: Confirmed carcinogen. Poison by inhalation.
PROP: A liquid. Mp: -33.4°, bp: 162° (decomp). Sol in H_2O.

PMU750 CAS: 79-09-4 $C_3H_6O_2$ HR: 3
PROPIONIC ACID
DOT: UN 1848
SYNS: ETHANECARBOXYLIC ACID ♦ METHYL ACETIC ACID ♦ PROPIONIC ACID (ACGIH,DOT,OSHA)
OSHA PEL: TWA 10 ppm
ACGIH TLV: TWA 10 ppm
DFG MAK: 10 ppm (31 mg/m^3)
DOT Classification: 8; Label: Corrosive
SAFETY PROFILE: Moderately toxic by ingestion and skin contact. A corrosive irritant to eyes, skin, and mucous membranes. Flammable liquid.
PROP: Oily liquid; pungent, disagreeable, rancid odor. D: 0.998 @ 15°/4°, mp: -21.5°, bp: 141.1°, vap press: 10 mm @ 39.7°, vap d: 2.56, autoign temp: 955°F. Misc in water, alc, ether, chloroform.

PMV750 CAS: 107-12-0 C_3H_5N HR: 3
PROPIONONITRILE
DOT: UN 2404
SYNS: ETHYL CYANIDE ♦ PROPANENITRILE ♦ PROPIONIC NITRILE
ACGIH TLV: CL 5 mg(CN)/m^3 (skin)
NIOSH REL: TWA (Nitriles) 14 mg/m^3
DOT Classification: 3; Label: Flammable Liquid, Poison
SAFETY PROFILE: Poison by ingestion and skin contact. Moderately toxic by inhalation. Experimental teratogenic effects. A skin and eye irritant. Dangerous fire hazard.
PROP: Colorless liquid; ethereal odor. Fp: -103.5°, bp: 97.1°, d: 0.783 @ 21°/4°, vap d: 1.9, flash p: 36°F, lel: 3.1%, mp: 91.8°. Misc with alc and ether.

PMY300 CAS: 114-26-1 $C_{11}H_{15}NO_3$ HR: 3
PROPOXUR

SYNS: o-ISOPROPOXYPHENYL METHYLCARBAMATE ♦ N-METHYL-2-ISOPROPOXYPHENYLCARBAMATE
OSHA PEL: TWA 0.5 mg/m^3
ACGIH TLV: TWA 0.5 mg/m^3; Animal Carcinogen
DFG MAK: 2 mg/m^3
SAFETY PROFILE: Confirmed carcinogen. A human poison by ingestion. Moderately toxic by inhalation and skin contact. An experimental teratogen.
PROP: A white to tan, crystalline solid or powder. Mp: 84–87°. Sltly sol in water; sol in all polar org solv.

PNC250 CAS: 109-60-4 $C_5H_{10}O_2$ HR: 3

n-PROPYL ACETATE

DOT: UN 1276
SYNS: 1-ACETOXYPROPANE
OSHA PEL: TWA 200 ppm; STEL 250 ppm
ACGIH TLV: TWA 200 ppm; STEL 250 ppm
DFG MAK: 200 ppm (850 mg/m^3)
DOT Classification: 3; Label: Flammable Liquid
SAFETY PROFILE: Mildly toxic by ingestion and inhalation. A skin irritant. A flammable liquid.
PROP: Clear, colorless liquid; pleasant odor. Mp: –92.5°, bp: 101.6°, flash p: 58°F, lel: 2.0%, uel: 8.0%, d: 0.887, autoign temp: 842°F, vap press: 40 mm @ 28.8°, vap d: 3.52. Misc with alc, ether; sol in water.

PND000 CAS: 71-23-8 C_3H_8O HR: 3

n-PROPYL ALCOHOL

DOT: UN 1274
SYNS: 1-HYDROXYPROPANE ♦ n-PROPANOL ♦ PROPYLIC ALCOHOL
OSHA PEL: TWA 200 ppm; STEL 250 ppm
ACGIH TLV: TWA 200 ppm; STEL 250 ppm (skin); (Proposed: TWA 200 ppm; STEL 250 ppm Confirmed Animal Carcinogen with Unknown Relevance to Humans)
DOT Classification: 3; Label: Flammable Liquid
SAFETY PROFILE: Moderately toxic by inhalation and ingestion. A skin and severe eye irritant. A flammable liquid.
PROP: Clear liquid; alcohol-like odor. Mp: –127°, bp: 97.19°, flash p: 59°F (CC), ULC: 55–60, d: 0.8044 @ 20°/4°, lel: 2.1%, uel: 13.5%, autoign temp: 824°F, vap press: 10 mm @ 14.7°, vap d: 2.07. Misc in water, alc, and ether.

PNJ400 CAS: 78-87-5 $C_3H_6Cl_2$ HR: 3

PROPYLENE DICHLORIDE

DOT: UN 1279
SYNS: 1,2-DICHLOROPROPANE ♦ PROPYLENE CHLORIDE
OSHA PEL: TWA 75 ppm; STEL 110 ppm
ACGIH TLV: TWA 75 ppm; STEL 110 ppm; Not Classifiable as a Human Carcinogen
DFG MAK: Confirmed Animal Carcinogen with Unknown Relevance to Humans

DOT Classification: 3; Label: Flammable Liquid
SAFETY PROFILE: Suspected carcinogen. Moderately toxic by inhalation and ingestion. Mildly toxic by skin contact. An eye irritant. A flammable liquid.
PROP: Colorless liquid. Bp: 96.8°, flash p: 60°F, d: 1.1593 @ 20°/20°, vap press: 40 mm @ 19.4°, vap d: 3.9, autoign temp: 1035°F, lel: 3.4%, uel: 14.5%.

PNL000 CAS: 6423-43-4 $C_3H_6N_2O_6$ HR: 3
PROPYLENE GLYCOL DINITRATE
SYNS: PGDN ♦ 1,2-PROPYLENE GLYCOL DINITRATE
OSHA PEL: TWA 0.05 ppm
ACGIH TLV: TWA 0.05 ppm (skin)
DFG MAK: 0.05 ppm (0.34 mg/m^3)
SAFETY PROFILE: Poison by ingestion. An eye irritant.

PNL250 CAS: 107-98-2 $C_4H_{10}O_2$ HR: 3
PROPYLENE GLYCOL MONOMETHYL ETHER
SYNS: METHOXY ETHER of PROPYLENE GLYCOL ♦ 1-METHOXY-2-PROPANOL ♦ PROPYLENE GLYCOL METHYL ETHER
OSHA PEL: TWA 100 ppm; STEL 150 ppm
ACGIH TLV: TWA 100 ppm; STEL 150 ppm
DFG MAK: 100 ppm (370 mg/m^3)
SAFETY PROFILE: Mildly toxic by ingestion, inhalation, and skin contact. A skin and eye irritant. An experimental teratogen. Very dangerous fire hazard.
PROP: Colorless liquid. Mp: –96.7°, bp: 126–127°, flash p: 100°F, d: 0.919 @ 25°/25°.

PNL400 CAS: 75-55-8 C_3H_7N HR: 3
PROPYLENE IMINE
DOT: UN 1921
SYNS: 2-METHYLAZACYCLOPROPANE ♦ 2-METHYLAZIRIDINE ♦ 2-METHYLETHYLENIMINE
OSHA PEL: TWA 2 ppm (skin)
ACGIH TLV: TWA 2 ppm (skin); Animal Carcinogen
DFG MAK: Animal Carcinogen, Suspected Human Carcinogen
DOT Classification: 3; Label: Flammable Liquid
SAFETY PROFILE: Confirmed carcinogen. Poison by ingestion and skin contact. Moderately toxic by inhalation. Severe eye irritant. A flammable liquid.
PROP: Liquid. Vap d: 2.0, flash p: 14°F.

PNL600 CAS: 75-56-9 C_3H_6O HR: 3
PROPYLENE OXIDE
DOT: UN 1280
SYNS: 1,2-EPOXYPROPANE ♦ METHYL ETHYLENE OXIDE ♦ METHYL OXIRANE
OSHA PEL: TWA 20 ppm
ACGIH TLV: TWA 20 ppm; Animal Carcinogen; (Proposed: TWA 5 ppm Confirmed Animal Carcinogen with Unknown Relevance to Humans)
DFG MAK: Animal Carcinogen, Suspected Human Carcinogen

DOT Classification: 3; Label: Flammable Liquid
SAFETY PROFILE: Confirmed carcinogen. Moderately toxic by ingestion, inhalation, and skin contact. An experimental teratogen. A severe skin and eye irritant. Flammable liquid. It can react violently with many substances.
PROP: Colorless liquid; ethereal odor. Bp: 33.9°, lel: 2.8%, uel: 37%, fp: –104.4°, flash p: –35°F (TOC), d: 0.8304 @ 20°/20°, vap press: 400 mm @ 17.8°, vap d: 2.0. Sol in water, alc, and ether.

PNQ500 CAS: 627-13-4 $C_3H_7NO_3$ HR: 3
n-PROPYL NITRATE
DOT: UN 1865
OSHA PEL: TWA 25 ppm; STEL 40 ppm
ACGIH TLV: TWA 25 ppm; STEL 40 ppm
DFG MAK: 25 ppm (110 mg/m^3)
DOT Classification: 3; Label: Flammable Liquid
SAFETY PROFILE: A flammable liquid and dangerous fire hazard A shock-sensitive explosive.
PROP: Pale-yellow liquid; sickly odor. Bp: 110.5°, d: 1.054 @ 20°/4°, flash p: 68°F, autoign temp: 347°F (in air), lel: 2%, uel: 100%. Very sltly sol in water; sol in alc, ether.

PON250 CAS: 129-00-0 $C_{16}H_{10}$ HR: 3
PYRENE
SYNS: BENZO(def)PHENANTHRENE
OSHA PEL: TWA 0.2 mg/m^3
SAFETY PROFILE: Poison by inhalation. Moderately toxic by ingestion. A skin irritant.
PROP: Pale-yellow plates by sublimation or colorless solid. Solutions have a slight blue color. Mp: 149–150°, d: 1.271 @ 23°, bp: 404°. Insol in water; sol in Et_2O, CS_2, C_6H_6, and toluene.

POO250 CAS: 8003-34-7 HR: 3
PYRETHRINS
SYNS: CINERIN I or II ♦ PYRETHRUM (ACGIH)
OSHA PEL: TWA 5 mg/m^3
ACGIH TLV: TWA 5 mg/m^3; Not Classifiable as a Human Carcinogen
DFG MAK: 5 mg/m^3
SAFETY PROFILE: Moderately toxic to humans by ingestion. An allergen. Combustible liquid.
PROP: Viscous liquid. Bp: 170° @ 0.1 mm (decomp).

POP250 CAS: 110-86-1 C_5H_5N HR: 3
PYRIDINE
DOT: UN 1282
SYNS: AZABENZENE ♦ AZINE

OSHA PEL: TWA 5 ppm
ACGIH TLV: TWA 5 ppm
DFG MAK: 5 ppm (16 mg/m^3)
DOT Classification: 3; Label: Flammable Liquid, Poison
SAFETY PROFILE: Moderately toxic by ingestion and skin contact. Mildly toxic by inhalation. A skin and severe eye irritant. A flammable liquid. It can react violently with many substances.
PROP: Colorless liquid; sharp, penetrating, empyreumatic odor; burning taste. Bp: 115.3°, lel: 1.8%, uel: 12.4%, fp: –42°, flash p: 68°F (CC), d: 0.982, autoign temp: 900°F, vap press: 10 mm @ 13.2°, vap d: 2.73. Volatile with steam; misc with water, alc, ether.

QQS200 CAS: 106-51-4 $C_6H_4O_2$ HR: 3
QUINONE
DOT: UN 2587
SYNS: BENZOQUINONE (DOT) ♦ 1,4-CYCLOHEXADIENEDIONE ♦ 1,4-DIOXYBENZENE
OSHA PEL: TWA 0.1 ppm
ACGIH TLV: TWA 0.1 ppm
DFG MAK: 0.1 ppm (0.45 mg/m^3)
DOT Classification: 6.1; Label: Poison
SAFETY PROFILE: Poison by ingestion. Causes severe damage to the skin and mucous membranes. The moist material self-heats and decomposes exothermically above 60°C.
PROP: Yellow crystals; characteristic chlorine-like odor from pet ether or water. Mp: 115.7°, bp: sublimes, d: 1.318 @ 20°/4°. Sol in EtOH and Et_2O; very spar sol in H_2O.

REA000 CAS: 108-46-3 $C_6H_6O_2$ HR: 3
RESORCINOL
DOT: UN 2876
SYNS: m-BENZENEDIOL ♦ C.I. 76505 ♦ 1,3-DIHYDROXYBENZENE ♦ m-HYDROQUINONE ♦ 3-HYDROXYPHENOL ♦ RESORCINE
OSHA PEL: TWA 10 ppm; STEL 20 ppm
ACGIH TLV: TWA 10 ppm; STEL 20 ppm; Not Classifiable as a Human Carcinogen
DOT Classification: 6.1; Label: KEEP AWAY FROM FOOD
SAFETY PROFILE: Human poison by ingestion. Moderately toxic by skin contact. A skin and severe eye irritant. Combustible.
PROP: Platelets from EtOH. Very white crystals, become pink on exposure to light when not perfectly pure; unpleasant sweet taste. Mp: 110°, bp: 280.5°, flash p: 261°F (CC), d: 1.285 @ 15°, autoign temp: 1126°F, vap press: 1 mm @ 108.4°, vap d: 3.79. Very sol in alc, ether, glycerin; sltly sol in chloroform; sol in water.

RHF000 CAS: 7440-16-6 Rh HR: 3
RHODIUM
OSHA PEL: TWA Metal, Fume, Insol Compounds: 0.1 mg(Rh)/m^3; Sol Compounds: 0.001 mg(Rh)/m^3

ACGIH TLV: TWA (Metal) 1 mg/m^3, (insoluble compounds as Rh) 1 mg/m^3, (soluble compounds as Rh) 0.01 mg/m^3
NIOSH REL: (Rhodium (as Rh)): TWA 0.1 mg/m^3
SAFETY PROFILE: May be a sensitizer. Most rhodium compounds have only moderate toxicity by ingestion. Flammable when exposed to heat or flame.
PROP: A hard, lustrous, silvery-white, metallic element. Mp: 1966°, bp: 3727°, d: 12.41 @ 20°. Oxidizes slowly at 6°.

RMA500 CAS: 299-84-3 $C_8H_8Cl_3O_3PS$ HR: 3
RONNEL
SYNS: O,O-DIMETHYL-O-2,4,5-TRICHLOROPHENYL PHOSPHOROTHIOATE ♦ 2,4,5-TRICHLOROPHENOL, O-ESTER with O,O-DIMETHYL PHOSPHOROTHIOATE
OSHA PEL: TWA 10 mg/m^3
ACGIH TLV: TWA 10 mg/m^3; Not Classifiable as a Human Carcinogen
SAFETY PROFILE: Poison by ingestion. Moderately toxic by skin contact. An experimental teratogen.
PROP: White or crystalline powder. Mp: 41°, vap press: 8 X 10^{-4} mm, bp: 97° @ 0.01 mm.

RNZ000 CAS: 83-79-4 $C_{23}H_{22}O_6$ HR: 3
ROTENONE
OSHA PEL: TWA 5 mg/m^3
ACGIH TLV: TWA 5 mg/m^3; Not Classifiable as a Human Carcinogen
DFG MAK: 5 mg/m^3
SAFETY PROFILE: Human poison by ingestion. Experimental poison by ingestion. A skin and eye irritant.
PROP: Orthorhombic plates or crystals from Me_2CO (aq). Mp: 163° (dimorphic form mp: 185–186°). D: 1.27 @ 20°. Almost insol in water; sol in alc, acetone, carbon tetrachloride, chloroform, ether, and other org solvs. Decomp on exposure to light and air.

SAM000 CAS: 4342-30-7 $C_{19}H_{32}O_3Sn$ HR: 3
SALICYLOYLOXYTRIBUTYLSTANNANE
SYNS: TRIBUTYLTIN SALICYLATE ♦ TRI-N-BUTYLTIN SALICYLATE
OSHA PEL: TWA 0.1 mg(Sn)/m^3 (skin)
ACGIH TLV: TWA 0.1 mg(Sn)/m^3 (skin) (Proposed: TWA 0.1 mg(Sn)/m^3; STEL 0.2 mg(Sn)/m^3 (skin))
NIOSH REL: (Organotin Compounds) TWA 0.1 mg(Sn)/m^3
SAFETY PROFILE: Poison by ingestion.

SBO500 CAS: 7782-49-2 Se HR: 3
SELENIUM

DOT: UN 2658
SYNS: C.I. 77805 ♦ SELENIUM METAL POWDER, NON-PYROPHORIC (DOT)
OSHA PEL: TWA 0.2 mg(Se)/m^3
ACGIH TLV: TWA 0.2 mg(Se)/m^3
DFG MAK: 0.1 mg(Se)/m^3
DOT Classification: 6.1; Label: KEEP AWAY FROM FOOD
SAFETY PROFILE: Poison by inhalation. Selenosis in humans has occurred from ingestion of 3.2 mg selenium per day. It can react violently with many substances.
PROP: Steel-gray, metalloid element. Viscosity of liquid decreases with temperature. Mp: 170–217°, bp: 690°, d: 4.26–4.81, vap press: 1 mm @ 356°. Insol in water and alc; very sltly sol in ether.

SBS000 CAS: 7783-79-1 F_6Se HR: 3

SELENIUM HEXAFLUORIDE

DOT: UN 2194
SYN: SELENIUM FLUORIDE
OSHA PEL: TWA 0.05 ppm (Se)
ACGIH TLV: TWA 0.05 ppm (Se)
DFG MAK: 0.1 mg(Se)/m^3
DOT Classification: 2.3; Label: Poison Gas
SAFETY PROFILE: Poison by inhalation.
PROP: Thermally, very stable, colorless gas. White solid at low temp. Sublimes before melting. Inert to water. Mp: –46.6°, d: 3.27 @ –46.6°. Sublimes @ –34°.

SBT500 CAS: 7791-23-3 Cl_2OSe HR: 3

SELENIUM OXYCHLORIDE

DOT: UN 2879
SYNS: SELENINYL CHLORIDE ♦ SELENIUM CHLORIDE OXIDE
OSHA PEL: TWA 0.2 mg(Se)/m^3
ACGIH TLV: TWA 0.2 mg(Se)/m^3
DFG MAK: 0.1 mg(Se)/m^3
DOT Classification: 8; Label: Corrosive, Poison
SAFETY PROFILE: Poison by skin contact.
PROP: Colorless to yellowish liquid with high dielectric construction. Mp: 10.9°, bp: 176.4°, d: 2.42 @ 22°, vap press: 1 mm @ 34.8°. Sol in org solvs.

SCH002 CAS: 112945-52-5 O_2Si HR: 2

SILICA, AMORPHOUS FUMED

SYNS: AMORPHOUS SILICA DUST ♦ COLLOIDAL SILICA ♦ SILICON DIOXIDE (FCC)
SAFETY PROFILE: Moderately toxic by ingestion. An inhalation hazard. Does not cause silicosis.

PROP: A finely powdered microcellular silica foam with minimum SiO_2 content of 89.5%. Insol in water; sol in hydrofluoric acid.

SCI000 CAS: 7631-86-9 O_2Si HR: 1

SILICA, AMORPHOUS HYDRATED

SYNS: SILICA GEL ♦ SILICIC ACID

OSHA PEL: TWA 6 mg/m^3

ACGIH TLV: TWA (nuisance particulate) 10 mg/m^3 of total dust (when toxic impurities are not present, e.g., quartz <1%)

DFG MAK: 4 mg/m^3 as total dust

SAFETY PROFILE: The pure unaltered form is considered a nuisance dust. Some deposits contain small amounts of crystalline quartz and are therefore fibrogenic. When diatomaceous earth is calcined (with or without fluxing agents) some silica is converted to cristobalite and is therefore fibrogenic.

PROP: Transparent, tasteless crystals or amorphous powder. Melts to a glass at ordinary temps. Chemically resistant to most reagents. Mp: 1716–1736°, bp: 2230°. Insol in H_2O; sol in HF (giving fluorosilicate ions).

SCI500 O_2Si HR: 3

SILICA, CRYSTALLINE

SYNS: CHALCEDONY ♦ CRISTOBALITE ♦ PURE QUARTZ ♦ SAND ♦ SILICON DIOXIDE ♦ TRIPOLI

OSHA PEL: Total Dust: TWA 30 $mg/m^3/2(\%SiO_2+2)$ Respirable Fraction: TWA 0.05 mg/m^3

ACGIH TLV: TWA Respirable Fraction: 0.05 mg/m^3

DFG MAK: 0.15 mg/m^3

NIOSH REL: (Silica, Crystalline) TWA 50 $\mu g/m^3$

SAFETY PROFILE: Moderately toxic as an acute irritating dust. The prolonged inhalation of dusts containing free silica may result in the development of a disabling pulmonary fibrosis known as silicosis.

PROP: Transparent, tasteless crystals or amorph powder. Mp: 1710°, bp: 2230°, d: (amorph) 2.2, d: (crystalline) 2.6, vap press: 10 mm @ 1732°. Practically insol in water or acids. Dissolves readily in HF, forming silicon tetrafluoride.

SCJ000 CAS: 14464-46-1 O_2Si HR: 3

SILICA, CRYSTALLINE–CRISTOBALITE

SYNS: CALCINED DIATOMITE ♦ CRISTOBALITE

OSHA PEL: Total Dust: TWA 30 $mg/m^3/2(\%SiO_2+2)$ TWA Respirable Fraction: 0.05 mg/m^3

ACGIH TLV: TWA Respirable Fraction: 0.05 mg/m^3

DFG MAK: 0.15 mg/m^3

NIOSH REL: (Silica, Crystalline) TWA 50 $\mu g/m^3$

SAFETY PROFILE: Confirmed carcinogen. An inhalation hazard.

PROP: White, cubic-system crystals formed from quartz at temperatures above 1000°C (NTIS** PB246–697).

SCJ500 CAS: 14808-60-7 O_2Si HR: 3

SILICA, CRYSTALLINE–QUARTZ

SYNS: CHALCEDONY ♦ PURE QUARTZ ♦ QUARTZ ♦ SAND ♦ SILICA FLOUR (powdered crystalline silica)

OSHA PEL: Total Dust: TWA 30 $mg/m^3/2(\%SiO_2+2)$ TWA Respirable Fraction: 0.1 mg/m^3

ACGIH TLV: TWA Respirable Fraction: 0.1 mg/m^3 (Proposed: 0.05 mg/m^3; Suspected Human Carcinogen)

DFG MAK: 0.15 mg/m^3 as fine dust

NIOSH REL: TWA 50 $\mu g/m^3$; 3,000,000 $fibers/m^3$

SAFETY PROFILE: Confirmed carcinogen. An inhalation hazard.

PROP: White to reddish crystals. Stable below 8°. Low (þ&□-) quartz, stable at room temp; transforms to high (β-) quartz at 5°; the two forms are related by small rotations of the SiO_4 tetrahedron. Piezoelectric and pyroelectric. Mp: 1710°, bp: 2230°, d: 2.6.

SCK000 CAS: 15468-32-3 O_2Si HR: 3

SILICA, CRYSTALLINE–TRIDYMITE

SYNS: CHRISTENSENITE ♦ TRIDYMITE

OSHA PEL: Total Dust: TWA 30 $mg/m^3/2(\%SiO_2+2)$ TWA 0.05 mg/m^3

ACGIH TLV: (Silica, Crystalline) TWA Respirable Fraction: 0.05 mg/m^3

DFG MAK: 0.15 mg/m^3

NIOSH REL: TWA 50 $\mu g/m^3$

SAFETY PROFILE: Confirmed carcinogen. About twice as toxic as silica in causing silicosis.

PROP: White or colorless platelets or orthorhombic crystals formed from quartz @ temperatures >870° (NTIS** PB246–697). Stable from 870–914° at atmospheric pressure, but persists as a metastable phase below 8° forming low tridymite below 1° and middle tridymite from 117–1°.

SCK600 CAS: 60676-86-0 O_2Si HR: 3

SILICA, FUSED

SYNS: AMORPHOUS SILICA ♦ FUSED SILICA ♦ SILICA, AMORPHOUS-FUSED (ACGIH) ♦ SILICON DIOXIDE

OSHA PEL: Total Dust: TWA 30 $mg/m^3/2(\%SiO_2+2)$ TWA 0.1 mg/m^3

ACGIH TLV: TWA 0.1 mg/m^3 (Respirable Fraction)

DFG MAK: 0.3 mg/m^3 as fine dust

NIOSH REL: (Silica, Crystalline) TWA 0.05 mg/m^3

SAFETY PROFILE: An inhalation hazard.

PROP: Made up of spherical submicroscopic particles under 0.1 micron in size (AMIHBC 9,389,54).

SCL000 CAS: 7699-41-4 H_2O_3Si HR: 3

SILICA, GEL and AMORPHOUS–PRECIPITATED

SYNS: METASILICIC ACID ♦ PRECIPITATED SILICA ♦ SILICA GEL ♦ SILICIC ACID

OSHA PEL: TWA 6 mg/m^3

ACGIH TLV: TWA (nuisance particulate) 10 mg/m^3 of total dust (when toxic impurities are not present, e.g., quartz <1%)

DFG MAK: 0.3 mg/m^3 as fine dust

SAFETY PROFILE: An inhalation hazard. An eye irritant and nuisance dust.

PROP: White amorphous powder. Insol in H_2O.

SCP000 CAS: 7440-21-3 Si HR: 3

SILICON

DOT: UN 1346

SYNS: SILICON POWDER, amorphous (DOT)

OSHA PEL: TWA Total Dust: 10 mg/m^3 of total; Respirable Fraction: 5 mg/m^3

ACGIH TLV: TWA (nuisance particulate) 10 mg/m^3 of total dust (when toxic impurities are not present, e.g., quartz <1%)

DOT Classification: 4.1; Label: Flammable Solid

SAFETY PROFILE: A nuisance dust. Moderately toxic by ingestion. An eye irritant. Elemental Si is flammable when exposed to flame or by chemical reaction with oxidizers. It can react violently with many substances.

PROP: Cubic, steel-gray crystals or black-brown amorphous powder. Bulk Si is unreactive to O_2, H_2O, H halides (except HF) but dissolves in hot aq alkalies. Reactive to halogens, e.g., F_2 at room temp, Cl_2 at 3°. Acted on by N_2 at 14°. S reacts at 6°, and P at 10°. Mp: 1410°, bp: 2355°, d: 2.42 or 2.3 @ 20°, vap press: 1 mm @ 1724°. Almost insol in water; sol in molten alkali oxides.

SCQ000 CAS: 409-21-2 CSi HR: 3

SILICON CARBIDE

SYNS: CARBORUNDUM ♦ SILICON MONOCARBIDE

OSHA PEL: TWA Total Dust: 10 mg/m^3; Respirable Fraction: 5 mg/m^3

ACGIH TLV: TWA (nuisance particulate) 10 mg/m^3 of total dust (when toxic impurities are not present, e.g., quartz <1%); Not Classifiable as a Human Carcinogen

DFG MAK: (with fibers) Animal Carcinogen, Suspected Human Carcinogen; (without fibers) 1.5 mg/m^3

NIOSH REL: (Silicon Carbide, Total Dust) TWA 10 mg/m^3

SAFETY PROFILE: Suspected carcinogen with experimental neoplastigenic data. A nuisance dust.

PROP: Bluish-black, iridescent, hexagonal or cubic crystals; colorless when pure. Mp: 2600°, bp: subl >2000°, decomp @ 2210°, d: 3.17. Insol in H_2O and acids; sol in fused KOH.

SDF650 CAS: 7783-61-1 F_4Si HR: 3

SILICON FLUORIDE

DOT: UN 1859

SYNS: SILICON TETRAFLUORIDE (DOT) ♦ TETRAFLUOROSILANE

OSHA PEL: TWA 2.5 mg(F)/m^3

ACGIH TLV: TWA 2.5 mg(F)/m^3; BEI: 3 mg/g creatinine of fluorides in urine prior to shift; 10 mg/g creatinine of fluorides in urine at end of shift.

NIOSH REL: (Inorganic Fluorides) TWA 2.5 mg(F)/m^3

DOT Classification: 2.3; Label: Poison Gas, Corrosive

SAFETY PROFILE: A poison. Moderately toxic by inhalation. A corrosive irritant to skin, eyes, and mucous membranes.

PROP: Colorless, fuming gas; very pungent odor. Rapidly hydrol by H_2O. Subl –95°, mp: –77°, bp: –65° @ 181 mm, d: 4.67.

SDH575 CAS: 7803-62-5 H_4Si HR: 3

SILICON TETRAHYDRIDE

DOT: UN 2203

SYNS: MONOSILANE ♦ SILANE

OSHA PEL: TWA 5 ppm

ACGIH TLV: TWA 5 ppm

DOT Classification: 2.1; Label: Flammable Gas

SAFETY PROFILE: Mildly toxic by inhalation. Silanes are irritating to skin, eyes, and mucous membranes. Easily ignited in air. It can react violently with many substances.

PROP: Colorless gas with repulsive odor; slowly decomp by water. D: 0.68 @ –185°, mp: –185°, bp: –112°, fp: –200°. Insol in Et_2O, C_6H_6, $CHCl_3$, and EtOH. Insol in H_2O; decomp in aq KOH.

SDI500 CAS: 7440-22-4 Ag HR: 2

SILVER

SYNS: C.I. 77820

OSHA PEL: Metal, Dust, and Fume: TWA 0.01 mg/m^3

ACGIH TLV: TWA (metal) 0.1 mg/m^3, (soluble compounds as Ag) 0.01 mg/m^3

DFG MAK: 0.1 mg/m^3; (salts) 0.01 mg/m^3

NIOSH REL: (Silver, metal and soluble compounds) TWA 0.01 mg/m^3

SAFETY PROFILE: Inhalation of dusts can cause argyrosis. Flammable in the form of dust

PROP: Soft, ductile, malleable, lustrous white metal. Tarnishes in air with formation of black sulfide. Physical properties dependent on mechanical treatment. Attacked by Cl_2, S, H_2S, metal cyanides (in air), chromic, nitric, and sulfuric acids. Mp: 961°, bp: 2163°, d: 10.50 @ 20°.

SDL500 CAS: 23606-32-8 $AgH_4N_3O_6$ HR: 2

SILVER AMMONIUM NITRATE

OSHA PEL: TWA 0.01 $mg(Ag)/m^3$
ACGIH TLV: TWA 0.01 $mg(Ag)/m^3$
SAFETY PROFILE: A severe eye irritant.
PROP: White crystals.

SDP000 CAS: 506-64-9 CAgN HR: 3
SILVER CYANIDE
DOT: UN 1684
DOT Classification: 6.1; Label: Poison
OSHA PEL: TWA 5 $mg(CN)/m^3$
ACGIH TLV: CL 5 $mg(CN)/m^3$ (skin)
DFG MAK: 5 mg/m^3
NIOSH REL: (Cyanide) CL 5 $mg(CN)/m^3$/10M
SAFETY PROFILE: Deadly poison by ingestion. A skin and severe eye irritant.
PROP: White, odorless, tasteless powder that darkens upon exposure to light. Stable in dry air. Mp: 320° (decomp), d: 3.95. Insol in H_2O, alcohols, and dil acids. Sol in aq alkali cyanides, boiling conc HNO_3.

SDS000 CAS: 7761-88-8 NO_3•Ag HR: 3
SILVER(I) NITRATE (1:1)
DOT: UN 1493
SYNS: NITRIC ACID, SILVER(1+) SALT ♦ SILVER NITRATE (DOT)
OSHA PEL: TWA 0.01 $mg(Ag)/m^3$
ACGIH TLV: TWA 0.01 $mg(Ag)/m^3$
DOT Classification: 5.1; Label: Oxidizer
SAFETY PROFILE: A human poison. Experimental poison by ingestion. A powerful caustic and irritant to skin, eyes, and mucous membranes. It can react violently with many substances.
PROP: Colorless, odorless, transparent, large or small white crystals. Not photosensitive when pure. Mp: 212°, bp: 444° (decomp), d: 4.352 @ 19°. Very sol in ammonia, water; sltly sol in ether.

SFA000 CAS: 26628-22-8 N_3Na HR: 3
SODIUM AZIDE
DOT: UN 1687
OSHA PEL: As NH_3: CL 0.1 ppm; As NaN_3: Cl 0.3 mg/m^3 (skin)
ACGIH TLV: CL 0.29 mg/m^3; Not Classifiable as a Human Carcinogen; CL 0.11 ppm (as hydrazoic acid vapor); Not Classifiable as a Human Carcinogen
DFG MAK: 0.2 mg/m^3
DOT Classification: 6.1; Label: Poison
SAFETY PROFILE: Poison by ingestion and skin contact. It can react violently with many substances.

PROP: Colorless, hexagonal crystals. Mp: decomp, d: 1.846. Insol in ether; sol in liquid ammonia.

SFE000 CAS: 7631-90-5 $HO_3S{\cdot}Na$ HR: 3
SODIUM BISULFITE
SYNS: SODIUM ACID SULFITE ♦ SODIUM HYDROGEN SULFITE ♦ SULFUROUS ACID, MONOSODIUM SALT
OSHA PEL: TWA 5 mg/m^3
ACGIH TLV: TWA 5 mg/m^3; Not Classifiable as a Human Carcinogen
DOT Classification: 8; Label: Corrosive
SAFETY PROFILE: Moderately toxic by ingestion. A corrosive irritant to skin, eyes, and mucous membranes. An allergen.
PROP: White, crystalline powder; odor of sulfur dioxide, disagreeable taste. Yellow in soln. D: 1.48, mp: decomp. Very sol in hot or cold water; sltly sol in alc.

SFE500 CAS: 1303-96-4 $B_4O_7{\cdot}2Na$ HR: D
SODIUM BORATE
SYNS: BORATES, TETRA, SODIUM SALT, anhydrous (OSHA, ACGIH) ♦ SODIUM BORATE anhydrous
OSHA PEL: 10 mg/m^3 (anhydrous, decahydrate, pentahydrate)
ACGIH TLV: TWA 1 mg/m^3
SAFETY PROFILE: An inhalation hazard.
PROP: White or colorless monoclinic crystals. Mp: 741°, bp: 1575° (decomp), d: 2.367. Slowly soluble in water, MeOH, polyols; spar sol in Me_2CO, EtOAc, EtOH.

SFF000 CAS: 1303-96-4 $B_4O_7{\cdot}2Na{\cdot}10H_2O$ HR: 3
SODIUM BORATE DECAHYDRATE
SYNS: BORATES, TETRA, SODIUM SALT, anhydrous (OSHA, ACGIH) ♦ BORAX (8CI) ♦ SODIUM TETRABORATE ♦ SODIUM TETRABORATE DECAHYDRATE
OSHA PEL: TWA 10 mg/m^3
ACGIH TLV: TWA 5 mg/m^3
SAFETY PROFILE: Moderately toxic to humans by ingestion.
PROP: Hard, odorless crystals, granules, or crystalline powder. D: 1.73, mp: 75° (when rapidly heated).

SGA500 CAS: 143-33-9 CNNa HR: 3
SODIUM CYANIDE
DOT: UN 1689
SYNS: HYDROCYANIC ACID, SODIUM SALT
OSHA PEL: TWA 5 $mg(CN)/m^3$ (skin)
ACGIH TLV: CL 5 $mg(CN)/m^3$ (skin)
DFG MAK: 5 $mg(CN)/m^3$
NIOSH REL: CL 5 $mg(CN)/m^3$/10M

DOT Classification: 6.1; Label: Poison
SAFETY PROFILE: A deadly human poison by ingestion. An experimental teratogen. Flammable by chemical reaction with heat, moisture, acid. It can react violently with many substances.
PROP: White or colorless, deliquescent, cubic crystals or powder. Undergoes cubic to hexagonal transition on cooling below 10°. Mp: 563.7°, bp: 1496°, vap press: 1 mm @ 817°. Very sol in H_2O; spar sol in EtOH.

SHF500 CAS: 7681-49-4 FNa HR: 3
SODIUM FLUORIDE
DOT: UN 1690
SYNS: DISODIUM DIFLUORIDE ♦ SODIUM HYDROFLUORIDE ♦ SODIUM MONOFLUORIDE
OSHA PEL: TWA 2.5 mg(F)/m^3
ACGIH TLV: TWA 2.5 mg(F)/m^3; BEI: 3 mg/g creatinine of fluorides in urine prior to shift; 10 mg/g creatinine of fluorides in urine at end of shift.
NIOSH REL: TWA (Inorganic Fluorides) 2.5 mg(F)/m^3
DOT Classification: 6.1; Label: KEEP AWAY FROM FOOD
SAFETY PROFILE: Poison by ingestion and skin contact. Experimental teratogenic effects. A corrosive irritant to skin, eyes, and mucous membranes.
PROP: Clear, lustrous, cubic crystals, or white powder or balls. Mp: 996°, bp: 1695°, d: 2 @ 41°, vap press: 1 mm @ 1077°. Mod sol in H_2O; spar sol in EtOH.

SHG500 CAS: 62-74-8 $C_2H_2FO_2{\cdot}Na$ HR: 3
SODIUM FLUOROACETATE
DOT: UN 2629
SYNS: 1080 ♦ COMPOUND No. 1080 ♦ FLUOROACETIC ACID, SODIUM SALT ♦ SODIUM MONOFLUOROACETATE
OSHA PEL: TWA 0.05 mg/m^3 (skin); STEL 0.15 mg/m^3 (skin)
ACGIH TLV: TWA 0.05 mg/m^3 (skin); STEL 0.15 mg/m^3 (skin) (Proposed: TWA 0.05 mg/m^3 (skin))
DFG MAK: 0.05 mg/m^3
DOT Classification: 6.1; Label: Poison
SAFETY PROFILE: A poison by ingestion and skin contact.
PROP: Fine, white powder, or monoclinic crystals. Mp: 200°. Sol in water; spar sol in EtOH, MeOH; prac insol in Me_2CO, CCl_4.

SHS000 CAS: 1310-73-2 HNaO HR: 3
SODIUM HYDROXIDE
DOT: UN 1823/UN 1824
SYNS: CAUSTIC SODA ♦ LYE (DOT) ♦ SODA LYE ♦ SODIUM HYDRATE (DOT)
OSHA PEL: CL 2 mg/m^3
ACGIH TLV: CL 2 mg/m^3

DFG MAK: 2 mg/m^3

NIOSH REL: (Sodium Hydroxide) CL 2 mg/m^3/15M
DOT Classification: 8; Label: Corrosive
SAFETY PROFILE: Moderately toxic by ingestion. A corrosive irritant to skin, eyes, and mucous membranes. It can react violently with many substances.
PROP: White, pieces, lumps, sticks or deliquescent, orthorhombic powder. Undergoes polymorphic transition at 2°. Readily reacts with atm CO_2 forming Na_2CO_3. Mp: 323°, bp: 1390°, d: 2.120 @ 20°/4°, vap press: 1 mm @ 739°. Very sol in water and alc; insol in Et_2O, Me_2CO.

SII000 CAS: 7681-57-4 $O_5S_2•2Na$ HR: 3
SODIUM METABISULFITE
SYNS: DISODIUM DISULFITE ♦ SODIUM DISULFITE ♦ SODIUM PYROSULFITE
OSHA PEL: TWA 5 mg/m^3
ACGIH TLV: TWA 5 mg/m^3; Not Classifiable as a Human Carcinogen
SAFETY PROFILE: An inhalation hazard.
PROP: Colorless crystals or white to yellowish powder; odor of sulfur dioxide. Decomp on heating with ultimate formation of Na_2SO_4. Very sol in H_2O, $NaHSO_3$; mod sol in EtOH.

SIY250 CAS: 13721-39-6 $O_4V•3Na$ HR: 3
SODIUM ORTHOVANADATE
SYNS: SODIUM VANADATE ♦ SODIUM VANADIUM OXIDE ♦ TRISODIUM ORTHOVANADATE
ACGIH TLV: TWA 0.05 $mg(V_2O_5)/m^3$
NIOSH REL: (Vanadium Compounds) CL 0.05 $mg(V)/m^3$/15M
SAFETY PROFILE: Poison by ingestion.
PROP: Colorless, hexagonal prisms. Mp: 850–866°.

SJT500 CAS: 10102-18-8 $O_3Se•2Na$ HR: 3
SODIUM SELENITE
DOT: UN 2630
SYNS: DISODIUM SELENITE ♦ SELENIOUS ACID, DISODIUM SALT
OSHA PEL: TWA 0.2 $mg(Se)/m^3$
ACGIH TLV: TWA 0.2 $mg(Se)/m^3$
DFG MAK: 0.1 $mg(Se)/m^3$
DOT Classification: 6.1; Label: Poison
SAFETY PROFILE: Poison by ingestion. Experimental teratogenic effects.
PROP: White crystals or powder. Undergoes monoclinic to hexagonal transformation at 6°. Decomp on heating to form Na_2O and SeO_2 which decomp further to selenium and O_2. Mp: 710°. Very sol in H_2O.

SKC000 CAS: 10101-83-4 $O_4Te \cdot 2Na$ HR: 3

SODIUM TELLURATE

OSHA PEL: TWA 0.1 mg(Te)/m^3
ACGIH TLV: TWA 0.1 mg(Te)/m^3
SAFETY PROFILE: Poison by ingestion.
PROP: White powder or colorless monoclinic crystals.

SKN500 CAS: 13472-45-2 $O_4W \cdot 2Na$ HR: 3

SODIUM TUNGSTATE

SYNS: DISODIUM TETRAOXATUNGSTATE (2-) ♦ DISODIUM TUNGSTATE ♦ SODIUM WOLFRAMATE
ACGIH TLV: TWA 5 mg(W)/m^3
NIOSH REL: (Tungsten) TWA 1 mg(W)/m^3
SAFETY PROFILE: Poison by ingestion.
PROP: White, rhombic crystals. Mp: 695°, d: 4.179. Sol in water.

SKP000 CAS: 13718-26-8 $O_3V \cdot Na$ HR: 3

SODIUM VANADATE

SYNS: SODIUM METAVANADATE ♦ VANADIC ACID, MONOSODIUM SALT
ACGIH TLV: TWA 0.05 mg(V_2O_5)/m^3
NIOSH REL: (Vanadium Compounds) CL 0.05 mg(V)/m^3/15M
SAFETY PROFILE: Poison by ingestion.

SLQ000 CAS: 7803-52-3 H_3Sb HR: 3

STIBINE

DOT: UN 2676
SYNS: ANTIMONY HYDRIDE ♦ ANTIMONY TRIHYDRIDE ♦ HYDROGEN ANTIMONIDE
OSHA PEL: TWA 0.1 ppm
ACGIH TLV: TWA 0.1 ppm
DFG MAK: 0.1 ppm (0.52 mg/m^3)
DOT Classification: 2.3; Label: Poison Gas, Flammable Gas
SAFETY PROFILE: Poison by inhalation. Potentially explosive decomposition at 200°C. Flammable when exposed to heat or flame.
PROP: Thermally unstable gas, forming colorless liquid and crystals at liquid air temp. Disagreeable odor. Mp: –88°, bp: –18.4°, d: 2.26 @ –25°. Gas is sltly sol in water; very sol in alc, carbon disulfide, and org solvs.

SLU500 CAS: 8052-41-3 HR: 3

STODDARD SOLVENT

SYNS: NAPHTHA SAFETY SOLVENT ♦ WHITE SPIRITS
OSHA PEL: TWA 100 ppm
ACGIH TLV: TWA 100 ppm
NIOSH REL: (Refined Petroleum Solvents) TWA 350 mg/m^3; CL 1800 mg/m^3/15M

SAFETY PROFILE: Mildly toxic by inhalation. A human eye irritant. Flammable liquid.
PROP: Clear, colorless liquid. Composed of 85% nonane and 15% trimethyl benzene. Bp: 220–300°, flash p: 100–110°F, lel: 1.1%, uel: 6%, autoign temp: 450°F, d: 1.0. Insol in water; misc with abs alc, benzene, ether, chloroform, carbon tetrachloride, carbon disulfide, and some oils (not castor oil).

SMH000 CAS: 7789-06-2 $CrO_4 \cdot Sr$ HR: 3
STRONTIUM CHROMATE (1:1)
OSHA PEL: CL 0.1 mg(CrO_3)/m^3
ACGIH TLV: TWA 0.0005 ppm; Suspected Human Carcinogen
DFG TRK: 0.1 mg/m^3; Animal Carcinogen, Suspected Human Carcinogen
NIOSH REL: TWA 0.0001 mg(Cr(VI))/m^3
SAFETY PROFILE: Confirmed human carcinogen. Moderately toxic by ingestion.
PROP: Monoclinic, yellow crystals. D: 3.895 @ 15°. Sol in HCl, HNO_3, and AcOH.

SMN500 CAS: 57-24-9 $C_{21}H_{22}N_2O_2$ HR: 3
STRYCHNINE
DOT: UN 1692
OSHA PEL: TWA 0.15 mg/m^3
ACGIH TLV: TWA 0.15 mg/m^3
DFG MAK: 0.15 mg/m^3
DOT Classification: 6.1; Label: Poison
SAFETY PROFILE: Human poison by ingestion. An allergen.
PROP: Hard, white, crystalline alkaloid; very bitter taste. Mp: 268°, bp: 270°, d: 1.359 @ 18°.

SMQ000 CAS: 100-42-5 C_8H_8 HR: 3
STYRENE
DOT: UN 2055
SYNS: ETHENYLBENZENE ♦ STYRENE MONOMER, inhibited (DOT) ♦ VINYLBENZENE ♦ VINYLBENZOL
OSHA PEL: TWA 50 ppm; STEL 100 ppm
ACGIH TLV: TWA 20 ppm; STEL 40 ppm; Not Classifiable as a Human Carcinogen; BEI: 800 mg(mandelic acid)/L in urine at end of shift; 0.55 mg/L styrene in blood at end of shift; 0.02 mg/L styrene in blood prior to next shift
DFG MAK: 20 ppm (86 mg/m^3); BAT: 2 g/L of mandelic acid in urine at end of shift
NIOSH REL: (Styrene) TWA 50 ppm; CL 100 ppm
DOT Classification: 3; Label: Flammable Liquid
SAFETY PROFILE: Confirmed carcinogen. Poison by ingestion and inhalation. Mildly toxic to humans by inhalation. An experimental teratogen. A skin and severe eye irritant. A very dangerous fire hazard. It can react violently with many substances.

PROP: Colorless, refractive, oily liquid with penetrating odor. Mp: –33°, bp: 146°, lel: 1.1%, uel: 6.1%, flash p: 88°F, d: 0.9074 @ 20°/4°, autoign temp: 914°F, vap d: 3.6, fp: –33°, ULC: 40–50. Very sltly sol in water; misc in alc and ether.

SNE000 CAS: 110-61-2 $C_4H_4N_2$ HR: 3

SUCCINONITRILE

SYNS: 1,4-BUTANEDINITRILE ♦ ETHYLENE CYANIDE ♦ SUCCINIC ACID DINITRILE

ACGIH TLV: CL 5 mg(CN)/m^3 (skin)

NIOSH REL: (Nitriles) TWA 20 mg/m^3

SAFETY PROFILE: Poison by ingestion. An experimental teratogen. Combustible.

PROP: Colorless, odorless, waxy material. Mp: 58.1°, bp: 267°, flash p: 270°F, d: 1.022 @ 25°, vap press: 2 mm @ 100°, vap d: 2.1. Sltly sol in ether, water, alc; sol in acetone.

SOD100 CAS: 3689-24-5 $C_8H_{20}O_5P_2S_2$ HR: 3

SULFOTEP

DOT: UN 1704

SYNS: BIS-O,O-DIETHYLPHOSPHOROTHIONIC ANHYDRIDE ♦ ETHYL THIOPYROPHOSPHATE ♦ SULFOTEPP ♦ TEDP ♦ TEDP (OSHA) ♦ TEDTP ♦ TETRAETHYL DITHIONOPYROPHOSPHATE ♦ TETRAETHYL DITHIOPYROPHOSPHATE

OSHA PEL: TWA 0.2 mg/m^3 (skin)

ACGIH TLV: TWA 0.2 mg/m^3 (skin); Not Classifiable as a Human Carcinogen

DFG MAK: 0.0075 ppm (0.1 mg/m^3)

DOT Classification: 6.1; Label: Poison

SAFETY PROFILE: Poison by ingestion, skin contact, and inhalation.

PROP: A pale-yellow, mobile liquid. D: 1.19 @ 25°/4°, bp: 110–113° @ 0.2 mm. Almost insol in water.

SOH500 CAS: 7446-09-5 O_2S HR: 3

SULFUR DIOXIDE

DOT: UN 1079

SYNS: SULFUROUS ACID ANHYDRIDE ♦ SULFUROUS ANHYDRIDE ♦ SULFUROUS OXIDE ♦ SULFUR OXIDE

OSHA PEL: TWA 2 ppm; STEL 5 ppm

ACGIH TLV: TWA 2 ppm; STEL 5 ppm; Not Classifiable as a Human Carcinogen

DFG MAK: 0.5 pm (1.3mg/m^3)

NIOSH REL: (Sulfur Dioxide) TWA 0.5 ppm

DOT Classification: 2.3; Label: Poison Gas

SAFETY PROFILE: A poison gas. A corrosive irritant to eyes, skin, and mucous membranes. A nonflammable gas. It can react violently with many substances.

PROP: Colorless, nonflammable gas or liquid under pressure; pungent odor. Catalytically oxidized by air to SO_3. Mp: –75.5°, bp: –10.0°, d: (liquid) 1.434 @ 0°, vap d: 2.264 @ 0°, vap press: 2538 mm @ 21.1°. Sol in water, decreases with temp.

SOI000 CAS: 2551-62-4 F_6S HR: 1

SULFUR HEXAFLUORIDE
DOT: UN 1080
SYNS: SULFUR FLUORIDE
OSHA PEL: TWA 1000 ppm
ACGIH TLV: TWA 1000 ppm
DFG MAK: 1000 ppm (6100 mg/m^3)
DOT Classification: 2.2; Label: Nonflammable Gas
SAFETY PROFILE: In ert itself but can contain low-sulfur fluorides. Some of these are toxic, very reactive chemically, and corrosive in nature. In high concentrations and when pure it may act as a simple asphyxiant.
PROP: Colorless, odorless gas of high chemical stability and inertness. Nonflammable. White sublimable solid at low temps. Stable to H_2O and to glass. Mp: –51° (subl @ –64°), vap d: 6.602, d (liquid): 1.67 @ –100°. Very insol in H_2O; slightly sol in EtOH.

SOI500 CAS: 7664-93-9 H_2O_4S HR: 3
SULFURIC ACID
DOT: UN 1830/UN 1832
SYNS: MATTING ACID (DOT) ♦ NORDHAUSEN ACID (DOT) ♦ OIL OF VITRIOL (DOT) ♦ SPENT SULFURIC ACID (DOT) ♦ VITRIOL, OIL OF (DOT)
OSHA PEL: TWA 1 mg/m^3
ACGIH TLV: TWA 1 mg/m^3; STEL 3 ppm; Suspected Human Carcinogen (Contained in strong inorganic mists)
DFG MAK: 1 mg/m^3
NIOSH REL: (Sulfuric Acid) TWA 1 mg/m^3
DOT Classification: 8; Label: Corrosive
SAFETY PROFILE: A poison by inhalation. Suspected human carcinogen when contained in strong inorganic mists. Moderately toxic by ingestion. A severe eye irritant. Extremely irritating, corrosive, and toxic to skin and eyes. An experimental teratogen. It can react violently with many substances.
PROP: Viscous, colorless oily liquid; odorless. Mp: 10.49°, d: 1.834, vap press: 1 mm @ 145.8°, bp: 290°, decomp @ 340°. Misc with water and alc (liberating great heat).

SON510 CAS: 10025-67-9 Cl_2S_2 HR: 3
SULFUR MONOCHLORIDE
SYNS: CHLORIDE of SULFUR (DOT) ♦ DISULFUR DICHLORIDE ♦ SULFUR SUBCHLORIDE ♦ THIOSULFUROUS DICHLORIDE
OSHA PEL: CL 1 ppm
ACGIH TLV: CL 1 ppm
DFG MAK: 1 ppm (5.6 mg/m^3)
NIOSH REL: (Sulfur Monochloride) CL 1 ppm
SAFETY PROFILE: Poison by ingestion and inhalation. A fuming, corrosive liquid very irritating to skin, eyes, and mucous membranes. It can react violently with many substances.

PROP: Amber to yellowish-red, oily, fuming liquid; penetrating odor. Hydrolyzes to $HCl + SO_2 + H_2S$. Mp: –77°, bp: 138.0°, flash p: 245°F (CC), d: 1.6885 @ 15.5°/15.5°, autoign temp: 453°F, vap press: 10 mm @ 27.5°, vap d: 4.66. Decomp in water. Sol in CS_2 and org solvs.

SOQ450 CAS: 5714-22-7 $F_{10}S_2$ HR: 3

SULFUR PENTAFLUORIDE

SYNS: SULFUR DECAFLUORIDE

OSHA PEL: CL 0.01 ppm

ACGIH TLV: CL 0.01 ppm

DFG MAK: 0.025 ppm (0.26 mg/m^3)

NIOSH REL: (Sulfur Pentafluoride) CL 0.01 ppm

SAFETY PROFILE: Moderately toxic by inhalation.

PROP: Colorless liquid. Stable to H_2O in the presence of acid or base. Disproportionates at 1° to $SF_6 + SF_4$. Mp: –53°, bp: 26.7°.

SOR000 CAS: 7783-60-0 F_4S HR: 3

SULFUR TETRAFLUORIDE

DOT: UN 2418

SYN: TETRAFLUOROSULFURANE

OSHA PEL: CL 0.1 ppm

ACGIH TLV: CL 0.1 ppm; BEI: 3 mg/g creatinine of fluorides in urine prior to shift; 10 mg/g creatinine of fluorides in urine at end of shift.

NIOSH REL: (Inorganic Fluorides) TWA 2.5 mg(F)/m^3

DOT Classification: 2.3; Label: Poison Gas

SAFETY PROFILE: Poison by inhalation. A powerful irritant.

PROP: Colorless gas. Readily hydrolyzes. Stable in dry Pyrex glass vessels. Bp: –38°, mp: –121°, d: 1.95° @ –78°. Very sol in C_6H_6.

SOT500 CAS: 13637-84-8 $ClFO_2S$ HR: 3

SULFURYL CHLORIDE FLUORIDE

SYNS: CHLORO FLUORO SULFONE ♦ CHLOROSULFONYL FLUORIDE ♦ FLUOROSULFONYL CHLORIDE ♦ SULFURYL FLUOROCHLORIDE ♦ TL 212

OSHA PEL: TWA 2.5 mg(F)/m^3

ACGIH TLV: TWA 2.5 mg(F)/m^3; BEI: 3 mg/g creatinine of fluorides in urine prior to shift; 10 mg/g creatinine of fluorides in urine at end of shift.

NIOSH REL: (Inorganic Fluorides) TWA 2.5 mg(F)/m^3

SAFETY PROFILE: Poison by inhalation.

PROP: Colorless gas with pungent smell. Hydrolyzes by H_2O to $HCl + HF + H_2SO_4$. Mp: –124.7°, bp: 7.1°, d: 1.623 @ 0°.

SOU500 CAS: 2699-79-8 F_2O_2S HR: 3

SULFURYL FLUORIDE

DOT: UN 2191
SYNS: SULFURIC OXYFLUORIDE
OSHA PEL: TWA 5 ppm; STEL 10 ppm
ACGIH TLV: TWA 5 ppm; STEL 10 ppm
DOT Classification: 2.3; Label: Poison Gas
SAFETY PROFILE: Poison by ingestion. Mildly toxic by inhalation. Can react with water.
PROP: Odorless, colorless, fairly inert gas. Stable to 4°. Mp: –137°, bp: –55°, d: 1.7 @ –55° (approx). Sol in water.

TAB750 CAS: 14807-96-6 $H_2O_3Si \cdot 3/4Mg$ HR: 2
TALC
SYNS: C.I. 77718 ♦ TALCUM
OSHA PEL: TWA 2 mg/m^3
ACGIH TLV: TWA 2 mg/m^3, respirable dust (use asbestos TLV if asbestos fibers are present); Not Classifiable as a Human Carcinogen
DFG MAK: 2 mg/m^3
NIOSH REL: Talc (containing no asbestos) 2 mg/m^3
SAFETY PROFILE: The talc with less than 1 percent asbestos is a nuisance dust. Talc with greater percentage of asbestos may be a human carcinogen. A human skin irritant.
PROP: White to grayish-white, fine powder; odorless and tasteless. Powdered native hydrous magnesium silicate. Insol in water, cold acids, or alkalies. Containing less than 1% crystalline silica. Insol in H_2O.

TAE750 CAS: 7440-25-7 Ta HR: 2
TANTALUM
OSHA PEL: TWA 5 mg/m^3
ACGIH TLV: TWA 5 mg/m^3
DFG MAK: 4 mg/m^3
SAFETY PROFILE: An inhalation hazard. The dry powder ignites spontaneously in air.
PROP: Gray, malleable, soft ductile metal when pure. Very air and H_2O corrosion resistant (oxide film). Attacked by HF, and by fused alkalies, as well as by fuming H_2SO_4. Mp: 2996°, bp: 5429°, d: 16.69. Insol in water.

TAI750 CAS: 7803-68-1 H_6O_6Te HR: 3
TELLURIC ACID
SYNS: ORTHOTELLURIC ACID ♦ TELLURATE ♦ TELLURIUM HYDROXIDE
OSHA PEL: TWA 0.1 mg(Te)/m^3
ACGIH TLV: TWA 0.1 mg(Te)/m^3
SAFETY PROFILE: Poison by ingestion.
PROP: White crystals or solid; stable to 1°. D: (monoclinic) 3.068, (cubic) 3.163. Mp: 136°. Sltly sol in concentrated nitric acid; very sol in water.

TAJ000 CAS: 13494-80-9 Te HR: 3
TELLURIUM
OSHA PEL: TWA 0.1 mg(Te)/m^3
ACGIH TLV: TWA 0.1 mg(Te)/m^3
DFG MAK: 0.1 mg/m^3
NIOSH REL: (Tellurium and Compounds) TWA 0.1 mg/m^3
SAFETY PROFILE: Poison by ingestion. An experimental teratogen.
PROP: Silvery-white, metallic, lustrous element; quite brittle. Semiconductor; poor conductor of heat. Forms golden-yellow vapor. Red when colloidal. Mp: 449.5°, bp: 989.8°, d: 6.24 @ 20°, vap press: 1 mm @ 520°. Insol in water, benzene, carbon disulfide, and all solvs with which it does not react.

TAK250 CAS: 7783-80-4 F_6Te HR: 3
TELLURIUM HEXAFLUORIDE
DOT: UN 2195
OSHA PEL: TWA 0.02 ppm (Te)
ACGIH TLV: TWA 0.02 ppm (Te)
DOT Classification: 2.3; Label: Poison Gas
SAFETY PROFILE: Poison by inhalation.
PROP: Colorless gas; repulsive odor. White solid at low temp; subl before melting. Hydrol by H_2O and aq KOH. Mp: –37.6°, bp: –38.9° (subl), d: (solid) 4.006 @ –191°, (liquid) 2.499 @ –10°.

TBC640 CAS: 84-15-1 $C_{18}H_{14}$ HR: 2
o-TERPHENYL
SYN: 1,2-DIPHENYLBENZENE
OSHA PEL: CL 0.5 ppm
ACGIH TLV: TWA CL 0.5 ppm
SAFETY PROFILE: Moderately toxic by ingestion. Slightly combustible.
PROP: Prisms from MeOH. Mp: 58–59°, bp: 337°, flash p: >230°F.

TBD000 CAS: 26140-60-3 $C_{18}H_{14}$ HR: 2
TERPHENYLS
SYNS: DIPHENYLBENZENE ♦ TRIPHENYL
OSHA PEL: CL 0.5 ppm
ACGIH TLV: TWA CL 0.5 ppm
NIOSH REL: (Terphenyls) CL 0.5 ppm
SAFETY PROFILE: Moderately toxic by ingestion. Slightly combustible when exposed to heat or flame.

TBP000 CAS: 76-11-9 $C_2Cl_4F_2$ HR: 1
1,1,1,2-TETRACHLORO-2,2-DIFLUOROETHANE
SYNS: CFC 112a ♦ 1,1-DIFLUORO-1,2,2,2-TETRACHLOROETHANE ♦ REFRIGERANT 112a

OSHA PEL: TWA 500 ppm
ACGIH TLV: TWA 500 ppm
DFG MAK: 1000 ppm (8500 mg/m^3)
NIOSH REL: (1,1,1,2-Tetrachloro-2,2-difluoroethane) TWA 500 ppm
SAFETY PROFILE: Mildly toxic by inhalation.
PROP: D: 1.65 @ 20°/4°, mp: 38–40°, bp: 91°.

TBP050 CAS: 76-12-0 $C_2Cl_4F_2$ HR: 3
1,1,2,2-TETRACHLORO-1,2-DIFLUOROETHANE
SYNS: CFC-112 ♦ 1,2-DIFLUORO-1,1,2,2-TETRACHLOROETHANE ♦ FREON 112
OSHA PEL: TWA 500 ppm
ACGIH TLV: TWA 500 ppm
DFG MAK: 200 ppm (1700 mg/m^3)
NIOSH REL: (1,1,2,2-Tetrachloro-1,2-difluoroethane) TWA 500 ppm
SAFETY PROFILE: A poison by inhalation. Moderately toxic by ingestion. A skin and eye irritant.
PROP: Liquid or crystals. Mp: 24.65°, bp: 92.8°, d: 1.6447 @ 25°, vap d: 7.03.

TBP250 CAS: 31242-94-1 $C_{12}H_6Cl_4O$ HR: 3
TETRACHLORODIPHENYL OXIDE
SYNS: PHENYL ETHER TETRACHLORO ♦ TETRACHLOROPHENYL ETHER
OSHA PEL: TWA 0.5 mg/m^3
SAFETY PROFILE: Poison by ingestion.

TBP750 CAS: 25322-20-7 $C_2H_2Cl_4$ HR: 3
TETRACHLOROETHANE
DOT: UN 1702
NIOSH REL: (Tetrachloroethane) Reduce to lowest feasible level
DOT Classification: 6.1; Label: Poison
SAFETY PROFILE: Poison by ingestion and inhalation. Mildly toxic by skin contact.

TBQ100 CAS: 79-34-5 $C_2H_2Cl_4$ HR: 3
1,1,2,2-TETRACHLOROETHANE
SYNS: ACETYLENE TETRACHLORIDE ♦ 1,1-DICHLORO-2,2-DICHLOROETHANE ♦ TCE
OSHA PEL: TWA 1 ppm (skin)
ACGIH TLV: TWA 1 ppm (skin); Animal Carcinogen
DFG MAK: 1 ppm (7 mg/m^3); Confirmed Animal Carcinogen with Unknown Relevance to Humans
NIOSH REL: (1,1,2,2-Tetrachlorethane) Reduce to lowest level
SAFETY PROFILE: Suspected carcinogen. Poison by inhalation and ingestion. A strong irritant of eyes and mucous membranes. It can react violently with many substances.
PROP: Heavy, colorless, mobile liquid; chloroform-like odor. Mp: –43.8°, bp: 146.4°, d: 1.600 @ 20°/4°, fp: –43.8°.

TBR000 CAS: 1335-88-2 $C_{10}H_4Cl_4$ HR: 3
TETRACHLORONAPHTHALENE
SYN: HALOWAX
OSHA PEL: TWA 2 mg/m^3
ACGIH TLV: TWA 2 mg/m^3
SAFETY PROFILE: A poison.
PROP: Colorless to pale-yellow solid; aromatic odor. Mw: 265.96, specific gravity: 1.59–1.65, mp: 115°, bp: 311.5–360°, flash p: 410°F. Insol in water.

TCF000 CAS: 78-00-2 $C_8H_{20}Pb$ HR: 3
TETRAETHYL LEAD
DOT: NA 1649
SYNS: TEL ♦ TETRAETHYLPLUMBANE
OSHA PEL: TWA 0.075 $mg(Pb)/m^3$ (skin)
ACGIH TLV: TWA 0.1 $mg(Pb)/m^3$ (skin); Not Classifiable as a Human Carcinogen
DFG MAK: 0.05 mg/m^3
DOT Classification: 6.1; Label: Poison, Flammable Liquid
SAFETY PROFILE: Poison by ingestion. Moderately toxic by inhalation and skin contact. Experimental teratogenic effects. A combustible liquid.
PROP: Colorless, oily liquid; pleasant characteristic odor. Mp: 125–150°, bp: 198–202° with decomp, d: 1.659 @ 18°, vap press: 1 mm @ 38.4°, flash p: 200°F.

TCF250 CAS: 107-49-3 $C_8H_{20}O_7P_2$ HR: 3
TETRAETHYL PYROPHOSPHATE
DOT: NA 2783/NA 3018
SYNS: BIS-O,O-DIETHYLPHOSPHORIC ANHYDRIDE ♦ TEPP (ACGIH)
OSHA PEL: TWA 0.05 mg/m^3 (skin)
ACGIH TLV: TWA 0.004 mg/m^3 (skin)
DFG MAK: 0.005 ppm (0.060 mg/m^3)
DOT Classification: 6.1; Label: Poison
SAFETY PROFILE: A poison by ingestion and skin contact.
PROP: Water-white to amber hygroscopic liquid. D: 1.20, bp: 104–110°. Misc in H_2O and most org solvs. Spar sol in pet ether.

TCF750 CAS: 597-64-8 $C_8H_{20}Sn$ HR: 3
TETRAETHYLSTANNANE
SYN: TETRAETHYL TIN
OSHA PEL: TWA 0.1 $mg(Sn)/m^3$ (skin)
ACGIH TLV: TWA 0.1 $mg(Sn)/m^3$ (skin) (Proposed: TWA 0.1 $mg(Sn)/m^3$; STEL 0.2 $mg(Sn)/m^3$ (skin))
NIOSH REL: (Organotin Compounds) TWA 0.1 $mg(Sn)/m^3$
SAFETY PROFILE: Poison by ingestion.

PROP: Colorless liquid. D: 1.187 @ 23°, mp: −112°, bp: 181°. Insol in water; sol in org solvs.

TCR750 CAS: 109-99-9 C_4H_8O HR: 3

TETRAHYDROFURAN

DOT: UN 2056

SYNS: BUTYLENE OXIDE ♦ CYCLOTETRAMETHYLENE OXIDE ♦ 1,4-EPOXYBUTANE ♦ OXACYCLOPENTANE ♦ TETRAMETHYLENE OXIDE ♦ THF

OSHA PEL: TWA 200 ppm; STEL 250 ppm

ACGIH TLV: TWA 200 ppm; STEL 250 ppm; (Proposed: BEI: 8 mg/L tetrahydrofuran in urine at end of shift)

DFG MAK: 50 ppm (150 mg/m^3)

DOT Classification: 3; Label: Flammable Liquid

SAFETY PROFILE: Moderately toxic by ingestion. Mildly toxic by inhalation. Irritant to eyes and mucous membranes. A flammable liquid. It can react violently with many substances.

PROP: Colorless, mobile liquid; ether-like odor. Bp: 65.4°, flash p: 1.4°F (TCC), lel: 1.8%, uel: 11.8%, fp: −65°, d: 0.888 @ 21°/4°, vap press: 114 mm @ 15°, vap d: 2.5, autoign temp: 610°F. Misc with water, alc, ketones, esters, ethers, and hydrocarbons.

TDR500 CAS: 75-74-1 $C_4H_{12}Pb$ HR: 3

TETRAMETHYL LEAD

SYNS: TML

OSHA PEL: TWA 0.075 $mg(Pb)/m^3$ (skin)

ACGIH TLV: TWA 0.15 $mg(Pb)/m^3$ (skin)

DFG MAK: 0.05 mg/m^3

SAFETY PROFILE: Poison by ingestion. Moderately toxic by skin contact. An experimental teratogen. A flammable liquid.

PROP: Colorless liquid. Mp: −27.5°, lel: 1.8%, bp: 110°, d: 1.99, vap d: 9.2, flash p: 100°F.

TDW250 CAS: 3333-52-6 $C_8H_{12}N_2$ HR: 3

TETRAMETHYLSUCCINONITRILE

SYN: TMSN

OSHA PEL: TWA 0.5 ppm (skin)

ACGIH TLV: TWA 0.5 ppm (skin)

DFG MAK: 0.5 ppm (2.8 mg/m^3)

NIOSH REL: (Nitriles) CL 6 mg/m^3/15M

SAFETY PROFILE: Poison by ingestion. An experimental teratogen. A human skin irritant and allergen.

PROP: Crystallizes in plates; almost no odor. Mp: 172° (sublimes).

TDY250 CAS: 509-14-8 CN_4O_8 HR: 3

TETRANITROMETHANE

DOT: UN 1510
SYNS: TNM
OSHA PEL: TWA 1 ppm
ACGIH TLV: TWA 0.005 ppm; Animal Carcinogen
DFG MAK: Confirmed Animal Carcinogen, Suspected Human Carcinogen
DOT Classification: 5.1; Label: Oxidizer, Poison
SAFETY PROFILE: Confirmed carcinogen. Poison by ingestion. Irritating to the skin, eyes, mucous membranes. A very dangerous fire and severe explosion. It can react violently with many substances.
PROP: Colorless or yellow liquid. Mp: 13°, bp: 125.7°, d: 1.650 @ 13°, vap press: 10 mm @ 22.7°. Insol in water; very sol in alc, ether.

TEE500 CAS: 7722-88-5 $O_7P_2{\bullet}4Na$ HR: 3
TETRASODIUM PYROPHOSPHATE
SYNS: SODIUM PYROPHOSPHATE (FCC) ♦ TETRASODIUM DIPHOSPHATE ♦ TSPP
OSHA PEL: TWA 5 mg/m^3
ACGIH TLV: TWA 5 mg/m^3
SAFETY PROFILE: Poison by ingestion.
PROP: White orthorhombic crystalline powder. Mp: 993°, d: 2.534. Sol in water; insol in alc.

TEG250 CAS: 479-45-8 $C_7H_5N_5O_8$ HR: 3
TETRYL
DOT: UN 0208
SYNS: N-METHYL-N,2,4,6-TETRANITROANILINE ♦ NITRAMINE ♦ PICRYLNITROMETHYLAMINE ♦ N,2,4,5-TETRANITRO-N-METHYLANILINE ♦ TETRIL ♦ 2,4,6-TRINITROPHENYLMETHYLNITRAMINE
OSHA PEL: TWA 0.1 mg/m^3 (skin)
ACGIH TLV: TWA 1.5 mg/m^3
DFG MAK: Confirmed Animal Carcinogen with Unknown Relevance to Humans
DOT Classification: EXPLOSIVE 1.1D; Label: EXPLOSIVE 1.1D
SAFETY PROFILE: An irritant, sensitizer, and allergen. A dangerous fire and explosion hazard.
PROP: Yellow, monoclinic crystals or prisms from EtOH. Mp: 131–132°, bp: explodes @ 187°, d: 1.57 @ 19°. Spar sol in EtOH.

TEI000 CAS: 7440-28-0 Tl HR: 3
THALLIUM
OSHA PEL: TWA 0.1 $mg(Tl)/m^3$ (skin)
ACGIH TLV: TWA 0.1 $mg(Tl)/m^3$ (skin)
DFG MAK: 0.1 mg/m^3
SAFETY PROFILE: Human poison. Flammable in the form of dust.
PROP: Bluish-white, soft, malleable metal. Tarnishes readily in air. Moderately strong base; reacts with steam or moist air releasing TlOH. Mp: 303.5°, bp: 1457°, d: 11.85 @

20°, vap press: 1 mm @ 825°. Readily dissolves in dil H_2SO_4 and HNO_3; does not dissolve in basic solns.

TEI250 CAS: 563-68-8 $C_2H_3O_2 \cdot Tl$ HR: 3
THALLIUM ACETATE
SYNS: THALLIUM MONOACETATE ♦ THALLOUS ACETATE
OSHA PEL: TWA 0.1 mg(Tl)/m^3 (skin)
ACGIH TLV: TWA 0.1 mg(Tl)/m^3 (skin)
SAFETY PROFILE: Human poison by ingestion. An experimental teratogen.
PROP: Silk-white crystals, or hygroscopic, colorless, light-sensitive monoclinic crystals from EtOH. Mp: 131°, d: 3.68. Sol in water, alc.

TEI750 CAS: 7789-40-4 BrTl HR: 3
THALLIUM BROMIDE
OSHA PEL: TWA 0.1 mg(Tl)/m^3 (skin)
ACGIH TLV: TWA 0.1 mg(Tl)/m^3 (skin)
SAFETY PROFILE: Poison by ingestion.
PROP: Air-stable, light-sensitive, yellowish-white powder, solids, or crystals. Mp: 460–480° (approx), bp: 815°, d: 7.557, vap press: 10 mm @ 522°.

TEJ000 CAS: 6533-73-9 $CO_3 \cdot 2Tl$ HR: 3
THALLIUM(I) CARBONATE (2:1)
SYNS: DITHALLIUM CARBONATE ♦ THALLOUS CARBONATE
OSHA PEL: TWA 0.1 mg(Tl)/m^3 (skin)
ACGIH TLV: TWA 0.1 mg(Tl)/m^3 (skin)
SAFETY PROFILE: Poison by ingestion and skin contact.
PROP: Monoclinic, colorless crystals stable in air to 1°; loses CO_2 on further heating; dissolves acidic oxides. Mp: 273°, d: 7.11. Sol in H_2O; insol in EtOH.

TEJ250 CAS: 7791-12-0 ClTl HR: 3
THALLIUM CHLORIDE
SYNS: THALLIUM MONOCHLORIDE ♦ THALLOUS CHLORIDE
OSHA PEL: TWA 0.1 mg(Tl)/m^3 (skin)
ACGIH TLV: TWA 0.1 mg(Tl)/m^3 (skin)
SAFETY PROFILE: Poison by ingestion. An experimental teratogen.
PROP: Colorless or white powder, or photosensitive crystals from water. Insulator, wide gap (3eV); metallic at high pressures. Mp: 430°, bp: 720°, d: 7.00, vap press: 10 mm @ 517°. Sol in 260 parts cold water, 70 parts boiling water; insol in alc.

TEJ500 HR: 3
THALLIUM COMPOUNDS
SAFETY PROFILE: Extremely toxic. Reproductive organs and the fetus are highly susceptible. Thallium is an experimental teratogen.

TEK750 CAS: 10102-45-1 $NO_3 \cdot Tl$ HR: 3

THALLIUM NITRATE

DOT: UN 2727

SYNS: THALLIUM MONONITRATE ♦ THALLOUS NITRATE

OSHA PEL: TWA 0.1 mg(Tl)/m^3 (skin)

ACGIH TLV: TWA 0.1 mg(Tl)/m^3 (skin)

DOT Classification: 6.1; Label: Poison, Oxidizer

SAFETY PROFILE: Poison by ingestion.

PROP: Cubic crystals or white solids from water. Mp: 206°, bp: 430°, d: 5.55. Decomp @ 450°; decomp at 800–809° *in vacuo* to give Tl_2O, NO, NO_2.

TEL050 CAS: 1314-32-5 O_3Tl_2 HR: 3

THALLIUM(III) OXIDE

SYNS: THALLIC OXIDE ♦ THALLIUM PEROXIDE ♦ THALLIUM SESQUIOXIDE

OSHA PEL: TWA 0.1 mg(Tl)/m^3 (skin)

ACGIH TLV: TWA 0.1 mg(Tl)/m^3 (skin)

SAFETY PROFILE: Poison by ingestion. Combustible by chemical reaction.

PROP: Hexagonal brown-black crystals, amorph prisms, powder, or solid which reacts with $AlCl_3$ to give TlCl; vaporizes to give Tl_2O and O_2. Mp: 717° ± 5°, bp: 1196°, d(amorph): 9.65 @ 21°, d(hexagonal): 10.19 @ 22°. Insol in H_2O; decomp in acids.

TFC600 CAS: 96-69-5 $C_{22}H_{30}O_2S$ HR: 3

4,4′-THIOBIS(6-tert-BUTYL-m-CRESOL)

SYNS: BIS(3-tert-BUTYL-4-HYDROXY-6-METHYLPHENYL) SULFIDE ♦ 4,4′-THIOBIS(2-tert-BUTYL-5-METHYLPHENOL)

OSHA PEL: TWA Total Dust: 10 mg/m^3; Respirable Fraction: 5 mg/m^3

ACGIH TLV: TWA 10 mg/m^3; Not Classifiable as a Human Carcinogen

NIOSH REL: (4,4′-Thiobis(6-t-butyl-m-cresol), Dust): TWA 10 mg/m^3; air: TWA 5 mg/m^3

SAFETY PROFILE: Poison by ingestion and inhalation.

PROP: Light-gray to tan powder. Mp: 150°, d: 1.10.

TFD750 CAS: 91-71-4 $C_{21}H_{17}AsN_2O_5S_2$ HR: 3

THIOCARBAMIZINE

SYNS: 4-CARBAMIDOPHENYL BIS(o-CARBOXYPHENYLTHIO)ARSENITE ♦ (p-UREIDOBENZENEARSYLENEDITHIO)DI-o-BENZOIC ACID

OSHA PEL: TWA 0.5 mg(As)/m^3

ACGIH TLV: BEI: 35 μ (As)/L inorganic arsenic and methylated metabolites in urine; (Proposed: BEI: 50 μ (As)/L creatinine) inorganic arsenic and methylated metabolites in urine)

SAFETY PROFILE: Moderately toxic by ingestion.

PROP: White crystal powder. Sltly sol in water, alc, and alkali; insol in acids.

TFI000 CAS: 139-65-1 $C_{12}H_{12}N_2S$ HR: 3

4,4'-THIODIANILINE

SYNS: BIS(p-AMINOPHENYL)SULFIDE ♦ p,p'-DIAMINODIPHENYL SULFIDE ♦ DI(p-AMINOPHENYL)SULPHIDE ♦ 4,4'-THIOBISBENZENAMINE ♦ THIODI-p-PHENYLENEDIAMINE

DFG MAK: Animal Carcinogen, Suspected Human Carcinogen

SAFETY PROFILE: Confirmed carcinogen. Moderately toxic by ingestion.

PROP: Needles from water. Mp: 108°. Sol in aq HCl; insol in H_2O.

TFJ100 CAS: 68-11-1 $C_2H_4O_2S$ HR: 3

THIOGLYCOLIC ACID

DOT: UN 1940

SYNS: MERCAPTOACETATE ♦ 2-MERCAPTOACETIC ACID ♦ THIOGLYCOLLIC ACID

OSHA PEL: TWA 1 ppm (skin)

ACGIH TLV: TWA 1 ppm (skin)

DOT Classification: 8; Label: Corrosive

SAFETY PROFILE: Poison by ingestion and skin contact. A corrosive irritant to skin, eyes, and mucous membranes.

PROP: Liquid, strong odor. Mp: –16.5°, bp: 108° @ 15 mm. Misc with water, alc, ether, chloroform, and benzene.

TFL000 CAS: 7719-09-7 Cl_2OS HR: 3

THIONYL CHLORIDE

DOT: UN 1836

SYNS: SULFINYL CHLORIDE ♦ SULFUR CHLORIDE OXIDE ♦ SULFUROUS OXYCHLORIDE

OSHA PEL: CL 1 ppm

ACGIH TLV: CL 1 ppm

DOT Classification: 8; Label: Corrosive, Poison

SAFETY PROFILE: Moderately toxic by inhalation. Violent reaction with water releases hydrogen chloride and sulfur dioxide. A corrosive irritant that causes burns to the skin and eyes. It can react violently with many substances.

PROP: Colorless to yellow to red liquid; suffocating odor. Mp: –105°, bp: 78.8° @ 746 mm, d: 1.640 @ 15.5°/15.5°, vap press: 100 mm @ 21.4°. Misc with benzene, chloroform, carbon tetrachloride. Sol in $CHCl_3$, C_6H_6, and CCl_4.

TGB250 CAS: 7440-31-5 Sn HR: 2

TIN

OSHA PEL: Organic Compounds: TWA 0.1 mg(Sn)/m^3 (skin); Inorganic Compounds (except oxides): TWA 2 mg/m^3

ACGIH TLV: TWA metal, oxide, and inorganic compounds (except SnH_4) as Sn 2 mg/m^3; organic compounds TWA 0.1 mg(Sn)/m^3; STEL 0.2 mg(Sn)/m^3 (skin)

DFG MAK: Inorganic 2 mg/m^3, organic 0.1 mg/m^3
NIOSH REL: (Organotin Compounds) TWA 0.1 $mg(Sn)/m^3$
SAFETY PROFILE: An inhalation hazard. Combustible in the form of dust.
PROP: Cubic, gray, crystalline metallic element. Mp: 231.9°, stabilizes @ <18°, d: 7.31, vap press: 1 mm @ 1492°, bp: 2625°.

TGC000 CAS: 7772-99-8 Cl_2Sn HR: 3
TIN(II) CHLORIDE (1:2)
SYNS: C.I. 77864 ♦ STANNOUS CHLORIDE (FCC) ♦ TIN DICHLORIDE
OSHA PEL: TWA 2 $mg(Sn)/m^3$
ACGIH TLV: TWA 2 $mg(Sn)/m^3$
SAFETY PROFILE: Poison by ingestion.
PROP: Colorless or white crystals. D: 2.71, mp: 246°, bp: 623°. Sol in less than its own weight of water; very sol in hydrochloric acid (dilute or conc); sol in alc, ethyl acetate, glacial acetic acid, sodium hydroxide solution.

TGC500 HR: D
TIN COMPOUNDS
OSHA PEL: Organic Compounds: TWA 0.1 $mg(Sn)/m^3$ (skin); Inorganic Compounds (except oxides): TWA 2 mg/m^3
ACGIH TLV: TWA metal, oxide, and inorganic compounds (except SnH_4) as Sn 2 mg/m^3; organic compounds 0.1 mg/m^3 (skin) (Proposed: TWA 0.1 $mg(Sn)/m^3$; STEL 0.2 $mg(Sn)/m^3$ (skin))
DFG MAK: Inorganic 2 mg/m^3, organic 0.1 mg/m^3
NIOSH REL: (Organotin Compounds) TWA 0.1 $mg(Sn)/m^3$
SAFETY PROFILE: Elemental tin and inorganic tin compounds have low toxicity. Some of the organic tin compounds are strong poisons.

TGD100 CAS: 7783-47-3 F_2Sn HR: 3
TIN FLUORIDE
SYNS: STANNOUS FLUORIDE ♦ TIN DIFLUORIDE
OSHA PEL: TWA 2 $mg(Sn)/m^3$; 2.5 $mg(F)/m^3$
ACGIH TLV: TWA 2 $mg(Sn)/m^3$; TWA 2.5 $mg(F)/m^3$; BEI: 3 mg/g creatinine of fluorides in urine prior to shift; 10 mg/g creatinine of fluorides in urine at end of shift.
NIOSH REL: TWA 2.5 $mg(F)/m^3$
SAFETY PROFILE: Poison by ingestion.
PROP: Monoclinic, lamellar plates or colorless hygroscopic prisms. Mp: 213°, d: (25°) 4.57, sublimes @ 7°. Sol in water (about 30%). Forms an oxyfluoride, $SnOF_2$, on exposure to air. Prac insol in MeOH, Et_2O, and $CHCl_3$.

TGG760 CAS: 13463-67-7 O_2Ti HR: 1

TITANIUM DIOXIDE
SYNS: C.I. 77891 ♦ TITANIUM OXIDE
OSHA PEL: TWA Total Dust: 10 mg/m^3; Respirable Fraction: 5 mg/m^3
ACGIH TLV: TWA (nuisance particulate) 10 mg/m^3 of total dust (when toxic impurities are not present, e.g., quartz <1%); Not Classifiable as a Human Carcinogen
DFG MAK: 1.5 mg/m^3
SAFETY PROFILE: A nuisance dust. A human skin irritant.
PROP: White amorphous powder or white solid with very high refractive index. Loses O_2 in air to form $TiO_{1.985}$ which melts at *ca.* 18°. Mp: 1860° (decomp), d: 4.26. Insol in water, hydrochloric acid, dil sulfuric acid, and alc; sol in hot concentrated H_2SO_4 and HF.

TGJ750 CAS: 119-93-7 $C_{14}H_{16}N_2$ HR: 3
o-TOLIDINE
SYNS: 4,4'-BI-o-TOLUIDINE ♦ C.I. 37230 ♦ 3,3'-DIMETHYLBENZIDIN ♦ o-TOLIDIN
ACGIH TLV: Animal Carcinogen
DFG MAK: Animal Carcinogen, Suspected Human Carcinogen
NIOSH REL: (o-Toluidine) CL 0.02 mg/m^3/60M; avoid skin contact
SAFETY PROFILE: Confirmed carcinogen. Moderately toxic by ingestion.
PROP: White to reddish crystals or leaflets. Mp: 129–131°. Very sltly sol in water; sol in alc, ether, acetic acid.

TGK750 CAS: 108-88-3 C_7H_8 HR: 3
TOLUENE
DOT: UN 1294
SYNS: METHYLBENZENE ♦ PHENYLMETHANE ♦ TOLUOL (DOT)
OSHA PEL: TWA 100 ppm; STEL 150 ppm
ACGIH TLV: TWA 50 ppm (skin); Not Classifiable as a Human Carcinogen; BEI: 1.6 g/g creatinine of hippuric acid in urine at end of shift; 0.05 mg/L toluene in venous blood prior to last shift of workweek; (Proposed: BEI: BEI: 0.05 mg/L toluene in venous blood prior to last shift of work ek; 0.5 mg/L o-cresol in urine at end of shift)
DFG MAK: 50 ppm (190 mg/m^3); BAT: 340 µg/dL in blood at end of shift
NIOSH REL: (Toluene) TWA 100 ppm; CL 200 ppm/10M
DOT Classification: 3; Label: Flammable Liquid
SAFETY PROFILE: Mildly toxic by inhalation. An experimental teratogen. A skin and severe eye irritant. A flammable liquid.
PROP: Colorless liquid; benzol-like odor. Mp: –95 to –94.5°, fp: –95°, bp: 110.4°, flash p: 40°F (CC), ULC: 75–80, lel: 1.27%, uel: 7%, d: 0.866 @ 20°/4°, autoign temp: 996°F, vap press: 36.7 mm @ 30°, vap d: 3.14. Insol in water; sol in acetone; misc in abs alc, ether, chloroform.

TGL750 CAS: 95-80-7 $C_7H_{10}N_2$ HR: 3
TOLUENE-2,4-DIAMINE
DOT: UN 1709

SYNS: 3-AMINO-p-TOLUIDINE ♦ C.I. 76035 ♦ 4-METHYL-m-PHENYLENEDIAMINE ♦ MTD ♦ 2,4-TOLUENEDIAMINE (DOT) ♦ 2,4-TOLUYLENEDIAMINE (DOT)
DFG MAK: Animal Carcinogen, Suspected Human Carcinogen
DOT Classification: 6.1; Label: KEEP AWAY FROM FOOD
SAFETY PROFILE: Confirmed carcinogen. A skin and eye irritant.
PROP: Needles from water or prisms from alc. Mp: 99°, bp: 292°, bp: 148–150° @ 8 mm, vap press: 1 mm @ 106.5°.

TGM750 CAS: 584-84-9 $C_9H_6N_2O_2$ HR: 3
TOLUENE-2,4-DIISOCYANATE
DOT: UN 2206/UN 2207/UN 2478/UN 3080
SYNS: 2,4-DIISOCYANATO-1-METHYLBENZENE (9CI) ♦ 4-METHYL-PHENYLENE ISOCYANATE ♦ TDI (OSHA) ♦ 2,4-TOLYLENEDIISOCYANATE
OSHA PEL: TWA 0.005 ppm; STEL 0.02 ppm
ACGIH TLV: TWA 0.005 ppm; STEL 0.02 ppm; Not Classifiable as a Human Carcinogen (Proposed: 0.005 ppm; STEL 0.02 ppm (skin, sensitizer); Not Classifiable as a Human Carcinogen)
DFG MAK: 0.01 ppm (0.072 mg/m^3)
NIOSH REL: (Diisocyanates) TWA 0.005 ppm; CL 0.02 ppm/10M
DOT Classification: 6.1; Label: KEEP AWAY FROM FOOD (UN 2207)
SAFETY PROFILE: Confirmed carcinogen. Poison by ingestion and inhalation. A severe skin and eye irritant. Combustible liquid.
PROP: Clear, faintly yellow liquid; sharp, pungent odor. Mp: 19.5–21.5°, d: (liquid) 1.2244 @ 20°/4°, bp: 124–126° @ 18 mm, flash p: 270°F (OC), vap d: 6.0, lel: 0.9%, uel: 9.5%. Misc with alc (decomp), ether, acetone, carbon tetrachloride, benzene, chlorobenzene, kerosene, olive oil.

TGM800 CAS: 91-08-7 $C_9H_6N_2O_2$ HR: 3
TOLUENE-2,6-DIISOCYANATE
DOT: UN 2207
SYNS: 2,6-DIISOCYANATO-1-METHYLBENZENE ♦ 2-METHYL-m-PHENYLENE ISOCYANATE ♦ 2,6-TDI ♦ 2,6-TOLUENE DIISOCYANATE
ACGIH TLV: TWA 0.005 ppm; STEL 0.02 ppm; Not Classifiable as a Human Carcinogen (Proposed: 0.005 ppm; STEL 0.02 ppm (skin, sensitizer); Not Classifiable as a Human Carcinogen)
DFG MAK: 0.01 ppm (0.072 mg/m^3)
NIOSH REL: (Diisocyanates) TWA 0.005 ppm; CL 0.02 ppm/10M
DOT Classification: 6.1; Label: KEEP AWAY FROM FOOD (UN 2207); DOT Class: 6.1; Label: Poison (UN 2206); DOT Class: 6.1; Label: Poison, Flammable Liquid (UN 3080); DOT Class: 3; Label: Flammable Liquid, Poison (UN 2478)
SAFETY PROFILE: Confirmed carcinogen. Poison by ingestion and inhalation. Flammable liquid.
PROP: A liquid. Bp: 129–133° @ 18 mm.

TGQ500 CAS: 108-44-1 C_7H_9N HR: 3
m-TOLUIDINE

DOT: UN 1708
SYNS: 3-AMINOPHENYLMETHANE ♦ m-METHYLANILINE ♦ 3-METHYLBENZENAMINE ♦ m-TOLYLAMINE
OSHA PEL: TWA 2 ppm (skin)
ACGIH TLV: TWA 2 ppm (skin); Not Classifiable as a Human Carcinogen
DOT Classification: 6.1; Label: Poison
SAFETY PROFILE: Poison by ingestion. A skin and eye irritant. Flammable liquid.
PROP: Colorless liquid. Mp: –43.6°, bp: 203.3°, d: 0.989 @ 20°/4°, vap press: 1 mm @ 41°, vap d: 3.90. Sltly sol in water; sol in alc, ether.

TGQ750 CAS: 95-53-4 C_7H_9N HR: 3
o-TOLUIDINE
DOT: UN 1708
SYNS: 1-AMINO-2-METHYLBENZENE ♦ o-AMINOTOLUENE ♦ C.I. 37077 ♦ 2-METHYL-1-AMINOBENZENE ♦ o-METHYLBENZENAMINE
OSHA PEL: TWA 5 ppm (skin)
ACGIH TLV: TWA 2 ppm (skin); Animal Carcinogen
DFG MAK: Animal Carcinogen, Suspected Human Carcinogen
DOT Classification: 6.1; Label: Poison
SAFETY PROFILE: Confirmed carcinogen. Poison by ingestion. Moderately toxic by skin contact. An experimental teratogen. A skin and severe eye irritant. A flammable liquid.
PROP: Colorless liquid. Mp: –16.3° (dimorph), bp: 200–202°, ULC: 20–25, flash p: 185° (CC), d: 1.004 @ 20°/4°, autoign temp: 900°F, vap press: 1 mm @ 44°, vap d: 3.69. Sltly sol in water, dil acid; sol in alc and ether.

TGR000 CAS: 106-49-0 C_7H_9N HR: 3
p-TOLUIDINE
DOT: UN 1708
SYNS: 4-AMINO-1-METHYLBENZENE ♦ p-AMINOTOLUENE ♦ C.I. 37107 ♦ 4-METHYLANILINE
OSHA PEL: TWA 2 ppm (skin)
ACGIH TLV: TWA 2 ppm (skin); Animal Carcinogen
DFG MAK: Confirmed Animal Carcinogen with Unknown Relevance to Humans
DOT Classification: 6.1; Label: Poison
SAFETY PROFILE: Confirmed carcinogen. Poison by ingestion. A severe skin and eye irritant. Flammable.
PROP: Colorless leaflets. Mp: 44.5° (anhyd), mp: 42°, bp: 82.2° @ 10 mm, flash p: 188°F (CC), d: 1.046 @ 20°/4°, autoign temp: 900°F, vap press: 1 mm @ 42°, vap d: 3.90. Sol in water, dil acid, CS_2; very sol in alc, ether.

THZ000 CAS: 2155-70-6 $C_{16}H_{32}O_2Sn$ HR: 3
TRIBUTYL(METHACRYLOXY)STANNANE
SYNS: TRIBUTYLSTANNYL METHACRYLATE ♦ TRIBUTYLTIN METHACRYLATE
OSHA PEL: TWA 0.1 mg(Sn)/m^3 (skin)

ACGIH TLV: TWA 0.1 mg(Sn)/m^3 (skin) (Proposed: TWA 0.1 mg(Sn)/m^3; STEL 0.2 mg(Sn)/m^3 (skin))
DFG MAK: 0.0021 ppm (0.05 mg/m^3)
NIOSH REL: (Organotin Compounds) TWA 0.1 mg(Sn)/m^3
SAFETY PROFILE: Poison by ingestion.

TIA250 CAS: 126-73-8 $C_{12}H_{27}O_4P$ HR: 3
TRIBUTYL PHOSPHATE
SYNS: TBP ♦ TRI-n-BUTYL PHOSPHATE
OSHA PEL: TWA 0.2 ppm
ACGIH TLV: TWA 0.2 ppm
SAFETY PROFILE: Moderately toxic by ingestion and inhalation. A skin, eye, and mucous membrane irritant. Combustible.
PROP: Colorless odorless liquid. Bp: 289° (decomp), mp: <–80°, flash p: 295°F (COC), d: 0.982 @ 20°, vap d: 9.20. Sol in water; misc in alc and ether.

TII250 CAS: 76-03-9 $C_2HCl_3O_2$ HR: 3
TRICHLOROACETIC ACID
DOT: UN 1839/UN 2564
SYNS: SODIUM TCA SOLUTION ♦ TCA ♦ TRICHLOROETHANOIC ACID
OSHA PEL: TWA 1 ppm
ACGIH TLV: TWA 1 ppm; Confirmed Animal Carcinogen with Unknown Relevance to Humans
DOT Classification: 8; Label: Corrosive
SAFETY PROFILE: Poison by ingestion. A corrosive irritant to skin, eyes, and mucous membranes.
PROP: Colorless, rhombic, deliq crystals. Bp: 197.5°, fp: 57.7°, flash p: none, mp: 57–58°, d: 1.6298 @ 61°/4°, vap press: 1 mm @ 51.0°. Sol in water and alc.

TIK250 CAS: 120-82-1 $C_6H_3Cl_3$ HR: 3
1,2,4-TRICHLOROBENZENE
OSHA PEL: CL 5 ppm
ACGIH TLV: CL 5 ppm
DFG MAK: Confirmed Animal Carcinogen with Unknown Relevance to Humans
SAFETY PROFILE: Poison by ingestion. An experimental teratogen. A skin irritant. Combustible.
PROP: Colorless liquid. Mp: 17°, bp: 213°, flash p: 230°F (CC), d: 1.454 @ 25°/25°, vap press: 1 mm @ 38.4°, vap d: 6.26. Sol in water.

TIL360 CAS: 2431-50-7 $C_4H_5Cl_3$ HR: 3
2,3,4-TRICHLOROBUTENE-1
DFG MAK: Animal Carcinogen; Suspected Human Carcinogen
SAFETY PROFILE: Confirmed carcinogen. Poison by ingestion.

TIN000 CAS: 79-00-5 $C_2H_3Cl_3$ HR: 3
1,1,2-TRICHLOROETHANE
SYNS: ETHANE TRICHLORIDE ♦ VINYL TRICHLORIDE
OSHA PEL: TWA 10 ppm (skin)
ACGIH TLV: TWA 10 ppm (skin); Not Classifiable as a Human Carcinogen
DFG MAK: 10 ppm (55 mg/m^3); Confirmed Animal Carcinogen with Unknown Relevance to Humans
SAFETY PROFILE: Suspected carcinogen. Poison by ingestion. Moderately toxic by inhalation and skin contact. An eye and severe skin irritant.
PROP: Nonflammable, mobile liquid with a pleasant odor. Bp: 114°, fp: –35°, d: 1.4416 @ 20°/4°, vap press: 40 mm @ 35.2°. Insol in H_2O; misc in most org solvs.

TIO750 CAS: 79-01-6 C_2HCl_3 HR: 3
TRICHLOROETHYLENE
DOT: UN 1710
SYNS: ACETYLENE TRICHLORIDE ♦ 1-CHLORO-2,2-DICHLOROETHYLENE ♦ 1,1-DICHLORO-2-CHLOROETHYLENE ♦ 1,2,2-TRICHLOROETHYLENE
OSHA PEL: TWA 50 ppm; STEL 200 ppm
ACGIH TLV: TWA 50 ppm; STEL 200 ppm (Proposed: TWA 50 ppm; 100 STEL; Not Suspected as a Human Carcinogen); BEI: 100 mg(trichloroacetic acid)/g creatinine in urine at end of workweek
DFG MAK: Confirmed Human Carcinogen; BAT: 500 µg/dL in blood at end of shift or workweek
NIOSH REL: (Trichloroethylene) TWA 250 ppm; (Waste Anesthetic Gases) CL 2 ppm/1H
DOT Classification: 6.1; Label: KEEP AWAY FROM FOOD
SAFETY PROFILE: Confirmed carcinogen. An experimental teratogen. Mildly toxic to humans by ingestion and inhalation. An eye and severe skin irritant. Nonflammable.
PROP: Clear, colorless, nonflammable, mobile liquid; characteristic sweet odor of chloroform. D: 1.4649 @ 20°/4°, bp: 86.7°, mp: –84°, fp: –86.8°, autoign temp: 788°F, vap press: 100 mm @ 32°, vap d: 4.53, refr index: 1.477 @ 20°. Immisc with water; misc with alc, ether, acetone, carbon tetrachloride. Insol in H_2O; sol in most org solvs.

TIP500 CAS: 75-69-4 CCl_3F HR: 2
TRICHLOROFLUOROMETHANE
SYNS: FLUOROTRICHLOROMETHANE (OSHA) ♦ FREON 11 ♦ MONOFLUOROTRICHLOROMETHANE
OSHA PEL: CL 1000 ppm
ACGIH TLV: CL 1000 ppm; Not Classifiable as a Human Carcinogen
DFG MAK: 1000 ppm (5700 mg/m^3)
SAFETY PROFILE: High concentrations cause narcosis and anesthesia in humans.
PROP: Colorless liquid. Mp: –111°, bp: 24.1°, d: 1.484 @ 17.2°.

TIT500 CAS: 1321-65-9 $C_{10}H_5Cl_3$ HR: 3
TRICHLORONAPHTHALENE

OSHA PEL: TWA 5 mg/m^3 (skin)
ACGIH TLV: TWA 5 mg/m^3 (skin)
DFG MAK: 5 mg/m^3
SAFETY PROFILE: A poison. The chlorinated naphthalenes have toxic effects on the skin and liver.
PROP: A white solid.

TJB600 CAS: 96-18-4 $C_3H_5Cl_3$ HR: 3
1,2,3-TRICHLOROPROPANE
SYNS: ALLYL TRICHLORIDE ♦ GLYCEROL TRICHLOROHYDRIN ♦ TRICHLOROHYDRIN
OSHA PEL: TWA 10 ppm
ACGIH TLV: TWA 10 ppm (skin); Animal Carcinogen
DFG MAK: Confirmed Animal Carcinogen, Suspected Human Carcinogen
SAFETY PROFILE: Confirmed carcinogen. Poison by ingestion. Moderately toxic by inhalation and skin contact. A skin and severe eye irritant. Moderately flammable.
PROP: Bp: 158°, d: 1.414 @ 20°/20°, flash p: 180°F (OC).

TJO000 CAS: 121-44-8 $C_6H_{15}N$ HR: 3
TRIETHYLAMINE
DOT: UN 1296
SYNS: N,N-DIETHYLETHANAMINE ♦ TEN
OSHA PEL: TWA 10 ppm; STEL 15 ppm
ACGIH TLV: TWA 10 ppm; STEL 15 ppm (Proposed: 1 ppm; STEL 5 ppm)
DFG MAK: 1 ppm (4.2 mg/m^3)
DOT Classification: 3; Label: Flammable Liquid
SAFETY PROFILE: Moderately toxic by ingestion and skin contact. Mildly toxic by inhalation. A skin and severe eye irritant. A very dangerous fire hazard.
PROP: Colorless liquid with fishy or ammonia odor. Mp: –114.8°, bp: 89.5°, flash p: 20°F (OC), d: 0.7255 @ 25°/4°, vap d: 3.48, lel: 1.2%, uel: 8.0%. Misc in water, alc, ether.

TJY100 CAS: 75-63-8 $CBrF_3$ HR: 1
TRIFLUOROBROMOMETHANE
DOT: UN 1009
SYNS: BROMOFLUOROFORM ♦ BROMOTRIFLUOROMETHANE ♦ FREON 13B1 ♦ R13B1 (DOT) ♦ TRIFLUOROMONOBROMOMETHANE
OSHA PEL: TWA 1000 ppm
ACGIH TLV: TWA 1000 ppm
DFG MAK: 1000 ppm (6200 mg/m^3)
DOT Classification: 2.2; Label: Nonflammable Gas
SAFETY PROFILE: Mildly toxic by inhalation.
PROP: A gas. D: 1.58, fp: –168°, bp: –57.8°.

TKB250 CAS: 406-90-6 $C_4H_5F_3O$ HR: 3

2,2,2-TRIFLUOROETHYL VINYL ETHER
SYNS: (2,2,2-TRIFLUOROETHOXY)ETHENE
NIOSH REL: (Waste Anesthetic Gases) CL 2 ppm/1H
SAFETY PROFILE: Poison by inhalation. An experimental teratogen.
PROP: A liquid. D: 1.14 @ 20°/4°, bp: 42.5° @ 751 mm.

TKV000 CAS: 552-30-7 $C_9H_4O_5$ HR: 2
TRIMELLITIC ANHYDRIDE
SYNS: 1,2,4-BENZENETRICARBOXYLIC ACID ANHYDRIDE ♦ 4-CARBOXYPHTHALIC ANHYDRIDE ♦ TMA ♦ TRIMELLIC ACID ANHYDRIDE
OSHA PEL: TWA 0.005 ppm
ACGIH TLV: TWA 0.005 ppm (Proposed: CL 0.04 mg/m^3)
DFG MAK: 0.04 mg/m^3
NIOSH REL: (Trimellitic Anhydride): handle as extremely toxic
SAFETY PROFILE: Moderately toxic by ingestion. Irritant to lungs and air passages.
PROP: Crystals or needles. Mp: 162°, bp: 240–245° @ 14 mm. Sol in acetone, ethyl acetate, dimethylformamide.

TLD500 CAS: 75-50-3 C_3H_9N HR: 3
TRIMETHYLAMINE
DOT: UN 1083/UN 1297
SYNS: TMA
OSHA PEL: TWA 10 ppm; STEL 15 ppm
ACGIH TLV: TWA 5 ppm; STEL 15 ppm
DOT Classification: 2.1; Label: Flammable Gas (UN 1083); DOT Class: 3; Label: Flammable Liquid (UN 1297)
SAFETY PROFILE: Mildly toxic by inhalation. A very dangerous fire hazard.
PROP: Volatile liquid with fishy odor, or colorless gas with pungent, ammonia-like odor; saline taste. Bp: 2.87°, lel: 2%, uel: 11.6%, fp: –117.1°, mp: –117.2°, d: 0.662 @ –5°, autoign temp: 374°F, vap d: 2.0, flash p: 20°F (CC). Misc with alc; sol in ether, benzene, toluene, xylene, chloroform.

TLG250 CAS: 137-17-7 $C_9H_{13}N$ HR: 3
2,4,5-TRIMETHYLANILINE
SYNS: 1-AMINO-2,4,5-TRIMETHYLBENZENE ♦ PSEUDOCUMIDINE ♦ 1,2,4-TRIMETHYL-5-AMINOBENZENE ♦ 2,4,5-TRIMETHYLBENZENAMINE
DFG MAK: Animal Carcinogen, Suspected Human Carcinogen
SAFETY PROFILE: Confirmed carcinogen. Moderately toxic by ingestion.
PROP: Needles in H_2O. Mp: 68°, bp: 234–235°.

TLL250 CAS: 25551-13-7 C_9H_{12} HR: 3
TRIMETHYL BENZENE
OSHA PEL: TWA 25 ppm
ACGIH TLV: TWA 25 ppm
SAFETY PROFILE: Mildly toxic by ingestion. A skin and eye irritant. Flammable.

TLY500 CAS: 540-84-1 C_8H_{18} HR: 3

2,2,4-TRIMETHYLPENTANE

SYNS: ISOBUTYLTRIMETHYLETHANE ♦ ISOOCTANE (DOT)

NIOSH REL: TWA (Alkanes) 350 mg/m^3

SAFETY PROFILE: High concentrations can cause narcosis. A very dangerous fire hazard.

PROP: Clear liquid; odor of gasoline. Bp: 99.2°, mp: –107.5°, fp: –116°, flash p: 10°F, d: 0.692 @ 20°/4°, autoign temp: 779°F, vap press: 40.6 mm @ 21°, vap d: 3.93, lel: 1.1%, uel: 6.0%.

TMD250 CAS: 512-56-1 $C_3H_9O_4P$ HR: 3

TRIMETHYL PHOSPHATE

SYNS: METHYL PHOSPHATE ♦ PHOSPHORIC ACID, TRIMETHYL ESTER ♦ TMP

DFG MAK: Confirmed Animal Carcinogen with Unknown Relevance to Humans

SAFETY PROFILE: Suspected carcinogen. An experimental teratogen. Moderately toxic by ingestion and skin contact.

PROP: Pleasant smelling liquid. D: 1.97 @ 19.5°/0°, bp: 197.2°. Sol in alc, water, ether, and org solvs.

TMD500 CAS: 121-45-9 $C_3H_9O_3P$ HR: 3

TRIMETHYL PHOSPHITE

DOT: UN 2329

SYNS: METHYL PHOSPHITE ♦ PHOSPHORUS ACID, TRIMETHYL ESTER

OSHA PEL: TWA 2 ppm

ACGIH TLV: TWA 2 ppm

DOT Classification: 3; Label: Flammable Liquid

SAFETY PROFILE: Moderately toxic by ingestion and skin contact. An experimental teratogen. A severe skin and eye irritant. Flammable liquid.

PROP: Air-sensitive, colorless liquid with powerful sickly odor. D: 1.046 @ 20°/4°, mp: –78°, bp: 111–112°, vap d: 4.3, bp: 232–234°F, flash p: 130°F (OC). Insol in water; sol in hexane, benzene, acetone, alc, ether, carbon tetrachloride, kerosene.

TMM250 CAS: 129-79-3 $C_{13}H_5N_3O_7$ HR: 3

2,4,7-TRINITROFLUOREN-9-ONE

SYNS: 2,4,7-TRINITROFLUORENONE (MAK)

DFG MAK: Confirmed Animal Carcinogen with Unknown Relevance to Humans

SAFETY PROFILE: Suspected carcinogen. Mildly toxic by ingestion. A skin and eye irritant.

PROP: Pale-yellow needles from AcOH or C_6H_6. Mp: 176°.

TMN000 CAS: 75321-19-6 $C_{16}H_7N_3O_6$ HR: 3

1,3,6-TRINITROPYRENE

DFG MAK: Confirmed Animal Carcinogen with Unknown Relevance to Humans

SAFETY PROFILE: Suspected carcinogen.

TMN490 CAS: 118-96-7 $C_7H_5N_3O_6$ HR: 3
2,4,6-TRINITROTOLUENE
DOT: UN 0209/UN 1356
SYNS: TNT (OSHA) ♦ TRINITROTOLUENE
OSHA PEL: TWA 0.5 mg/m^3 (skin)
ACGIH TLV: TWA 0.1 ppm
DFG MAK: 0.011 ppm (0.1 mg/m^3); Confirmed Animal Carcinogen with Unknown Relevance to Humans
DOT Classification: EXPLOSIVE 1.1D; Label: EXPLOSIVE 1.1D (UN 0209); DOT Class: 4.1; Label: Flammable Solid (UN 1356)
SAFETY PROFILE: Suspected carcinogen. Moderately toxic by ingestion. A skin irritant. Flammable or explosive when exposed to heat or flame.
PROP: Colorless, monoclinic, rhombohedral crystals from EtOH. Mp: 82°, bp: 240° (explodes), flash p: explodes, d: 1.654. Sol in hot water, alc, ether.

TMO600 CAS: 78-30-8 $C_{21}H_{21}O_4P$ HR: 3
TRIORTHOCRESYL PHOSPHATE
SYNS: o-CRESYL PHOSPHATE ♦ TOCP ♦ TOFK ♦ o-TOLYL PHOSPHATE ♦ TOTP ♦ TRI-o-CRESYL PHOSPHATE ♦ TRI-o-TOLYL PHOSPHATE
OSHA PEL: TWA 0.1 mg/m^3 (skin)
ACGIH TLV: TWA 0.1 mg/m^3 (skin); Not Classifiable as a Human Carcinogen
SAFETY PROFILE: Moderately toxic by ingestion. Combustible.
PROP: Colorless liquid. Mp: –25 to –30°, bp: 410° (slt decomp), flash p: 437°F, d: 1.17, autoign temp: 725°F, vap d: 12.7. Insol in water; sol in alc and ether.

TMQ500 CAS: 603-34-9 $C_{18}H_{15}N$ HR: 2
TRIPHENYLAMINE
SYN: N,N-DIPHENYLANILINE
OSHA PEL: TWA 5 mg/m^3
ACGIH TLV: TWA 5 mg/m^3
SAFETY PROFILE: Moderately toxic by ingestion.
PROP: Monoclinic crystals from EtOAc. D: 0.774 @ 0°/0°, mp: 127°, bp: 195–205° @ 10–22 mm.

TMT750 CAS: 115-86-6 $C_{18}H_{15}O_4P$ HR: 3
TRIPHENYL PHOSPHATE
SYNS: PHOSPHORIC ACID, TRIPHENYL ESTER ♦ TPP
OSHA PEL: TWA 3 mg/m^3
ACGIH TLV: TWA 3 mg/m^3; Not Classifiable as a Human Carcinogen
SAFETY PROFILE: Moderately toxic by ingestion. Combustible.
PROP: Colorless, odorless, crystalline solid or prisms from EtOH or EtOH/pet ether. Mp: 49–50°, bp: 245° @ 11 mm, flash p: 428°F (CC), d: 1.268 @ 60°, vap press: 1 mm @ 193.5°. Insol in water; sol in alc, benzene, ether, chloroform, and acetone.

TMX500 CAS: 1317-95-9 HR: 3
TRIPOLI
OSHA PEL: TWA 0.1 mg/m^3
ACGIH TLV: TWA 0.1 mg/m^3 (of contained respirable quartz dust)
NIOSH REL: (Silica, Crystalline) 10H TWA 0.05 mg/m^3
SAFETY PROFILE: The prolonged inhalation of dusts containing free silica may result in the development of a disabling pulmonary fibrosis known as silicosis.
PROP: Finely granulated white or gray siliceous rock. A form of crystalline silica.

TOA750 CAS: 7440-33-7 W HR: 3
TUNGSTEN
SYN: WOLFRAM
OSHA PEL: TWA (Insoluble compounds) 5 $mg(W)/m^3$; STEL 10 $mg(W)/m^3$; (soluble compounds) 1 $mg(W)/m^3$; STEL 3 $mg(W)/m^3$
ACGIH TLV: TWA (Insoluble compounds) 5 $mg(W)/m^3$; STEL 10 $mg(W)/m^3$; (soluble compounds) 1 $mg(W)/m^3$; STEL 3 $mg(W)/m^3$
NIOSH REL: (Tungsten, Insoluble) TWA 5 $mg(W)/m^3$
SAFETY PROFILE: An inhalation hazard. An experimental teratogen. A skin and eye irritant. Flammable in the form of dust.
PROP: A steely-gray to white, cuttable, forgeable, and spinnable metallic element. Fairly soft when pure. Mp: 3410°, d: 19.3 @ 20°, bp: 5900°.

TOC500 HR: 2
TUNGSTEN COMPOUNDS
OSHA PEL: TWA (insoluble compounds) 5 $mg(W)/m^3$; STEL 10 $mg(W)/m^3$; (soluble compounds) 1 $mg(W)/m^3$; STEL 3 $mg(W)/m^3$
ACGIH TLV: TWA (insoluble compounds) 5 $mg(W)/m^3$; STEL 10 $mg(W)/m^3$; (soluble compounds) 1 $mg(W)/m^3$; STEL 3 $mg(W)/m^3$
SAFETY PROFILE: Tungsten compounds are considered somewhat more toxic than those of molybdenum. Sodium tungstate (Na_2WO_4), the most soluble salt, is moderately toxic by ingestion.

TOD750 CAS: 8006-64-2 HR: 3
TURPENTINE
DOT: UN 1299/UN 1300
SYNS: OIL of TURPENTINE ♦ SPIRIT of TURPENTINE ♦ TURPENTINE STEAM DISTILLED
OSHA PEL: TWA 100 ppm
ACGIH TLV: TWA 100 ppm; (Proposed: TWA 100 ppm (sensitizer))
DFG MAK: 100 ppm (560 mg/m^3)
DOT Classification: 3; Label: Flammable Liquid

SAFETY PROFILE: Moderately toxic to humans by ingestion. A skin ans eye irritant. A very dangerous fire hazard.
PROP: Colorless liquid; characteristic odor. Bp: 154–170°, lel: 0.8%, flash p: 95°F (CC), d: 0.854–0.868 @ 25°/25°, autoign temp: 488°F, vap d: 4.84, ULC: 40–50.

UNS000 CAS: 7440-61-1 U HR: 3
URANIUM
DOT: UN 2979
OSHA PEL: TWA Soluble Compounds: 0.05 $mg(U)/m^3$; Insoluble Compounds 0.2 $mg(U)/m^3$; STEL 0.6 $mg(U)/m^3$
ACGIH TLV: TWA 0.2 $mg(U)/m^3$; STEL 0.6 $mg(U)/m^3$
DFG MAK: 0.25 mg/m^3
DOT Classification: 7; Label: RADIOACTIVE, SPONT Combustible
SAFETY PROFILE: A highly toxic element on an acute basis. The permissible levels for soluble compounds are based on chemical toxicity, whereas the permissible body level for insoluble compounds is based on radiotoxicity. A very dangerous fire hazard in the form of a solid or dust.
PROP: A heavy, silvery-white, malleable, ductile, softer-than-steel, metallic element. Tarnishes in air. þ&□- and β-forms are brittle, the γ-form softer and more malleable. Mp: 1132°, bp: 3818°, d: 18.95. Radioactive material.

UVA000 CAS: 51-79-6 $C_3H_7NO_2$ HR: 3
URETHANE
SYNS: ETHYL CARBAMATE ♦ ETHYL URETHANE
DFG MAK: Animal Carcinogen, Suspected Human Carcinogen
SAFETY PROFILE: Confirmed carcinogen. Moderately toxic by ingestion. An experimental teratogen.
PROP: Colorless, odorless crystals, prisms from C_6H_6 or toluene with cooling, saline taste. Mp: 49°, bp: 103° @ 54 mm, d: 1.107, vap press: 10 mm @ 77.8°, vap d: 3.07. Very sol in H_2O, EtOH, Et_2O, $CHCl_3$, and C_6H_6; spar sol in ligroin.

VAG000 CAS: 110-62-3 $C_5H_{10}O$ HR: 3
n-VALERALDEHYDE
DOT: UN 2058
SYNS: BUTYL FORMAL ♦ n-PENTANAL ♦ VALERIC ACID ALDEHYDE ♦ VALERIC ALDEHYDE
OSHA PEL: TWA 50 ppm
ACGIH TLV: TWA 50 ppm
DOT Classification: 3; Label: Flammable Liquid
SAFETY PROFILE: Moderately toxic by ingestion. Mildly toxic by inhalation and skin contact. A severe eye and skin irritant. A very dangerous fire hazard when exposed to heat or flame.
PROP: Liquid. Flash p: 53.6°F, fp: –92°, bp: 102–103°, d: 0.8095 @ 20°/4°. Very sltly sol in water; misc with org solvs.

VCP000 CAS: 7440-62-2 V HR: 3
VANADIUM
OSHA PEL: Respirable Dust and Fume: TWA 0.05 mg(V_2O_5)/m^3
NIOSH REL: TWA 1.0 mg(V)/m^3
SAFETY PROFILE: An inhalation hazard. Flammable in dust form from heat, flame, or sparks.
PROP: A bright, white, soft, ductile metal; sltly radioactive. Corrosion resistant (oxide film). Resistant to fused alkalies, attacked by hot concentrated mineral acids. Bp: 3380°, d: 6.11 @ 18.7°, mp: 1917°. Insol in water.

VCZ000 HR: D
VANADIUM COMPOUNDS
NIOSH REL: (Vanadium Compounds) CL 0.05 mg(V)/m^3/15M
SAFETY PROFILE: Variable toxicity. Vanadium compounds act chiefly as an irritant to the conjunctiva and respiratory tract. The absorption of V_2O_5 by inhalation is nearly 100%. VF_5 and the oxyhalogenides of pentavalent vanadium (VOF_3, $VOCl_3$, $VOBr_3$) are volatile. The fumes are highly toxic.

VDP000 CAS: 7727-18-6 Cl_3OV HR: 3
VANADIUM OXYTRICHLORIDE
DOT: UN 2443
SYNS: TRICHLOROOXOVANADIUM ♦ VANADIUM TRICHLORIDE OXIDE
ACGIH TLV: TWA 0.05 mg(V_2O_5)/m^3
NIOSH REL: (Vanadium Compounds) CL 0.05 mg(V)/m^3/15M
DOT Classification: 8; Label: Corrosive
SAFETY PROFILE: Poison by ingestion. A corrosive irritant to skin, eyes, and mucous membranes. Violently hygroscopic.
PROP: Lemon yellow liquid, freezing to deep orange solid. Mp: −77°, bp: 126.7°, d: 1.811 @ 32°.

VDU000 CAS: 1314-62-1 O_5V_2 HR: 3
VANADIUM PENTOXIDE (dust)
DOT: UN 2862
SYNS: C.I. 77938 ♦ VANADIC ANHYDRIDE
OSHA PEL: Respirable Dust and Fume: TWA 0.05 mg(V_2O_5)/m^3
ACGIH TLV: TWA 0.05 mg(V_2O_5)/m^3; Not Classifiable as a Human Carcinogen; BEI: 50 µg/g creatinine of vanadium in urine at end of shift at end of workweek.
DFG MAK: (fine dust) 0.05 mg/m^3
NIOSH REL: (Vanadium Compounds) CL 0.05 mg(V)/m^3/15M
DOT Classification: 6.1; Label: Poison
SAFETY PROFILE: Poison by ingestion and inhalation. An experimental teratogen.

PROP: Yellow to red crystalline powder or orange solid. Loses oxygen reversibly on heating. Amphoteric. Mild oxidizing agent. Mp: 677°, bp: decomp @ 1750°, d: 3.357 @ 18°. Insol.

VEA000 CAS: 1314-34-7 O_3V_2 HR: 3

VANADIUM SESQUIOXIDE

SYNS: VANADIC OXIDE ♦ VANADIUM OXIDE ♦ VANADIUM TRIOXIDE

ACGIH TLV: TWA 0.05 mg(V_2O_5)/m^3

NIOSH REL: (Vanadium Compound) CL 0.05 mg(V)/m^3/15M

SAFETY PROFILE: Poison by ingestion. Ignites when heated in air.

PROP: Black crystals or solid. Not amphoteric. Shows unusual electrical properties with a tenfold change in resistance between 225 and 4°. There is a metal-insulator transition at 155K. Mp: 1970°, d: 4.87 @ 18°.

VEF000 CAS: 7632-51-1 Cl_4V HR: 3

VANADIUM TETRACHLORIDE

DOT: UN 2444

SYN: VANADIUM CHLORIDE

ACGIH TLV: TWA 0.05 mg(V_2O_5)/m^3

NIOSH REL: (Vanadium Compounds) CL 0.05 mg(V)/m^3/15M

DOT Classification: 6.1; Label: Poison

SAFETY PROFILE: Poison by ingestion. A corrosive irritant to skin, eyes, and mucous membranes.

PROP: Reddish-brown liquid. Readily hydrolyzes; decomp slowly to $VCl_3 + Cl_2$ at room temp. Mp: –28°, bp: 148.5°, d: 1.816 @ 30°. Sol in CCl_4 and donor solvs.

VEP000 CAS: 7718-98-1 Cl_3V HR: 3

VANADIUM TRICHLORIDE

DOT: UN 2475

ACGIH TLV: TWA 0.05 mg(V_2O_5)/m^3

NIOSH REL: (Vanadium Compounds) CL 0.05 mg(V)/m^3/15M

DOT Classification: 8; Label: Corrosive

SAFETY PROFILE: Poison by ingestion. A corrosive irritant to skin, eyes, and mucous membranes.

PROP: Pink crystals or violet very hygroscopic solid; disproportionates @ >4° to VCl_4 + VCl_2. Mp: decomp, d: 3.00 @ 18°. Sol in aq HCl and most donor solvs.

VEZ000 CAS: 27774-13-6 O_5SV HR: 3

VANADYL SULFATE

DOT: UN 2931

SYNS: C.I. 77940

ACGIH TLV: TWA 0.05 mg(V_2O_5)/m^3

NIOSH REL: (Vanadium Compounds) CL 0.05 mg(V)/m^3/15M
DOT Classification: 6.1; Label: Poison
SAFETY PROFILE: A poison and an inhalation hazard.
PROP: Blue crystals or solid.

VLU250 CAS: 108-05-4 $C_4H_6O_2$ HR: 3
VINYL ACETATE
DOT: UN 1301
SYNS: ACETIC ACID VINYL ESTER ♦ VAC
OSHA PEL: TWA 10 ppm; STEL 20 ppm
ACGIH TLV: 10 ppm, STEL: 15 ppm; Animal Carcinogen
DFG MAK: 10 ppm (35 mg/m^3); Confirmed Animal Carcinogen with Unknown Relevance to Humans
NIOSH REL: (Vinyl Acetate) CL 15 mg/m^3/15M
DOT Classification: 3; Label: Flammable Liquid
SAFETY PROFILE: Confirmed carcinogen. Moderately toxic by ingestion and inhalation. A skin and eye irritant. Highly dangerous fire hazard. It can react violently with many substances.
PROP: Colorless, mobile liquid; polymerizes to solid on exposure to light. Mp: –92.8°, fp: –100°, bp: 73°, flash p: 18°F, d: 0.9335 @ 20°, autoign temp: 800°F, vap press: 100 mm @ 21.5°, lel: 2.6%, uel: 13.4%, vap d: 3.0. Misc in alc, ether. Somewhat sol in water.

VMP000 CAS: 593-60-2 C_2H_3Br HR: 3
VINYL BROMIDE
DOT: UN 1085
SYNS: BROMOETHENE ♦ BROMOETHYLENE
OSHA PEL: TWA 5 ppm
ACGIH TLV: TWA 0.5 ppm; Suspected Human Carcinogen
DFG MAK: Human Carcinogen
NIOSH REL: (Vinyl Bromide) Lowest Detectable Level
DOT Classification: 2.1; Label: Flammable Gas
SAFETY PROFILE: Confirmed carcinogen. Moderately toxic by ingestion. A very dangerous fire hazard.
PROP: A gas or liquid. Mp: –138°, bp: 15.6°, d: 1.51. Insol in water; misc in alc, ether.

VNP000 CAS: 75-01-4 C_2H_3Cl HR: 3
VINYL CHLORIDE
DOT: UN 1086
SYNS: CHLORETHENE ♦ CHLORETHYLENE ♦ CHLOROETHENE ♦ ETHYLENE MONOCHLORIDE ♦ MONOCHLOROETHYLENE (DOT) ♦ VCM ♦ VINYL CHLORIDE MONOMER ♦ VINYL C MONOMER
OSHA PEL: Cancer Suspect Agent
ACGIH TLV: TWA 1 ppm; Confirmed Human Carcinogen
DFG TRK: Confirmed Human Carcinogen
NIOSH REL: (Vinyl Chloride) Lowest Detectable Level
DOT Classification: 2.1; Label: Flammable Gas

SAFETY PROFILE: Confirmed human carcinogen. Moderately toxic by ingestion. Experimental teratogenic data. A severe irritant to skin, eyes, and mucous membranes. A very dangerous fire hazard.
PROP: Colorless liquid or gas (when inhibited); faintly sweet odor. Mp: –160°, bp: –13.9°, lel: 4%, uel: 22%, flash p: 17.6°F (COC), fp: –159.7°, d (liquid): 0.9195 @ 15°/4°, vap press: 2600 mm @ 25°, vap d: 2.15, autoign temp: 882°F. Sltly sol in water; sol in alc; very sol in ether.

VOA000 CAS: 106-87-6 $C_8H_{12}O_2$ HR: 3
VINYL CYCLOHEXENE DIOXIDE
SYNS: 1,2-EPOXY-4-(EPOXYETHYL)CYCLOHEXANE ♦ 4-(1,2-EPOXYETHYL)-7-OXABICYCLO(4.1.0)HEPTANE ♦ 1-ETHYLENEOXY-3,4-EPOXYCYCLOHEXANE ♦ VINYL CYCLOHEXENE DIEPOXIDE ♦ 4-VINYL-1-CYCLOHEXENE DIOXIDE (MAK)
OSHA PEL: TWA 10 ppm (skin)
ACGIH TLV: TWA 0.1 ppm; Animal Carcinogen
DFG MAK: Animal Carcinogen, Suspected Human Carcinogen
SAFETY PROFILE: Confirmed carcinogen. Moderately toxic by ingestion and skin contact. Mildly toxic by inhalation. A severe skin irritant. Combustible.
PROP: Colorless liquid. D: 1.098 @ 20°/20°, bp: 227°, flash p: 230°F, mp: –55°. Very sol in water.

VPA000 CAS: 75-02-5 $CH_2:CHF$ HR: 3
VINYL FLUORIDE
DOT: UN 1860
SYNS: FLUOROETHENE ♦ FLUOROETHYLENE ♦ MONOFLUOROETHYLENE
ACGIH TLV: TWA 1 ppm; Suspected Human Carcinogen
NIOSH REL: (Vinyl Chloride) TWA 1 ppm; CL 5 ppm/15M
DOT Classification: 2.1; Label: Flammable Gas
SAFETY PROFILE: A poison. A very dangerous fire hazard.
PROP: Colorless gas. Mp: –160.5°, bp: –51°, fp: –160.5°, lel: 2.6%, uel: 21.7%. Insol in water; sol in alc, ether.

VPK000 CAS: 75-35-4 $C_2H_2Cl_2$ HR: 3
VINYLIDENE CHLORIDE
DOT: UN 1303
SYNS: 1-1-DCE ♦ 1,1-DICHLOROETHYLENE ♦ VINYLIDENE DICHLORIDE ♦ VINYLIDINE CHLORIDE
OSHA PEL: TWA 1 ppm
ACGIH TLV: TWA 5 ppm; Not Classifiable as a Human Carcinogen
DFG MAK: 2 ppm (8 mg/m^3); Confirmed Animal Carcinogen with Unknown Relevance to Humans
NIOSH REL: (Vinyl Halides) TWA reduce to lowest detectable level
DOT Classification: 3; Label: Flammable Liquid
SAFETY PROFILE: Suspected carcinogen. An experimental teratoge. Poison by inhalation and ingestion. A very dangerous fire hazard.

PROP: Colorless, volatile liquid. Bp: 31.6°, lel: 7.3%, uel: 16.0%, fp: –122°, flash p: 0°F (OC), d: 1.213 @ 20°/4°, autoign temp: 1058°F.

VPP000 CAS: 75-38-7 $C_2H_2F_2$ HR: 3

VINYLIDENE FLUORIDE

DOT: UN 1959

SYNS: 1,1-DIFLUOROETHYLENE (DOT, MAK) ♦ R1132a (DOT) ♦ VDF ♦ VINYLIDENE DIFLUORIDE

ACGIH TLV: TWA 500 ppm; Not Classifiable as a Human Carcinogen

DFG MAK: Confirmed Animal Carcinogen with Unknown Relevance to Humans

DOT Classification: 2.1; Label: Flammable Gas

NIOSH REL: (Vinyl Halides) TWA reduce to lowest detectable level

SAFETY PROFILE: Suspected carcinogen. Mildly toxic by inhalation. A very dangerous fire hazard.

PROP: Odorless, colorless gas. Bp: <–70°, fp: –144°, lel: 5.5%, uel: 21.3%.

VQK650 CAS: 25013-15-4 C_9H_{10} HR: 3

VINYL TOLUENE

DOT: UN 2618

SYNS: METHYLSTYRENE

OSHA PEL: TWA 100 ppm

ACGIH TLV: TWA 50 ppm; STEL 100 ppm; Not Classifiable as a Human Carcinogen

DFG MAK: 100 ppm (490 mg/m^3)

DOT Classification: 3; Label: Flammable Liquid

SAFETY PROFILE: Moderately toxic by ingestion and inhalation. An experimental teratogen. A skin and eye irritant. Flammable.

WAT200 CAS: 81-81-2 $C_{19}H_{16}O_4$ HR: 3

WARFARIN

SYNS: COUMADIN ♦ 3-(1′-PHENYL-2′-ACETYLETHYL)-4-HYDROXYCOUMARIN

OSHA PEL: TWA 0.1 mg/m^3

ACGIH TLV: TWA 0.1 mg/m^3

DFG MAK: 0.5 mg/m^3

SAFETY PROFILE: A poison by ingestion and inhalation. Moderately toxic by skin contact. Human teratogenic effects. An experimental teratogen.

PROP: Colorless, odorless, tasteless crystals. Mp: 161°. Sol in acetone, dioxane; sltly sol in methanol, ethanol; very sol in alkaline aqueous sol; insol in water and benzene.

WBJ000 HR: 3

WELDING FUMES

OSHA PEL: TWA 5 $mg(Cd)/m^3$

ACGIH TLV: TWA 5 $mg(Cd)/m^3$

SAFETY PROFILE: When welding is done on a surface coated with cadmium, toxic and carcinogenic fumes of cadmium are evolved. When zinc-coated surfaces are welded, toxic quantities of zinc oxide may be liberated. When painted surfaces are welded, lead or other pigment fumes may be liberated. And when fluoride fluxes are used in welding, very toxic fluoride fumes are evolved. When oily surfaces are welded, offensive and toxic fumes can be liberated, and, when the welding torch is improperly ignited, carbon monoxide, which is very toxic, may be evolved.

XGS000 CAS: 1330-20-7 C_8H_{10} HR: 3

XYLENES

DOT: UN 1307
SYNS: DIMETHYLBENZENE ♦ METHYL TOLUENE ♦ XYLOL (DOT)
OSHA PEL: TWA 100 ppm; STEL 150 ppm
ACGIH TLV: TWA 100 ppm; STEL 150 ppm; BEI: methyl hippuric acids in urine at end of shift 1.5 g/g creatinine; Not Classifiable as a Human Carcinogen
DFG MAK: (all isomers) 100 ppm (440 mg/m^3); BAT: 150 µg/dL in blood at end of shift
NIOSH REL: (Xylene) TWA 100 ppm; CL 200 ppm/10M
DOT Classification: 3; Label: Flammable Liquid
SAFETY PROFILE: Mildly toxic by ingestion and inhalation. An experimental teratogen. An skin and severe eye irritant. A very dangerous fire hazard.
PROP: A clear liquid. Bp: 138.5°, flash p: 100°F (TOC), d: 0.864 @ 20°/4°, vap press: 6.72 mm @ 21°. Composition: as nonaromatics 0.07%, toluene 14%, ethyl benzene 19.27%, p-xylene 7.84%, m-xylene 65.01%, o-xylene 7.63%, C9 and aromatics 0.04% (TXAPA9 33,543,75).

XIJ000 CAS: 3634-83-1 $C_{10}H_8N_2O_2$ HR: 3

m-XYLENE DIISOCYANATE

DOT: UN 2207/UN 3080
SYNS: 1,3-BIS(ISOCYANATOMETHYL)BENZENE ♦ m-PHENYLENEDIMETHYLENE ISOCYANATE ♦ M-XDI
NIOSH REL: (Diisocyanates) TWA 0.005 ppm; CL 0.02 ppm/10M
DOT Classification: 6.1; Label: KEEP AWAY FROM FOOD (UN 2207); DOT Class: 6.1; Label: Poison (UN 2206); DOT Class: 6.1; Label: Poison, Flammable Liquid (UN 3080); DOT Class: 3; Label: Flammable Liquid, Poison (UN 2478)
SAFETY PROFILE: Moderately toxic by ingestion. A severe skin and eye irritant. A sensitizer. A flammable liquid.

XMA000 CAS: 1300-73-8 $C_8H_{11}N$ HR: 3

XYLIDINE

SYNS: DIMETHYLANILINE ♦ DIMETHYLPHENYLAMINE
OSHA PEL: TWA 0.2 ppm (skin)
ACGIH TLV: TWA 0.5 ppm (skin); Animal Carcinogen
DFG MAK: (all isomers except 2,4-xylidene) 5 ppm (25 mg/m^3)
SAFETY PROFILE: Confirmed carcinogen. Moderately toxic by ingestion. Combustible.
PROP: Usually liquid (except for o-4-xylidine). Bp: 213–226°, flash p: 206° (CC), d: 0.97–0.99, vap d: 4.17. Sltly sol in water; sol in alc.

XMS000 CAS: 95-68-1 $C_8H_{11}N$ HR: 3

2,4-XYLIDINE

SYNS: 1-AMINO-2,4-DIMETHYLBENZENE ♦ 2,4-DIMETHYLANILINE ♦ 2-METHYL-p-TOLUIDINE ♦ m-XYLIDINE

DFG MAK: Animal Carcinogen, Suspected Human Carcinogen

SAFETY PROFILE: Suspected carcinogen. Poison by ingestion.

PROP: Liquid. Bp: 214°, mp: 16°, d: 0.978 @ 19.6°/4°. Very sltly sol in water.

YEJ000 CAS: 7440-65-5 Y HR: 3

YTTRIUM

OSHA PEL: TWA 1 mg(Y)/m^3

ACGIH TLV: TWA 1 mg(Y)/m^3

DFG MAK: 5 mg(Y)/m^3

SAFETY PROFILE: It may have an anticoagulant effect on the blood. Flammable in the form of dust when reacted with air or halogens.

PROP: Hexagonal, silvery-metallic colored element. Reasonably air-stable; gray-black when finely divided. Brittle, rather harder than zinc. Reacts slowly with cold H_2O, rapidly with dil acid. Burns easily. Reacts with Cl_2 at 2° and O_2 at 4°. Mp: 1522°, bp: 3338°, d: 4.469.

ZFA000 CAS: 7646-85-7 Cl_2Zn HR: 3

ZINC CHLORIDE

DOT: UN 1840/UN 2331

SYNS: ZINC BUTTER ♦ ZINC DICHLORIDE

OSHA PEL: Fume: TWA 1 mg/m^3; STEL 2 mg/m^3

ACGIH TLV: TWA 1 mg/m^3; STEL 2 mg/m^3 (fume)

DOT Classification: 8; Label: Corrosive

SAFETY PROFILE: Poison by ingestion. An experimental teratogen. A corrosive irritant to skin, eyes, and mucous membranes. Mixtures of the powdered chloride and powdered zinc are flammable.

PROP: Odorless, colorless, cubic, white, highly deliq crystals. Mp: 290°, bp: 732°, d: 2.91 @ 25°, vap press: 1 mm @ 428°. Sol in MeOH, EtOH, Et_2O, and Me_2O; very sol in H_2O.

ZFJ100 CAS: 13530-65-9 $CrH_2O_4 \cdot Zn$ HR: 3

ZINC CHROMATE

SYNS: BASIC ZINC CHROMATE ♦ CHROMIC ACID, ZINC SALT ♦ C.I. 77955

OSHA PEL: CL 0.1 mg(CrO_3)/m^3

ACGIH TLV: TWA 0.01 mg(Cr)/M^3; Confirmed Human Carcinogen

DFG TRK: 0.1 mg/m^3; Human Carcinogen

NIOSH REL: (Chromium(VI)) TWA 0.001 mg(Cr(VI))/m^3

SAFETY PROFILE: Confirmed human carcinogen.

PROP: Lemon-yellow prisms. Mp: 316°. Sol in H_2O.

ZFS000 HR: D
ZINC COMPOUNDS
SAFETY PROFILE: Variable toxicity, but generally of low toxicity. However, zinc salts, such as chromates and arsenates, are experimental carcinogens.

ZIA000 CAS: 16871-71-9 $F_6Si \cdot Zn$ HR: 3
ZINC FLUOSILICATE
DOT: UN 2855 SYNS: ZINC FLUOROSILICATE (DOT) ♦ ZINC HEXAFLUOROSILICATE
OSHA PEL: TWA 2.5 mg(F)/m^3
ACGIH TLV: TWA 2.5 mg(F)/m^3; BEI: 3 mg/g creatinine of fluorides in urine prior to shift; 10 mg/g creatinine of fluorides in urine at end of shift.
NIOSH REL: TWA 2.5 mg(F)/m^3
DOT Classification: 6.1; Label: KEEP AWAY FROM FOOD
SAFETY PROFILE: Poison by ingestion.

ZKA000 CAS: 1314-13-2 OZn HR: 3
ZINC OXIDE
SYNS: C.I. 77947 ♦ ZINC OXIDE FUME (MAK)
OSHA PEL: Fume: TWA 5 mg/m^3; STEL 10 mg/m^3; Dust: TWA Total Dust: 10 mg/m^3; Respirable Fraction: 5 mg/m^3
ACGIH TLV: Fume: TWA 5 mg/m^3; STEL 10 mg/m^3; Dust: 10 mg/m^3 of total dust (when toxic impurities are not present, e.g., quartz <1%)
DFG MAK: 1.5 mg/m^3
NIOSH REL: TWA (Zinc Oxide) 5 mg/m^3; CL 15 mg/m^3/15M
SAFETY PROFILE: Moderately toxic to humans by ingestion. An experimental teratogen. A skin and eye irritant.
PROP: Odorless, white or yellowish powder. Hexagonal white crystals. Mp: >1800°, d: 5.47. Insol in water and alc; sol in dil acetic or mineral acids, ammonia.

ZMS000 CAS: 557-05-1 $Zn(C_{18}H_{35}O_2)_2$ HR: 3
ZINC STEARATE
SYNS: DIBASIC ZINC STEARATE ♦ STEARIC ACID, ZINC SALT ♦ ZINC DISTEARATE ♦ ZINC OCTADECANOATE
OSHA PEL: TWA Total Dust: 10 mg/m^3; Respirable Fraction: 5 mg/m^3
ACGIH TLV: TWA 10 mg/m^3 of total dust when toxic impurities are not present, e.g., quartz <1%
SAFETY PROFILE: Inhalation of zinc stearate has been reported as causing pulmonary fibrosis. A nuisance dust. Combustible.
PROP: White powder. Mp: 130°, flash p: 530°F (OC), autoign temp: 790°F. Insol in water, alc, ether; sol in benzene. Decomp in dil acids.

ZOA000 CAS: 7440-67-7 Zr HR: 3

ZIRCONIUM

DOT: UN 1358/UN 1932/UN 2008/UN 2009/UN 2858

OSHA PEL: TWA 5 mg(Zr)/m^3; STEL 10 mg(Zr)/m^3; Not Classifiable as a Human Carcinogen

ACGIH TLV: TWA 5 mg(Zr)/m^3; STEL 10 mg(Zr)/m^3; Not Classifiable as a Human Carcinogen

DFG MAK: 1 mg(Zr)/m^3

DOT Classification: 4.1; Label: Flammable Solid (UN 2858, UN 1358); DOT Class: 4.2; Label: Spontaneously Combustible (UN 2008, UN 2009, UN 1932)

SAFETY PROFILE: A very dangerous fire hazard in the form of dust when exposed to heat or flame or by chemical reaction with oxidizers. May ignite spontaneously. A dangerous explosion hazard in the form of dust by chemical reaction with air.

PROP: A grayish-white, lustrous, metallic element; very sltly radioactive. Very resistant to corrosion but embrittled by N, O, and C. Oxidizes rapidly at 6°. Nitrided slowly at 700°. Mp: 1852°, bp: 4200°, d: 6.506 @ 20°.

ZQA000 HR: 2

ZIRCONIUM COMPOUNDS

OSHA PEL: TWA 5 mg(Zr)/m^3; STEL 10 mg(Zr)/m^3

ACGIH TLV: TWA 5 mg(Zr)/m^3; STEL 10 mg(Zr)/m^3; Not Classifiable as a Human Carcinogen

DFG MAK: (insoluble compounds) 1 mg(Zr)/m^3

SAFETY PROFILE: Poisoning may occur due to excessive exposure to zirconium salts. Most zirconium compounds in common use are insoluble and considered inert.

Appendix I
Synonym Cross-Reference

1080 see SHG500
2-AAF see FDR000
ACETALDEHYDE see AAG250
ACETAMIDE see AAI000
2-ACETAMINOFLUORENE see FDR000
ACETANHYDRIDE see AAX500
ACETDIMETHYLAMIDE see DOO800
ACETIC ACID see AAT250
ACETIC ACID (aqueous solution) (DOT) see AAT250
ACETIC ACID AMIDE see AAI000
ACETIC ACID, AMYL ESTER see AOD725
ACETIC ACID, ANHYDRIDE (9CI) see AAX500
ACETIC ACID BENZYL ESTER see BDX000
ACETIC ACID n-BUTYL ESTER see BPU750
ACETIC ACID-tert-BUTYL ESTER see BPV100
ACETIC ACID DIMETHYLAMIDE see DOO800
ACETIC ACID-1,1-DIMETHYLETHYL ESTER see BPV100
ACETIC ACID, ISOPENTYL ESTER see IHO850
ACETIC ACID METHYL ESTER see MFW100
ACETIC ACID-1-METHYLETHYL ESTER (9CI) see INE100
ACETIC ACID-1-METHYLPROPYL ESTER (9CI) see BPV000
ACETIC ACID-2-METHYLPROPYL ESTER see IIJ000
ACETIC ACID VINYL ESTER see VLU250
ACETIC ALDEHYDE see AAG250
ACETIC ANHYDRIDE see AAX500
ACETIC PEROXIDE see PCL500
ACETONE see ABC750
ACETONE CYANOHYDRIN (ACGIH,DOT) see MLC750
ACETONE OILS (DOT) see ABC750
ACETONITRILE see ABE500
ACETONYL CHLORIDE see CDN200
2-ACETOXYBENZOIC ACID see ADA725
ACETOXYETHANE see EFR000
2-ACETOXYPENTANE see AOD735
1-ACETOXYPROPANE see PNC250
2-ACETOXYPROPANE see INE100
α-ACETOXYTOLUENE see BDX000
2-ACETYLAMINOFLUORENE (OSHA) see FDR000
ACETYL ANHYDRIDE see AAX500
ACETYLEN see ACI750
ACETYLENE see ACI750
ACETYLENE BLACK see CBT750
ACETYLENE DICHLORIDE see DFI210
ACETYLENE TETRABROMIDE see ACK250
ACETYLENE TETRACHLORIDE see TBQ100
ACETYLENE TRICHLORIDE see TIO750
ACETYL ETHYLENE see BOY500
ACETYL FLUORIDE see ACM000
ACETYL OXIDE see AAX500
ACETYLSALICYLIC ACID see ADA725
ACROLEIC ACID see ADS750
ACROLEIN see ADR000
ACRYLALDEHYDE see ADR000
ACRYLAMIDE see ADS250
ACRYLIC ACID see ADS750
ACRYLIC ACID BUTYL ESTER see BPW100
ACRYLIC ACID n-BUTYL ESTER (MAK) see BPW100
ACRYLIC ACID ETHYL ESTER see EFT000
ACRYLIC ACID-2-HYDROXYPROPYL ESTER see HNT600
ACRYLIC ACID METHYL ESTER (MAK) see MGA500
ACRYLIC ALDEHYDE see ADR000
ACRYLIC AMIDE see ADS250
ACRYLONITRILE see ADX500
ACTINOLITE see ARM250
ACTIVATED CARBON see CBT500
ADIPIC ACID DINITRILE see AER250
ADIPIC ACID NITRILE see AER250
ADIPONITRILE see AER250
AGE see AGH150
AGRICULTURAL LIMESTONE see CAO000
ALCOHOLS, n.o.s. (UN 1987) (DOT) see EFU000
ALLYL ALCOHOL see AFV500
ALLYL CHLORIDE see AGB250
ALLYL-2,3-EPOXYPROPYL ETHER see AGH150
ALLYL GLYCIDYL ETHER see AGH150
ALLYL PROPYL DISULFIDE see AGR500

ALLYL TRICHLORIDE see TJB600
ALUMINA see AHE250
α-ALUMINA (OSHA) see AHE250
ALUMINUM see AGX000
ALUMINUM BROMIDE see AGX750
ALUMINUM CHLORIDE see AGY750
ALUMINUM FLUORIDE see AHB000
ALUMINUM NITRATE (DOT) see AHD750
ALUMINUM(III) NITRATE (1:3) see AHD750
ALUMINUM OXIDE (2:3) see AHE250
ALUMINUM POWDER see AGX000
ALUMINUM PYRO POWDERS (OSHA) see AGX000
ALUMINUM TRIBROMIDE see AGX750
ALUMINUM TRICHLORIDE see AGY750
ALUMINUM TRIFLUORIDE see AHB000
ALUMINUM TRINITRATE see AHD750
ALUMINUM WELDING FUMES (OSHA) see AGX000
2-AMINOANILINE see PEY250
4-AMINOANILINE see PEY500
o-AMINOANISOLE see AOV900
p-AMINOANISOLE see AOW000
AMINOBENZENE see AOQ000
1-AMINOBUTANE see BPX750
2-AMINOBUTANE see BPY000
2-AMINO-4-CHLOROTOLUENE see CLK225
2-AMINO-5-CHLOROTOLUENE see CLK220
AMINOCYCLOHEXANE see CPF500
1-AMINO-2,4-DIMETHYLBENZENE see XMS000
1-AMINOETHANE see EFU400
2-AMINOETHANOL (MAK) see EEC600
β-AMINOETHYL ALCOHOL see EEC600
AMINOETHYLETHANEDIAMINE see DJG600
AMINOHEXAHYDROBENZENE see CPF500
6-AMINOHEXANOIC ACID CYCLIC LACTAM see CBF700
2-AMINOISOBUTANE see BPY250
1-AMINO-2-METHOXYBENZENE see AOV900
1-AMINO-2-METHYLBENZENE see TGQ750
4-AMINO-1-METHYLBENZENE see TGR000
1-AMINO-2-METHYL-5-NITROBENZENE see NMP500
1-AMINO-2-METHYLPROPANE see IIM000
1-AMINONAPHTHALENE see NBE700
2-AMINONAPHTHALENE see NBE500
4-AMINO-2-NITROANILINE see ALL750
1-AMINO-4-NITROBENZENE see NEO500
2-AMINOPENTANE see DHJ200
p-AMINOPHENYL ETHER see OPM000
3-AMINOPHENYLMETHANE see TGQ500
2-AMINOPROPANE see INK000
2-AMINOPYRIDINE see AMI000
AMINO-2-PYRIDINE see AMI000
o-AMINOPYRIDINE see AMI000
o-AMINOTOLUENE see TGQ750
p-AMINOTOLUENE see TGR000
3-AMINO-p-TOLUIDINE see TGL750
1-AMINO-2,4,5-TRIMETHYLBENZENE see TLG250
AMMONIA see AMY500
AMMONIA AQUEOUS see ANK250
AMMONIA GAS see AMY500
AMMONIA SOLUTIONS, with >10% but not >35% ammonia (UN 2672) (DOT) see ANK250
AMMONIA SOLUTIONS, with >35% but not >50% ammonia (UN 2073) (DOT) see ANK250
AMMONIUM AMIDOSULPHATE see ANU650
AMMONIUM BIFLUORIDE see ANJ000
AMMONIUM BOROFLUORIDE see ANH000
AMMONIUM CHLORIDE see ANE500
AMMONIUM CHLOROPLATINATE see ANF250
AMMONIUM DIFLUORIDE see ANJ000
AMMONIUM FLUOBORATE see ANH000
AMMONIUM FLUOROBORATE see ANH000
AMMONIUM HEXACHLOROPLATINATE(IV) see ANF250
AMMONIUM HYDROGEN DIFLUORIDE see ANJ000
AMMONIUM HYDROGEN FLUORIDE see ANJ000
AMMONIUM HYDROXIDE see ANK250
AMMONIUM METAVANADATE (DOT) see ANY250
AMMONIUM MOLYBDATE see ANM750
AMMONIUM MURIATE see ANE500
AMMONIUM PERFLUOROCAPRILATE see ANP625
AMMONIUM PERFLUOROOCTANOATE see ANP625
AMMONIUM PEROXYDISULFATE see ANR000
AMMONIUM PERSULFATE see ANR000
AMMONIUM PLATINIC CHLORIDE see ANF250

AMMONIUM SULFAMATE see ANU650
AMMONIUM SULPHAMATE see ANU650
AMMONIUM TETRAFLUOROBORATE see ANH000
AMMONIUM VANADATE see ANY250
AMORPHOUS SILICA see SCK600
AMORPHOUS SILICA see DCJ800
AMORPHOUS SILICA DUST see SCH002
AMOSITE see ARM250
AMPHIBOLE see ARM250
AMPROLENE see EJN500
n-AMYL ACETATE see AOD725
sec-AMYL ACETATE see AOD735
AMYL ACETIC ESTER see AOD725
AMYL ETHYL KETONE see ODI000
AMYL HYDRIDE (DOT) see PBK250
AMYL MERCAPTAN (DOT) see PBM000
AMYL METHYL KETONE (DOT) see MGN500
n-AMYL METHYL KETONE see MGN500
ANESTHETIC ETHER see EJU000
ANHYDROUS AMMONIA see AMY500
ANHYDROUS CALCIUM SULFATE see CAX500
ANHYDROUS HYDROBROMIC ACID see HHJ000
ANILINE see AOQ000
ANILINOBENZENE see DVX800
2-ANILINONAPHTHALENE see PFT500
2-ANISIDINE see AOV900
o-ANISIDINE see AOV900
p-ANISIDINE see AOW000
o-ANISYLAMINE see AOV900
p-ANISYLAMINE see AOW000
ANNALINE see CAX750
ANTHOPHYLITE see ARM250
ANTHRACENE see APG500
ANTHRACIN see APG500
ANTIMONOUS CHLORIDE (DOT) see AQC500
ANTIMONY see AQB750
ANTIMONY(III) CHLORIDE see AQC500
ANTIMONY(V) CHLORIDE see AQD000
ANTIMONY FLUORIDE see AQF250
ANTIMONY HYDRIDE see SLQ000
ANTIMONY OXIDE see AQF000
ANTIMONY PENTACHLORIDE (DOT) see AQD000
ANTIMONY(V) PENTAFLUORIDE see AQF250
ANTIMONY PERCHLORIDE see AQD000
ANTIMONY PEROXIDE see AQF000
ANTIMONY POWDER (DOT) see AQB750
ANTIMONY SESQUIOXIDE see AQF000
ANTIMONY TRICHLORIDE see AQC500
ANTIMONY TRIHYDRIDE see SLQ000
ANTIMONY TRIOXIDE see AQF000
ANTIMONY WHITE see AQF000
AROCLOR 1016 see PJL750
AROCLOR 1221 see PJL750
AROCLOR 1232 see PJL750
AROCLOR 1242 see PJL750
AROCLOR 1248 see PJL750
AROCLOR 1254 see PJL750
AROCLOR 1260 see PJL750
AROCLOR 1262 see PJL750
AROCLOR 1268 see PJL750
AROCLOR 2565 see PJL750
AROCLOR 4465 see PJL750
AROCLOR 5442 see PJL750
ARSENIC see ARA750
ARSENICALS see ARF750
ARSENIC COMPOUNDS see ARF750
ARSENIC HYDRIDE see ARK250
ARSENIC(III) OXIDE see ARI750
ARSENIC TRIHYDRIDE see ARK250
ARSENIC TRIOXIDE see ARI750
ARSENIOUS ACID see ARI750
ARSENOUS ACID see ARI750
ARSENOUS ACID ANHYDRIDE see ARI750
ARSINE see ARK250
A.S.A. see ADA725
ASBESTOS see ARM250
ASPHALT see ARO500
ASPHALT FUMES (ACGIH) see ARO500
AURAMINE HYDROCHLORIDE see IBA000
AURAMINE (MAK) see IBA000
AURAMINE (MAK) see IBB000
AZABENZENE see POP250
3-AZAPENTANE-1,5-DIAMINE see DJG600
AZIMETHYLENE see DCP800
AZINE see POP250
AZIRANE see EJM900
BACILLOMYCIN (8CI, 9CI) see BAB750
BACILLUS SUBTILIS BPN see BAB750
BANANA OIL see IHO850
BARIUM see BAH250
BARIUM AZIDE see BAI000
BARIUM AZIDE, dry or wetted with <50% water, by weight (UN 0224) (DOT) see BAI000
BARIUM AZIDE, wetted with not <50% water, by weight (UN 1571) (DOT) see BAI000
BARIUM CHROMATE(VI) see BAK250
BARIUM CHROMATE OXIDE see BAK250
BARIUM COMPOUNDS (soluble) see BAK500
BARIUM DINITRATE see BAN250
BARIUM FLUORIDE see BAM000

BARIUM NITRATE (DOT) see BAN250
BARIUM(II) NITRATE (1:2) see BAN250
BARIUM SULFATE see BAP000
BASIC ZINC CHROMATE see ZFJ100
BAYRITES see BAP000
BCME see BIK000
BEHP see DVL700
BENZAL CHLORIDE see BAY300
BENZENAMINE see AOQ000
BENZENE see BBL250
BENZENE CHLORIDE see CEJ125
1,2-BENZENEDIAMINE see PEY250
1,3-BENZENEDICARBONITRILE see PHX550
1,2-BENZENEDICARBOXYLIC ACID ANHYDRIDE see PHW750
1,2-BENZENEDICARBOXYLIC ACID DIMETHYL ESTER see DTR200
BENZENE-1,3-DIISOCYANATE see BBP000
1,2-BENZENEDIOL see CCP850
m-BENZENEDIOL see REA000
o-BENZENEDIOL see CCP850
p-BENZENEDIOL see HIH000
α-BENZENEHEXACHLORIDE see BBQ000
γ-BENZENE HEXACHLORIDE see BBQ500
trans-α-BENZENEHEXACHLORIDE see BBR000
BENZENE HEXACHLORIDE-α-isomer see BBQ000
BENZENE HEXACHLORIDE-γ-isomer see BBQ500
BENZENETETRAHYDRIDE see CPC579
1,2,4-BENZENETRICARBOXYLIC ACID ANHYDRIDE see TKV000
BENZENYL CHLORIDE see BFL250
BENZIDINE see BBX000
BENZINE (LIGHT PETROLEUM DISTILLATE) see PCT250
BENZO(d,e,f)CHRYSENE see BCS750
BENZOIC ACID, PEROXIDE see BDS000
BENZOIC TRICHLORIDE see BFL250
BENZOL (DOT) see BBL250
BENZO(def)PHENANTHRENE see PON250
BENZO(a)PYRENE see BCS750
3,4-BENZOPYRENE see BCS750
6,7-BENZOPYRENE see BCS750
BENZOQUINONE (DOT) see QQS200
BENZOTRICHLORIDE (DOT, MAK) see BFL250
BENZOYL PEROXIDE see BDS000
BENZYL ACETATE see BDX000
BENZYL CHLORIDE see BEE375
BENZYLENE CHLORIDE see BAY300
BENZYLIDENE CHLORIDE (DOT) see BAY300
BENZYLIDYNE CHLORIDE see BFL250
BENZYL TRICHLORIDE see BFL250
BERYLLIUM see BFO750
BERYLLIUM, powder (UN 1567) (DOT) see BFO750
BERYLLIUM COMPOUNDS see BFQ500
BGE see BRK750
BHC see BBQ500
α-BHC see BBQ000
β-BHC see BBR000
BHT (food grade) see BFW750
4,4′-BIANILINE see BBX000
BICYCLOPENTADIENE see DGW000
BINITROBENZENE see DUQ200
BIPHENYL see BGE000
4,4′-BIPHENYLDIAMINE see BBX000
4,4′-BIPHENYLENEDIAMINE see BBX000
BIPHENYL OXIDE see PFA850
BIS(ACETYLOXY)MERCURY see MCS750
BIS(2-AMINOETHYL)AMINE see DJG600
BIS(4-AMINOPHENYL)ETHER see OPM000
BIS(4-AMINOPHENYL)METHANE see MJQ000
BIS(p-AMINOPHENYL)SULFIDE see TFI000
BIS(3-tert-BUTYL-4-HYDROXY-6-METHYL PHENYL) SULFIDE see TFC600
BIS(2-CHLOROETHYL) ETHER see DFJ050
BIS(CHLOROMETHYL) ETHER see BIK000
BIS-CME see BIK000
BISCYCLOPENTADIENYLIRON see FBC000
BIS-O,O-DIETHYLPHOSPHORIC ANHYDRIDE see TCF250
BIS-O,O-DIETHYLPHOSPHOROTHIONIC ANHYDRIDE see SOD100
4,4′-BIS(DIMETHYLAMINO)BENZOPHENONE see MQS500
p,p′-BIS(DIMETHYLAMINO)DIPHENYLMETHANE see MJN000
BIS(2-DIMETHYLAMINOETHYL) ETHER see BJH750
BIS(p-DIMETHYLAMINOPHENYL)METHANE see MJN000
BIS(4-(DIMETHYLAMINO)PHENYL)METHANONE see MQS500
BIS(p-DIMETHYLAMINOPHENYL)METHYLENEIMINE see IBB000
1,1-BIS(p-DIMETHYLAMINOPHENYL)METHYLENIMINEHYDROCHLORIDE see IBA000BIS(2,3-EPOXYPROPYL)ETHER see DKM200

BIS(2-ETHYLHEXANOYLOXY)DIBUTYL STANNANE see BJQ250
BIS(2-ETHYLHEXYL)PHTHALATE see DVL700
BIS(2-HYDROXYETHYL)AMINE see DHF000
BIS(β-HYDROXYETHYL)NITROSAMINE see NKM000
2,2-BIS(HYDROXYMETHYL)-1,3-PROPANEDIOL see PBB750
BIS(4-ISOCYANATOCYCLOHEXYL)METHANE see MJM600
1,3-BIS(ISOCYANATOMETHYL)BENZENE see XIJ000
BIS(4-ISOCYANATOPHENYL)METHANE see MJP400
BISMUTH SESQUITELLURIDE see BKY000
BISMUTH TELLURIDE see BKY000
BIS(1-OXODODECYL)PEROXIDE see LBR000
4,4′-BI-o-TOLUIDINE see TGJ750
BITUMEN (MAK) see ARO500
BIVINYL see BOP500
BLASTING GELATIN (DOT) see NGY000
BORATES, TETRA, SODIUM SALT, anhydrous (OSHA, ACGIH) see SFE500
BORATES, TETRA, SODIUM SALT, anhydrous (OSHA, ACGIH) see SFF000
BORAX (8CI) see SFF000
BORIC ANHYDRIDE see BMG000
2-BORNANONE see CBA750
BOROFLUORIC ACID see FDD125
BORON BROMIDE see BMG400
BORON FLUORIDE see BMG700
BORON HYDRIDE see DDI450
BORON OXIDE see BMG000
BORON SESQUIOXIDE see BMG000
BORON TRIBROMIDE see BMG400
BORON TRIFLUORIDE see BMG700
BORON TRIOXIDE see BMG000
B(a)P see BCS750
BPL see PMT100
BROMIC ETHER see EGV400
BROMINE see BMP000
BROMINE, solution (DOT) see BMP000
BROMINE TRIFLUORIDE see BMQ325
BROMOCHLOROMETHANE see CES650
BROMOETHANE see EGV400
BROMOETHENE see VMP000
BROMOETHYLENE see VMP000
BROMOFLUOROFORM see TJY100
BROMOFORM see BNL000
BROMOMETHANE see MHR200
BROMOTRIFLUOROMETHANE see TJY100
BURNT LIME see CAU500
1,3-BUTADIENE see BOP500
BUTA-1,3-DIENE see BOP500
BUTADIENE DIMER see CPD750
BUTANE see BOR500
n-BUTANE (DOT) see BOR500
1,4-BUTANEDINITRILE see SNE000
BUTANE MIXTURES (DOT) see BOR500
n-BUTANENITRILE see BSX250
BUTANESULFONE see BOU250
BUTANE SULTONE see BOU250
1,4-BUTANESULTONE (MAK) see BOU250
BUTANETHIOL (OSHA) see BRR900
BUTANOL (DOT) see BPW500
2-BUTANOL see BPW750
n-BUTANOL see BPW500
sec-BUTANOL (DOT) see BPW750
tert-BUTANOL see BPX000
2-BUTANOL ACETATE see BPV000
2-BUTANONE (OSHA) see MKA400
2-BUTENAL see COB250
cis-BUTENEDIOIC ANHYDRIDE see MAM000
3-BUTENE-2-ONE see BOY500
3-BUTEN-2-ONE see BOY500
2-BUTOXYETHANOL see BPJ850
n-BUTOXYETHANOL see BPJ850
2-BUTOXYETHYL ACETATE see BPM000
2-BUTOXYETHYL ESTER ACETIC ACID see BPM000
BUTYL ACETATE see BPU750
1-BUTYL ACETATE see BPU750
2-BUTYL ACETATE see BPV000
n-BUTYL ACETATE see BPU750
sec-BUTYL ACETATE see BPV000
tert-BUTYL ACETATE see BPV100
n-BUTYL ACRYLATE see BPW100
BUTYLACRYLATE, INHIBITED (DOT) see BPW100
BUTYL ALCOHOL (DOT) see BPW500
n-BUTYL ALCOHOL see BPW500
sec-BUTYL ALCOHOL see BPW750
tert-BUTYL ALCOHOL see BPX000
sec-BUTYL ALCOHOL ACETATE see BPV000
n-BUTYLAMINE see BPX750
sec-BUTYLAMINE see BPY000
tert-BUTYLAMINE see BPY250
BUTYLATED HYDROXYTOLUENE see BFW750
BUTYL CELLOSOLVE see BPJ850
BUTYL CELLOSOLVE ACETATE see BPM000
BUTYLENE OXIDE see TCR750

1,4-BUTYLENE SULFONE see BOU250
n-BUTYL ETHYL KETONE see EHA600
BUTYL FORMAL see VAG000
n-BUTYL GLYCIDYL ETHER see BRK750
tert-BUTYLHYDROPEROXIDE see BRM250
BUTYL α-HYDROXYPROPIONATE see BRR600
n-BUTYL LACTATE see BRR600
BUTYL MERCAPTAN see BRR900
n-BUTYL MERCAPTAN see BRR900
BUTYL METHYL KETONE see HEV000
n-BUTYL-N-NITROSO-1-BUTAMINE see BRY500
tert-BUTYL PERACETATE see BSC250
tert-BUTYL PEROXIDE see BSC750
t-BUTYL PEROXYACETATE see BSC250
4-t-BUTYLPHENOL see BSE500
o-sec-BUTYLPHENOL see BSE000
p-tert-BUTYLPHENOL (MAK) see BSE500
BUTYL-2-PROPENOATE see BPW100
p-tert-BUTYLTOLUENE see BSP500
BUTYRIC ACID NITRILE see BSX250
BUTYRONE (DOT) see DWT600
BUTYRONITRILE see BSX250
BUTYRONITRILE (DOT) see BSX250
CADMIUM see CAD000
CADMIUM CHLORIDE see CAE250
CADMIUM COMPOUNDS see CAE750
CADMIUM DICHLORIDE see CAE250
CALCINED DIATOMITE see SCJ000
CALCINED MAGNESITE see MAH500
CALCITE see CAO000
CALCIUM CARBIMIDE see CAQ250
CALCIUM CARBONATE see CAO000
CALCIUM CYANAMID see CAQ250
CALCIUM CYANAMIDE see CAQ250
CALCIUM CYANIDE see CAQ500
CALCIUM HYDRATE see CAT225
CALCIUM HYDROSILICATE see CAW850
CALCIUM HYDROXIDE see CAT225
CALCIUM HYDROXIDE (ACGIH, OSHA) see CAT225
CALCIUM MONOSILICATE see CAW850
CALCIUM OXIDE see CAU500
CALCIUM SILICATE see CAW850
CALCIUM SILICATE, synthetic nonfibrous (ACGIH) see CAW850
CALCIUM SULFATE see CAX500
CALCIUM(II) SULFATE DIHYDRATE (1:1:2) see CAX750
CALCYANIDE see CAQ500
2-CAMPHANONE see CBA750
CAMPHECHLOR see CDV100
CAMPHOCHLOR see CDV100
CAMPHOR see CBA750
CAMPHOR, synthetic (ACGIH, DOT) see CBA750
CAPROLACTAM see CBF700
6-CAPROLACTAM see CBF700
ω-CAPROLACTAM (MAK) see CBF700
CARBAMALDEHYDE see FMY000
4-CARBAMIDOPHENYL BIS(o-CARBOXYPHENYLTHIO)ARSENITE see TFD750
CARBAMONITRILE see COH500
CARBOLIC ACID see PDN750
CARBOMETHENE see KEU000
CARBON see CBT500
CARBON, activated (DOT) see CBT500
CARBON BISULFIDE (DOT) see CBV500
CARBON BLACK see CBT750
CARBON BLACK, CHANNEL see CBT750
CARBON BLACK, FURNACE see CBT750
CARBON BLACK, LAMP see CBT750
CARBON BLACK, THERMAL see CBT750
CARBON DICHLORIDE see PCF275
CARBON DIFLUORIDE OXIDE see CCA500
CARBON DIOXIDE see CBU250
CARBON DIOXIDE, refrigerated liquid (UN 2187) (DOT) see CBU250
CARBON DIOXIDE, solid (UN 1845) (DOT) see CBU250
CARBON DISULFIDE see CBV500
CARBON FLUORIDE OXIDE see CCA500
CARBON HEXACHLORIDE see HCI000
CARBONIC ACID ANHYDRIDE see CBU250
CARBONIC ACID, CALCIUM SALT (1:1) see CAO000
CARBONIC ACID GAS see CBU250
CARBONIC ACID, NICKEL SALT (1:1) see NCY500
CARBON MONOXIDE see CBW750
CARBON MONOXIDE, refrigerated liquid (cryogenic liquid) (NA 9202) (DOT) see CBW750
CARBON NITRIDE see COO000
CARBON OXIDE (CO) see CBW750
CARBON OXYCHLORIDE see PGX000
CARBON OXYFLUORIDE see CCA500
CARBON SULPHIDE (DOT) see CBV500
CARBON TET see CBY000
CARBON TETRABROMIDE see CBX750
CARBON TETRACHLORIDE see CBY000
CARBONYL FLUORIDE see CCA500
CARBORUNDUM see SCQ000
o-CARBOXYPHENYL ACETATE see ADA725

4-CARBOXYPHTHALIC ANHYDRIDE see TKV000
CATECHOL see CCP850
CAUSTIC POTASH see PLJ500
CAUSTIC SODA see SHS000
CELLOSOLVE (DOT) see EES350
CELLOSOLVE ACETATE (DOT) see EES400
CEMENT, PORTLAND see PKS750
CESIUM HYDRATE see CDD750
CESIUM HYDROXIDE see CDD750
CFC 31 see CHI900
CFC-112 see TBP050
CFC-112a see TBP000
CHA see CPF500
CHALCEDONY see SCI500
CHALCEDONY see SCJ500
CHALK see CAO000
CHANNEL BLACK see CBT750
CHLORACETONE see CDN200
CHLORACETYL CHLORIDE see CEC250
CHLORETHENE see VNP000
CHLORETHYLENE see VNP000
CHLORIDE of SULFUR (DOT) see SON510
CHLORINATED CAMPHENE see CDV100
CHLORINATED DIPHENYL OXIDE see CDV175
CHLORINATED HYDROCHLORIC ETHER see DFF809
CHLORINE see CDV750
CHLORINE CYANIDE see COO750
CHLORINE DIOXIDE see CDW450
CHLORINE DIOXIDE, not hydrated (DOT) see CDW450
CHLORINE FLUORIDE see CDX750
CHLORINE FLUORIDE OXIDE see PCF750
CHLORINE OXIDE see CDW450
CHLORINE(IV) OXIDE see CDW450
CHLORINE OXYFLUORIDE see PCF750
CHLORINE TRIFLUORIDE see CDX750
CHLOROACETALDEHYDE see CDY500
2-CHLOROACETALDEHYDE see CDY500
CHLOROACETIC CHLORIDE see CEC250
CHLOROACETONE, stabilized (DOT) see CDN200
1-CHLOROACETOPHENONE see CEA750
α-CHLOROACETOPHENONE see CEA750
CHLOROACETYL CHLORIDE see CEC250
α-CHLOROALLYL CHLORIDE see DGG950
CHLOROALLYLENE see AGB250
CHLOROBEN see DEP600
CHLOROBENZAL see BAY300
o-CHLOROBENZAL MALONONITRILE see CEQ600
CHLOROBENZENE see CEJ125
CHLOROBENZOL (DOT) see CEJ125
2-CHLOROBENZYLIDENE MALONONITRILE see CEQ600
o-CHLOROBENZYLIDENE MALONONITRILE see CEQ600
CHLOROBROMOMETHANE see CES650
CHLOROBUTADIENE see NCI500
2-CHLORO-1,3-BUTADIENE see NCI500
3-CHLOROCHLORDENE see HAR000
CHLORO(CHLOROMETHOXY)METHANE see BIK000
CHLOROCYANIDE see COO750
CHLOROCYANOGEN see COO750
1-CHLORO-2,3-DIBROMOPROPANE see DDL800
1-CHLORO-2,2-DICHLOROETHYLENE see TIO750
CHLORODIFLUOROMETHANE see CFX500
1-CHLORO-2,3-EPOXYPROPANE see EAZ500
2-CHLOROETHANAL see CDY500
2-CHLORO-1-ETHANAL see CDY500
2-CHLOROETHANOL (MAK) see EIU800
CHLOROETHENE see VNP000
2-CHLOROETHYL ALCOHOL see EIU800
CHLOROETHYL ETHER see DFJ050
CHLOROFLUOROMETHANE see CHI900
CHLORO FLUORO SULFONE see SOT500
CHLOROFORM see CHJ500
CHLOROFORMIC ACID DIMETHYLAMIDE see DQY950
CHLOROFORMYL CHLORIDE see PGX000
CHLOROHYDRIC ACID see HHL000
CHLOROMETHANE see MIF765
4-CHLORO-2-METHYLANILINE see CLK220
CHLOROMETHYLBENZENE see BEE375
2-CHLORO-1-METHYLBENZENE (9CI) see CLK100
(CHLOROMETHYL)ETHYLENE OXIDE see EAZ500
CHLOROMETHYL METHYL ETHER see CIO250
CHLOROMETHYLOXIRANE see EAZ500
CHLOROMETHYL PHENYL KETONE see CEA750
4-CHLORONITROBENZENE see NFS525
4-CHLORO-1-NITROBENZENE see NFS525
CHLORONITROPROPANE see CJE000
1-CHLORO-1-NITROPROPANE see CJE000
CHLOROPENTAFLUOROETHANE see CJI500

2-CHLORO-1-PHENYLETHANONE see CEA750
CHLOROPHENYLMETHANE see BEE375
CHLOROPICRIN see CKN500
3-CHLOROPRENE see AGB250
β-CHLOROPRENE (OSHA, MAK) see NCI500
CHLOROPROPANONE see CDN200
1-CHLORO-2-PROPANONE see CDN200
3-CHLOROPROPENE see AGB250
α-CHLOROPROPIONIC ACID see CKS750
3-CHLORO-1-PROPYLENE see AGB250
o-CHLOROSTYRENE see CLE750
CHLOROSULFONYL FLUORIDE see SOT500
o-CHLOROTOLUENE see CLK100
α-CHLOROTOLUENE see BEE375
4-CHLORO-o-TOLUIDINE see CLK220
5-CHLORO-o-TOLUIDINE see CLK225
2-CHLORO-6-(TRICHLOROMETHYL)PYRIDINE see CLP750
2-CHLORO-1,1,2-TRIFLUOROETHYL DIFLUOROMETHYL ETHER see EAT900
CHRISTENSENITE see SCK000
CHROME see CMI750
CHROMIC(VI) ACID see CMH250
CHROMIC ACID, ZINC SALT see ZFJ100
CHROMIUM see CMI750
CHROMIUM CARBONYL (MAK) see HCB000
CHROMIUM HEXACARBONYL see HCB000
CHROMIUM METAL (OSHA) see CMI750
CHRYSOTILE see ARM250
C.I. 10305 see PID000
C.I. 10355 see DVX800
C.I. 11020 see DOT300
C.I. 23060 see DEQ600
C.I. 24110 see DCJ200
C.I. 37035 see NEO500
C.I. 37077 see TGQ750
C.I. 37105 see NMP500
C.I. 37107 see TGR000
C.I. 37230 see TGJ750
C.I. 37270 see NBE500
C.I. 41000 see IBA000
C.I. 41000B see IBB000
C.I. 76010 see PEY250
C.I. 76035 see TGL750
C.I. 76060 see PEY500
C.I. 76505 see REA000
C.I. 77491 see IHC450
C.I. 77575 see LCF000
C.I. 77600 see LCR000
C.I. 77718 see TAB750
C.I. 77755 see PLP000
C.I. 77775 see NCW500
C.I. 77777 see NDF500
C.I. 77779 see NCY500
C.I. 77795 see PJD500
C.I. 77805 see SBO500
C.I. 77820 see SDI500
C.I. 77864 see TGC000
C.I. 77891 see TGG760
C.I. 77938 see VDU000
C.I. 77940 see VEZ000
C.I. 77947 see ZKA000
C.I. 77955 see ZFJ100
CINERIN I or II see POO250
CMME see CIO250
COAL TAR see CMY800
COAL TAR CREOSOTE see CMY825
COAL TAR OIL see CMY825
COAL TAR OIL (DOT) see CMY825
COAL TAR PITCH VOLATILES see CMZ100
COAL TAR PITCH VOLATILES: PHENANTHRENE see PCW250
COBALT see CNA250
COBALT CARBONYL see CNB500
COBALT HYDROCARBONYL see CNC230
COBALT OCTACARBONYL see CNB500
COBALT TETRACARBONYL see CNB500
COLLOIDAL ARSENIC see ARA750
COLLOIDAL MANGANESE see MAP750
COLLOIDAL SILICA see SCH002
COMPOUND No. 1080 see SHG500
COPPER see CNI000
COPPER COMPOUNDS see CNK750
COPPER CYANIDE see CNL000
COPPER(I) CYANIDE see CNL000
COTTON DUST see CNT750
COUMADIN see WAT200
CREOSOTE see CMY825
CRESOL see CNW500
o-CRESOL see CNX000
CRESYLIC ACID see CNW500
o-CRESYLIC ACID see CNX000
CRESYLIC CREOSOTE see CMY825
o-CRESYL PHOSPHATE see TMO600
CRISTOBALITE see SCI500
CRISTOBALITE see SCJ000
CROTONALDEHYDE see COB250
CROTONIC ALDEHYDE see COB250
CSAC see EES400
CUMENE see COE750
CUMENE HYDROPEROXIDE (DOT) see IOB000
CUMENYL HYDROPEROXIDE see IOB000

CUPROUS CYANIDE see CNL000
CYANAMIDE see COH500
CYANIDE see COI500
CYANIDE ION see COI500
CYANIDE SOLUTIONS (DOT) see COI500
CYANIDES (OSHA) see PLC500
CYANOACETONITRILE see MAO250
2-CYANOACRYLIC ACID, METHYL ESTER see MIQ075
α-CYANOACRYLIC ACID METHYL ESTER see MIQ075
CYANOETHYLENE see ADX500
CYANOGAS see CAQ500
CYANOGEN see COO000
CYANOGENAMIDE see COH500
CYANOGEN CHLORIDE see COO750
CYANOGEN GAS (DOT) see COO000
CYANOGEN NITRIDE see COH500
CYANOMETHANE see ABE500
CYANOMETHANOL see HIM500
1-CYANOPROPANE see BSX250
2-CYANOPROPANE see IJX000
1,4-CYCLOHEXADIENEDIONE see QQS200
CYCLOHEXANAMINE see CPF500
CYCLOHEXANE see CPB000
CYCLOHEXANETHIOL see CPB625
CYCLOHEXANOL see CPB750
CYCLOHEXANONE see CPC000
CYCLOHEXENE see CPC579
CYCLOHEXENYLETHYLENE see CPD750
CYCLOHEXYL ALCOHOL see CPB750
CYCLOHEXYLAMINE see CPF500
CYCLOHEXYL MERCAPTAN (DOT) see CPB625
CYCLOHEXYLMETHANE see MIQ740
CYCLONITE see CPR800
CYCLONITE, desensitized (UN 0483) (DOT) see CPR800
CYCLONITE, wetted (UN 0072) (DOT) see CPR800
CYCLOPENTADIENE see CPU500
1,3-CYCLOPENTADIENE see CPU500
1,3-CYCLOPENTADIENE, DIMER see DGW000
CYCLOPENTADIENYLMANGANESE TRICARBONYL see CPV000
CYCLOPENTANE see CPV750
CYCLOTETRAMETHYLENE OXIDE see TCR750
CYCLOTRIMETHYLENETRINITRAMINE, desensitized (UN 0483) (DOT) see CPR800
CYCLOTRIMETHYLENETRINITRAMINE, wetted (UN 0072) (DOT) see CPR800
CYKLOHEXANTHIOL see CPB625
2,4-DAA see DBO000
DAB see DOT300
DALAPON (USDA) see DGI400
DAPM see MJQ000
DBCP see DDL800
DBE see EIY500
DBN see BRY500
DBNA see BRY500
DBOT see DEF400
DBP see DEH200
DBPC (technical grade) see BFW750
DCA see DFE200
DCB see DEP600
DCB see DEQ600
DCB see DEV000
1-1-DCE see VPK000
1,2-DCE see EIY600
DCEE see DFJ050
DDC see DQY950
DEA see DHF000
DEAE see DHO500
DECABORANE see DAE400
DECABORANE(14) see DAE400
DEHP see DVL700
DEK see DJN750
DENA see NJW500
DESMODUR H see DNJ800
DETA see DJG600
DFA see DVX800
DGE see DKM200
DIACETONE ALCOHOL see DBF750
DIACETOXYMERCURY see MCS750
2,4-DIAMINOANISOLE see DBO000
o-DIAMINOBENZENE see PEY250
p-DIAMINOBENZENE see PEY500
4,4′-DIAMINOBIPHENYL see BBX000
DI(-4-AMINO-3-CHLOROPHENYL)METHANE see MJM200
4,4′-DIAMINO-3,3′-DICHLORODIPHENYL see DEQ600
2,2′-DIAMINODIETHYLAMINE see DJG600
4,4-DIAMINODIPHENYL ETHER see OPM000
4,4′-DIAMINODIPHENYLMETHANE (DOT) see MJQ000
p,p′-DIAMINODIPHENYL SULFIDE see TFI000
1,2-DIAMINOETHANE see EEA500
1,6-DIAMINOHEXANE see HEO000
2,4-DIAMINO-1-METHOXYBENZENE see DBO000
1,4-DIAMINO-2-NITROBENZENE see

ALL750
DI-(4-AMINOPHENYL)METHANE see MJQ000
DI(p-AMINOPHENYL)SULPHIDE see TFI000
DIAMMONIUM MOLYBDATE see ANM750
o-DIANISIDINE see DCJ200
3,3′-DIANISIDINE see DCJ200
o-DIANISIDINE DIHYDROCHLORIDE see DOA800
DIANISIDINE DIISOCYANATE see DCJ400
DIATOMACEOUS EARTH see DCJ800
DIATOMACEOUS SILICA see DCJ800
DIAZIRINE see DCP800
DIAZOIMIDE see HHG500
DIAZOMETHANE see DCP800
DIBASIC ZINC STEARATE see ZMS000
DIBENZO-1,4-THIAZINE see PDP250
DIBENZOYL PEROXIDE (MAK) see BDS000
DIBORANE see DDI450
DIBORANE MIXTURES (NA 1911) see DDI450
DIBORON HEXAHYDRIDE see DDI450
1,2-DIBROMO-3-CHLOROPROPANE see DDL800
DIBROMODIFLUOROMETHANE see DKG850
1,2-DIBROMOETHANE (MAK) see EIY500
DIBUTYL ACID PHOSPHATE see DEG700
2-DIBUTYLAMINOETHANOL see DDU600
2-N-DIBUTYLAMINOETHANOL see DDU600
N,N-DI-n-BUTYLAMINOETHANOL (DOT) see DDU600
β-N-DIBUTYLAMINOETHYL ALCOHOL see DDU600
DIBU-
TYL-1,2-BENZENEDICARBOXYLATE see DEH200
2,6-DI-tert-BUTYL-p-CRESOL (OSHA, ACGIH) see BFW750
DIBUTYL HYDROGEN PHOSPHATE see DEG700
N,N-DIBUTYL-N-(2-HYDROXYETHYL)AMINE see DDU600
3,5-DI-tert-BUTYL-4-HYDROXYTOLUENE see BFW750
2,6-DI-tert-BUTYL-4-METHYLPHENOL see BFW750
DIBUTYLNITROSOAMINE see BRY500
N,N-DIBUTYLNITROSOAMINE see BRY500
DIBUTYLOXIDE of TIN see DEF400
DIBUTYLOXOSTANNANE see DEF400
DIBUTYLOXOTIN see DEF400
DI-tert-BUTYL PEROXIDE (MAK) see BSC750
DIBUTYL PHENYL PHOSPHATE see DEG600
DIBUTYL PHOSPHATE see DEG700
DI-n-BUTYL PHOSPHATE see DEG700
DIBUTYL PHTHALATE see DEH200
DIBUTYLSTANNANE OXIDE see DEF400
DIBUTYLTIN BIS(α-ETHYLHEXANOATE) see BJQ250
DIBUTYLTIN DI(2-ETHYLHEXANOATE) see BJQ250
DI-n-BUTYLTIN DI-2-ETHYLHEXANOATE see BJQ250
DIBUTYLTIN DI(2-ETHYLHEXOATE) see BJQ250
DICHLOROACETYLENE see DEN600
1,2-DICHLOROBENZENE see DEP600
1,4-DICHLOROBENZENE (MAK) see DEP800
o-DICHLOROBENZENE see DEP600
p-DICHLOROBENZENE see DEP800
DICHLOROBENZENE, PARA, solid (DOT) see DEP800
3′,3′-DICHLOROBENZIDINE see DEQ600
3,3′-DICHLOROBENZIDINE DIHYDROCHLORIDE see DEQ800
p-DICHLOROBENZOL see DEP800
3,3′-DICHLORO-4,4′-BIPHENYLDIAMINE see DEQ600
3,3′-DICHLORO-(1,1′-BIPHENYL)-4,4′-DIAMINE DIHYDROCHLORIDE see DEQ800
1,4-DICHLORO-2-BUTENE see DEV000
1,4-DICHLOROBUTENE-2 (MAK) see DEV000
1,1-DICHLORO-2-CHLOROETHYLENE see TIO750
1,1-DICHLORO-2,2-DICHLOROETHANE see TBQ100
DICHLORODIFLUOROMETHANE see DFA600
sym-DICHLORODIMETHYL ETHER (DOT) see BIK000
DICHLORODIMETHYLHYDANTOIN see DFE200
1,3-DICHLORO-5,5-DIMETHYL HYDANTOIN see DFE200
DICHLORODIPHENYL OXIDE see DFE800
1,1-DICHLOROETHANE see DFF809
1,2-DICHLOROETHANE see EIY600
1,1-DICHLOROETHYLENE see VPK000
1,2-DICHLOROETHYLENE see DFI200

1,2-DICHLOROETHYLENE see DFI210
cis-DICHLOROETHYLENE see DFI200
sym-DICHLOROETHYLENE see DFI210
DICHLOROETHYL ETHER see DFJ050
2,2'-DICHLOROETHYL ETHER (MAK) see DFJ050
DICHLOROETHYNE see DEN600
DICHLOROFLUOROMETHANE see DFL000
DICHLOROMETHANE (MAK, DOT) see MJP450
1,3-DICHLORO-5,5'-METHYLHYDANTOIN see DFE200
DICHLORONITROETHANE see DFU000
1,1-DICHLORO-1-NITROETHANE see DFU000
DICHLOROPHENYL ETHER see DFE800
1,2-DICHLOROPROPANE see PNJ400
1,3-DICHLOROPROPENE see DGG950
2,2-DICHLOROPROPIONIC ACID see DGI400
α-DICHLOROPROPIONIC ACID see DGI400
1,3-DICHLOROPROPYENE-1 see DGG950
1,3-DICHLOROPROPYLENE see DGG950
DICHLOROTETRAFLUOROETHANE see DGL600
DICHLOROTETRAFLUOROETHANE (OSHA, ACGIH) see FOO509
1,2-DICHLORO-1,1,2,2-TETRAFLUOROETHANE (MAK) see FOO509
DICOBALT CARBONYL see CNB500
DICOBALT OCTACARBONYL see CNB500
1,3-DICYANOBENZENE see PHX550
1,4-DICYANOBUTANE see AER250
DICYCLOHEXYL-METHANE-4,4'-DIISOCYANATE see MJM600
DICYCLOPENTADIENE see DGW000
DICYCLOPENTADIENYL IRON (OSHA, ACGIH) see FBC000
DI(2,3-EPOXYPROPYL) ETHER see DKM200
DIETHANOLAMINE see DHF000
DIETHYLAMINE see DHJ200
2-DIETHYLAMINOETHANOL see DHO500
β-DIETHYLAMINOETHANOL see DHO500
β-DIETHYLAMINOETHYL ALCOHOL see DHO500
DIETHYLCARBAMIDOYL CHLORIDE see DIW400
DIETHYLCARBAMOYL CHLORIDE see DIW400
N,N-DIETHYLCARBAMOYL CHLORIDE see DIW400
DIETHYL CARBINOL see IHP010
1,4-DIETHYLENE DIOXIDE see DVQ000
DIETHYLENE OXIMIDE see MRP750
DIETHYLENETRIAMINE see DJG600
DIETHYL ESTER SULFURIC ACID see DKB110
N,N-DIETHYLETHANAMINE see TJO000
DIETHYL ETHER (DOT) see EJU000
DI(2-ETHYLHEXYL)PHTHALATE see DVL700
DIETHYL KETONE see DJN750
DIETHYLNITROSAMINE see NJW500
DIETHYL OXIDE see EJU000
DIETHYL PHTHALATE see DJX000
DIETHYL-o-PHTHALATE see DJX000
DIETHYL SULFATE see DKB110
DIETHYL TETRAOXOSULFATE see DKB110
DIFLUOROCHLOROMETHANE see CFX500
DIFLUORODIBROMOMETHANE see DKG850
DIFLUORODICHLOROMETHANE see DFA600
1,1-DIFLUOROETHYLENE (DOT, MAK) see VPP000
DIFLUOROFORMALDEHYDE see CCA500
1,1-DIFLUORO-1,2,2,2-TETRACHLOROETHANE see TBP000
1,2-DIFLUORO-1,1,2,2-TETRACHLOROETHANE see TBP050
DIGLYCIDYL ETHER see DKM200
DIHYDROAZIRENE see EJM900
DIHYDROGEN DIOXIDE see HIB050
1,2-DIHYDRO-5-NITRO-ACENAPHTHYLENE see NEJ500
1,2-DIHYDROXYBENZENE see CCP850
1,3-DIHYDROXYBENZENE see REA000
DIHYDROXYBENZENE (OSHA) see HIH000
1,2-DIHYDROXYETHANE see EJC500
2,4-DIHYDROXY-2-METHYLPENTANE see HFP875
DIISOBUTYL KETONE see DNI800
1,3-DIISOCYANATOBENZENE see BBP000
1,6-DIISOCYANATOHEXANE see DNJ800
2,6-DIISOCYANATO-1-METHYLBENZENE see TGM800
2,4-DIISOCYANATO-1-METHYLBENZENE (9CI) see TGM750
1,5-DIISOCYANATONAPHTHALENE see NAM500
s-DIISOPROPYLACETONE see DNI800
DIISOPROPYLAMINE see DNM200
DIISOPROPYL ETHER see IOZ750
DIISOPROPYL OXIDE see IOZ750
DIKETONE ALCOHOL see DBF750

DILAUROYL PEROXIDE see LBR000
DIMAZINE see DSF400
3,3′-DIMETHOXYBENZIDINE see DCJ200
3,3′-DIMETHOXYBENZIDINE DIHYDRO-CHLORIDE see DOA800
3,3′-DIMETHOXYBENZIDINE-4,4′-DIISOCYANATE see DCJ400
3,3-DIMETHOXY-(1,1′-BIPHENYL)-4,4′-DIAMINE DIHYDROCHLORIDE see DOA800
3,3′-DIMETHOXY-4,4′-BIPHENYLENE DIISOCYANATE see DCJ400
N,N-DIMETHYLACETAMIDE see DOO800
DIMETHYLACETONE see DJN750
DIMETHYLACETONE AMIDE see DOO800
DIMETHYLAMIDE ACETATE see DOO800
DIMETHYLAMINE see DOQ800
4-DIMETHYLAMINOAZOBENZENE see DOT300
(DIMETHYLAMINO)CARBONYL CHLO-RIDE see DQY950
N,N-DIMETHYLAMINOCARBONYL CHLORIDE see DQY950
p,p-DIMETHYLAMINODIPHENYLMETHANE see MJN000
DIMETHYLANILINE see XMA000
2,4-DIMETHYLANILINE see XMS000
N,N-DIMETHYLANILINE see DQF800
N-DIMETHYL-ANILINE (OSHA) see DQF800
DIMETHYLBENZENE see XGS000
N,N-DIMETHYLBENZENEAMINE see DQF800
3,3′-DIMETHYLBENZIDIN see TGJ750
1,2-DIMETHYLBUTANE see IKS600
2,2-DIMETHYLBUTANE see DQT200
2,3-DIMETHYLBUTANE see DQT400
1,3-DIMETHYLBUTYL ACETATE see HFJ000
N,N-DIMETHYLCARBAMIDOYL CHLO-RIDE see DQY950
DIMETHYLCARBAMOYL CHLORIDE see DQY950
DIMETHYL CARBAMOYL CHLORIDE (ACGIH,DOT) see DQY950
DIMETHYLCARBINOL see INJ000
DIMETHYLCHLOROFORMAMIDE see DQY950
3,3′-DIMETHYLDIPHENYLMETHANE-4,4′-DIISOCYANATE see MJN750
DIMETHYLENEDIAMINE see EEA500
DIMETHYLENEIMINE see EJM900
DIMETHYLENE OXIDE see EJN500
1,1-DIMETHYLETHANOL see BPX000
1,1-DIMETHYLETHYLAMINE see BPY250
1,1-DIMETHYLETHYL HYDROPEROXIDE see BRM250
DIMETHYL FORMAL see MGA850
DIMETHYLFORMALDEHYDE see ABC750
DIMETHYLFORMAMIDE see DSB000
N,N-DIMETHYLFORMAMIDE (DOT) see DSB000
2,6-DIMETHYL-4-HEPTANONE see DNI800
1,1-DIMETHYLHYDRAZINE see DSF400
1,2-DIMETHYLHYDRAZINE see DSF600
DIMETHYLHYDRAZINE, symmetrical (DOT) see DSF600
DIMETHYLHYDRAZINE, unsymmetrical (DOT) see DSF400
DIMETHYL KETONE see ABC750
DIMETHYL MONOSULFATE see DUD100
N,N-DIMETHYLNITROSAMINE see NKA600
DIMETHYLPHENYLAMINE see XMA000
DIMETHYLPHENYLAMINE see DQF800
N,N-DIMETHYLPHENYLAMINE see DQF800
N,N-DIMETHYL-p-PHENYLAZOANILINE see DOT300
DIMETHYL PHTHALATE see DTR200
2,2-DIMETHYLPROPANE see NCH000
DIMETHYL SULFATE see DUD100
O,O-DIMETHYL-O-2,4,5-TRICHLOROPHENYL PHOSPHOROTHIOATE see RMA500
1,2-DINITROBENZENE see DUQ400
1,3-DINITROBENZENE see DUQ200
m-DINITROBENZENE see DUQ200
o-DINITROBENZENE see DUQ400
p-DINITROBENZENE see DUQ600
DINITRO-o-CRESOL see DUS700
2,4-DINITRO-o-CRESOL see DUS700
DINITROGEN MONOXIDE see NGU000
DINITROGLYCOL see EJG000
2,4-DINITRO-6-METHYLPHENOL see DUS700
DINITROPHENYLMETHANE see DVG600
DINITROTOLUENE see DVG600
DINOC see DUS700
DIOCTYL PHTHALATE see DVL700
DI-sec-OCTYL PHTHALATE see DVL700
2,3-p-DIOXANDITHIOL S,S-BIS(O,O-DIETHYL PHOSPHORODI-THIOATE) see DVQ709
DIOXANE see DVQ000
1,4-DIOXANE (MAK) see DVQ000
p-DIOXANE-2,3-DIYL ETHYL PHOS-PHORODITHIOATE see DVQ709
DIOXATHION see DVQ709
1,3-DIOXOPHTHALAN see PHW750

1,4-DIOXYBENZENE see QQS200
o-DIOXYBENZENE see CCP850
p-DIOXYBENZENE see HIH000
DIPA see DNM200
DIPHENYL (OSHA) see BGE000
DIPHENYLAMINE see DVX800
N,N-DIPHENYLANILINE see TMQ500
DIPHENYLBENZENE see TBD000
1,2-DIPHENYLBENZENE see TBC640
DIPHENYL mixed with DIPHENYL OXIDE see PFA860
DIPHENYL ETHER see PFA850
DIPHENYL OXIDE see PFA850
DIPHOSPHORUS PENTOXIDE see PHS250
DIPOTASSIUM PERSULFATE see DWQ000
DIPROPYLENE GLYCOL METHYL ETHER see DWT200
DIPROPYLENE GLYCOL MONOMETHYL ETHER see DWT200
DIPROPYL KETONE see DWT600
DIPROPYL METHANE see HBC500
DI-n-PROPYLNITROSAMINE see NKB700
DISODIUM DIFLUORIDE see SHF500
DISODIUM DISULFITE see SII000
DISODIUM SELENITE see SJT500
DISODIUM TETRAOXATUNGSTATE (2-) see SKN500
DISODIUM TUNGSTATE see SKN500
DISULFUR DICHLORIDE see SON510
DITHALLIUM CARBONATE see TEJ000
DITHIOCARBONIC ANHYDRIDE see CBV500
DIVINYL see BOP500
DIVINYLBENZENE see DXQ745
m-DIVINYLBENZENE see DXQ745
DMA see DOO800
DMA see DOQ800
DMAC see DOO800
DMAEE see BJH750
DMCC see DQY950
DMF see DSB000
DMFA see DSB000
DMH see DSF600
DMN see NKA600
DMNA see NKA600
DMP see DTR200
DMS see DUD100
1-DODECANETHIOL see LBX000
DODECANOYL PEROXIDE see LBR000
DODECYL MERCAPTAN see LBX000
DOLOMITE see CAO000
DOP see DVL700
DPN see NKB700
DPNA see NKB700
DRIERITE see CAX500
DRY ICE (UN 1845) (DOT) see CBU250
DTBP see BSC750
EAK see ODI000
ECH see EAZ500
EDC see EIY600
EGM see EJH500
EGME see EJH500
ENFLURANE see EAT900
EPICHLOROHYDRIN see EAZ500
1,4-EPOXYBUTANE see TCR750
1,2-EPOXY-4-(EPOXYETHYL)CYCLOHEXANE see VOA000
1,2-EPOXYETHANE see EJN500
4-(1,2-EPOXYETHYL)-7-OXABICYCLO(4.1.0)HEPTANE see VOA000
1,2-EPOXY-3-ISOPROPOXYPROPANE see IPD000
1,2-EPOXY-3-PHENOXYPROPANE see PFF360
1,2-EPOXYPROPANE see PNL600
2,3-EPOXY-1-PROPANOL (OSHA) see GGW500
2,3-EPOXYPROPYL BUTYL ETHER see BRK750
2,3-EPOXYPROPYL CHLORIDE see EAZ500
ETHANAL see AAG250
ETHANAMIDE see AAI000
ETHANECARBOXYLIC ACID see PMU750
1,2-ETHANEDIAMINE see EEA500
ETHANEDIOIC ACID see OLA000
1,2-ETHANEDIOL see EJC500
ETHANEDIONIC ACID see OLA000
ETHANE HEXACHLORIDE see HCI000
ETHANENITRILE see ABE500
ETHANE PENTACHLORIDE see PAW500
ETHANE TRICHLORIDE see TIN000
ETHANOIC ACID see AAT250
ETHANOL (MAK) see EFU000
ETHANOLAMINE see EEC600
ETHENE OXIDE see EJN500
ETHENONE see KEU000
ETHENYLBENZENE see SMQ000
4-ETHENYL-1-CYCLOHEXENE see CPD750
ETHINE see ACI750
2-ETHOXYETHANOL see EES350
2-ETHOXYETHANOL ACETATE see EES400
2-ETHOXYETHYL ACETATE see EES400
ETHYL ACETATE see EFR000
ETHYL ACETIC ESTER see EFR000
ETHYL ACETONE see PBN250
ETHYL ACRYLATE see EFT000
ETHYL ALCOHOL see EFU000

ETHYL ALDEHYDE see AAG250
ETHYLAMINE see EFU400
ETHYL AMYL KETONE see ODI000
ETHYL AMYL KETONE see EGI750
ETHYL BENZENE see EGP500
ETHYLBENZOL see EGP500
ETHYL BROMIDE see EGV400
ETHYL BUTYL KETONE see EHA600
ETHYL CARBAMATE see UVA000
ETHYL CHLORIDE see EHH000
ETHYL CYANIDE see PMV750
ETHYLDIMETHYLMETHANE see EIK000
ETHYLENE ALCOHOL see EJC500
ETHYLENE ALDEHYDE see ADR000
ETHYLENECARBOXYLIC ACID see ADS750
ETHYLENE CHLOROHYDRIN see EIU800
ETHYLENE CYANIDE see SNE000
ETHYLENEDIAMINE (OSHA) see EEA500
1,2-ETHYLENE DIBROMIDE see EIY500
ETHYLENE DICHLORIDE see EIY600
ETHYLENE GLYCOL see EJC500
ETHYLENE GLYCOL DINITRATE see EJG000
ETHYLENE GLYCOL ETHYL ETHER see EES350
ETHYLENE GLYCOL ISOPROPYL ETHER see INA500
ETHYLENE GLYCOL METHYL ETHER see EJH500
ETHYLENE GLYCOL METHYL ETHER ACETATE see EJJ500
ETHYLENE GLYCOL MONOBUTYL ETHER (MAK, DOT) see BPJ850
ETHYLENE GLYCOL MONOBUTYL ETHER ACETATE (MAK) see BPM000
ETHYLENE GLYCOL MONOETHYL ETHER (DOT) see EES350
ETHYLENE GLYCOL MONOETHYL ETHER ACETATE (MAK, DOT) see EES400
ETHYLENE GLYCOL, MONOISOPROPYL ETHER see INA500
ETHYLENE GLYCOL MONOMETHYL ETHER (MAK, DOT) see EJH500
ETHYLENE GLYCOL MONOMETHYL ETHER ACETATE see EJJ500
ETHYLENEIMINE see EJM900
ETHYLENE MONOCHLORIDE see VNP000
ETHYLENE OXIDE see EJN500
1-ETHYLENEOXY-3,4-EPOXYCYCLOHEXANE see VOA000
ETHYLENE TETRACHLORIDE see PCF275
ETHYLENE THIOUREA see IAQ000
N-ETHYL-ETHANAMINE see DHJ200
ETHYL ETHANOATE see EFR000
ETHYL ETHER see EJU000
ETHYL FORMATE see EKL000
ETHYL FORMIC ESTER see EKL000
2-ETHYLHEXYL PHTHALATE see DVL700
ETHYLIDENE CHLORIDE see DFF809
ETHYLIDENE DICHLORIDE see DFF809
ETHYLIDENE NORBORNENE see ELO500
5-ETHYLIDENE-2-NORBORNENE see ELO500
ETHYLIMINE see EJM900
ETHYL METHANOATE see EKL000
ETHYL METHYL KETONE (DOT) see MKA400
ETHYL METHYL KETONE PEROXIDE see MKA500
4-ETHYLMORPHOLINE see ENL000
N-ETHYLMORPHOLINE see ENL000
ETHYL NITRILE see ABE500
N-ETHYL-N-NITROSOBENZENAMINE see NKD000
N-ETHYL-N-NITROSO-ETHANAMINE see NJW500
ETHYL ORTHOSILICATE see EPF550
ETHYL PHTHALATE see DJX000
ETHYL PROPENOATE see EFT000
ETHYL SILICATE see EPF550
ETHYL SULFATE see DKB110
ETHYL THIOPYROPHOSPHATE see SOD100
ETHYL URETHANE see UVA000
ETHYNE see ACI750
ETO see EJN500
ETU see IAQ000
FAA see FDR000
FERRIC CHLORIDE see FAU000
FERRIC OXIDE see IHC450
FERROCENE see FBC000
FERROSULFATE see FBN100
FERROUS CHLORIDE see FBI000
FERROUS GLUCONATE see FBK000
FERROUS SULFATE see FBN100
FERROVANADIUM DUST see FBP000
FIBERGLASS see FBQ000
FIBROUS GLASS see FBQ000
FIBROUS GLASS DUST (ACGIH) see FBQ000
FLUE GAS see CBW750
FLUOBORIC ACID see FDD125
FLUOBORIC ACID (DOT) see FDD125
2-FLUORENYLACETAMIDE see FDR000
N-FLUOREN-2-YL ACETAMIDE see FDR000
N-2-FLUORENYLACETAMIDE see FDR000

FLUORIDES see FEY000
FLUORINE see FEZ000
FLUORINE MONOXIDE see ORA000
FLUORINE OXIDE see ORA000
FLUOROACETIC ACID, SODIUM SALT see SHG500
FLUOROCARBON-22 see CFX500
FLUOROCARBON 114 see FOO509
FLUOROETHENE see VPA000
FLUOROETHYLENE see VPA000
FLUOROFORMYL FLUORIDE see CCA500
FLUOROPHOSGENE see CCA500
FLUOROSULFONYL CHLORIDE see SOT500
FLUOROTRICHLOROMETHANE (OSHA) see TIP500
FORMAL see MGA850
FORMALDEHYDE see FMV000
FORMALDEHYDE CYANOHYDRIN see HIM500
FORMALIN (DOT) see FMV000
FORMAMIDE see FMY000
FORMIC ACID see FNA000
FORMIC ACID, ETHYL ESTER see EKL000
FORMIC ETHER see EKL000
N-FORMYLDIMETHYLAMINE see DSB000
FORMYLIC ACID see FNA000
FORMYL TRICHLORIDE see CHJ500
FREON 11 see TIP500
FREON 12 see DFA600
FREON 12-B2 see DKG850
FREON 13B1 see TJY100
FREON 21 see DFL000
FREON 22 see CFX500
FREON 30 see MJP450
FREON 31 see CHI900
FREON 112 see TBP050
FREON 113 see FOO000
FREON 114 see FOO509
FREON 115 see CJI500
2-FURALDEHYDE see FPQ875
2,5-FURANDIONE see MAM000
2-FURANMETHANOL see FPU000
FURFURAL see FPQ875
FURFURALDEHYDE see FPQ875
FURFURYL ALCOHOL see FPU000
FURRO L see DBO000
FURYL ALCOHOL see FPU000
2-FURYL-METHANAL see FPQ875
FUSED SILICA see SCK600
GASOLINE see GBY000
GERMANE (DOT) see GEI100
GERMANIUM HYDRIDE see GEI100
GERMANIUM TETRAHYDRIDE see GEI100
GLACIAL ACETIC ACID see AAT250
GLACIAL ACRYLIC ACID see ADS750
GLASS FIBERS see FBQ000
GLUTARALDEHYDE see GFQ000
GLUTARIC DIALDEHYDE see GFQ000
GLYCERIN see GGA000
GLYCERITOL see GGA000
GLYCEROL see GGA000
GLYCEROL EPICHLORHYDRIN see EAZ500
GLYCEROL TRICHLOROHYDRIN see TJB600
GLYCERYL TRINITRATE see NGY000
GLYCIDOL see GGW500
GLYCIDYL ALCOHOL see GGW500
GLYCIDYL BUTYL ETHER see BRK750
GLYCINOL see EEC600
GLYCOL DIBROMIDE see EIY500
GLYCOL DICHLORIDE see EIY600
GLYCOL DINITRATE see EJG000
GLYCOL ETHYLENE ETHER see DVQ000
GLYCOL ETHYL ETHER see EES350
GLYCOL MONOCHLOROHYDRIN see EIU800
GLYCOL MONOETHYL ETHER see EES350
GLYCOL MONOETHYL ETHER ACETATE see EES400
GLYCOL MONOMETHYL ETHER see EJH500
GLYCOL MONOMETHYL ETHER ACETATE see EJJ500
GLYCOLONITRILE see HIM500
GLYCYL ALCOHOL see GGA000
GRAIN ALCOHOL see EFU000
GRAPHITE SYNTHETIC (ACGIH,OSHA) see CBT500
GTN see NGY000
GYPSUM see CAX750
HALANE see DFE200
HALOCARBON 115 see CJI500
HALON 1011 see CES650
HALOWAX see TBR000
HCCPD see HCE500
HEMPA see HEK000
HEPTACHLOR see HAR000
3,4,5,6,7,8,8-HEPTACHLORODICYCLOPENTADIENE see HAR000
HEPTANE see HBC500
1-HEPTANETHIOL see HBD500
2-HEPTANONE see MGN500
3-HEPTANONE see EHA600
4-HEPTANONE see DWT600
HEPTYL HYDRIDE see HBC500
HEPTYL MERCAPTAN see HBD500

n-HEPTYLMERCAPTAN see HBD500
HEXACARBONYL CHROMIUM see HCB000
γ-HEXACHLORAN see BBQ500
HEXACHLOROBENZENE see HCC500
β-HEXACHLOROBENZENE see BBR000
HEXACHLOROBUTADIENE see HCD250
HEXACHLORO-1,3-BUTADIENE (MAK) see HCD250
α-HEXACHLOROCYCLOHEXANE see BBQ000
HEXACHLOROCYCLOPENTADIENE see HCE500
HEXACHLOROCYCLOPENTADIENE (ACGIH,DOT,OSHA) see HCE500
HEXACHLORODIPHENYL ETHER see CDV175
HEXACHLOROETHANE see HCI000
HEXACHLORONAPHTHALENE see HCK500
HEXAFLUOROACETONE see HCZ000
HEXAFLUOROPHOSPHORIC ACID see HDE000
HEXAHYDROANILINE see CPF500
HEXAHYDRO-2H-AZEPIN-2-ONE see CBF700
HEXAHYDROBENZENAMINE see CPF500
HEXAHYDROBENZENE see CPB000
HEXAHYDROCRESOL see MIQ745
HEXAHYDROMETHYLPHENOL see MIQ745
HEXAHYDROPHENOL see CPB750
HEXAHYDROTOLUENE see MIQ740
HEXALIN see CPB750
HEXAMETHYLENE see CPB000
HEXAMETHYLENEDIAMINE, solid (UN 2280) (DOT) see HEO000
HEXAMETHYLENE DIISOCYANATE (DOT) see DNJ800
1,6-HEXAMETHYLENE DIISOCYANATE see DNJ800
HEXAMETHYLPHOSPHORAMIDE see HEK000
HEXAMETHYLPHOSPHORIC ACID TRIAMIDE (MAK) see HEK000
HEXANAPHTHENE see CPB000
n-HEXANE see HEN000
1,6-HEXANEDIAMINE see HEO000
HEXANEDINITRILE see AER250
1,2-HEXANEDIOL see HFP875
1-HEXANETHIOL see HES000
2-HEXANONE see HEV000
HEXOGEN (Explosive) see CPR800
1,6-HEXOLACTAM see CBF700
HEXOLITE see CPR800
HEXONE see HFG500
sec-HEXYL ACETATE see HFJ000
HEXYLENE GLYCOL see HFP875
HEXYL MERCAPTAN see HES000
HI-FLASH NAPHTHA see NAH600
HMDA see HEO000
HMDI see DNJ800
HMPA see HEK000
HMPT see HEK000
HPT see HEK000
HYDRATED LIME see CAT225
HYDRAZINE see HGS000
HYDRAZINE AQUEOUS SOLUTIONS, with not >64% hydrazine, by weight (DOT) see HGU500
HYDRAZINE HYDRATE see HGU500
HYDRAZINE HYDRATE, with not >64% hydrazine, by weight (DOT) see HGU500
HYDRAZINE, HYDROCHLORIDE see HGV000
HYDRAZINE HYDROGEN SULFATE see HGW500
HYDRAZINE MONOCHLORIDE see HGV000
HYDRAZINE SULFATE (1:1) see HGW500
HYDRAZINOBENEZENE see PFI000
HYDRAZOIC ACID see HHG500
HYDRAZOMETHANE see MKN000
HYDRAZOMETHANE see DSF600
HYDRAZONIUM SULFATE see HGW500
HYDROBROMIC ACID see HHJ000
HYDROBROMIC ETHER see EGV400
HYDROCHLORIC ACID see HHL000
HYDROCHLORIC ETHER see EHH000
HYDROCHLORIDE see HHL000
HYDROCYANIC ACID see HHS000
HYDROCYANIC ACID, POTASSIUM SALT see PLC500
HYDROCYANIC ACID, SODIUM SALT see SGA500
HYDROFLUOBORIC ACID see FDD125
HYDROFLUORIC ACID see HHU500
HYDROGEN ANTIMONIDE see SLQ000
HYDROGENATED TERPHENYLS see HHW800
HYDROGEN AZIDE see HHG500
HYDROGEN BROMIDE (ACGIH,OSHA,MAK) see HHJ000
HYDROGEN CHLORIDE see HHX000
HYDROGEN CHLORIDE, anhydrous (UN 1050) (DOT) see HHL000
HYDROGEN CHLORIDE, refrigerated liquid (UN 2186) (DOT) see HHL000

HYDROGEN CYANAMIDE see COH500
HYDROGEN CYANIDE (ACGIH,OSHA) see HHS000
HYDROGEN FLUORIDE, anhydrous (UN 1052) (DOT) see HHU500
HYDROGEN HEXAFLUOROPHOSPHATE see HDE000
HYDROGEN PEROXIDE, 90% see HIB050
HYDROGEN PHOSPHIDE see PGY000
HYDROGEN POTASSIUM FLUORIDE see PKU250
HYDROGEN SELENIDE see HIC000
HYDROGEN SULFIDE see HIC500
HYDROGEN TETRAFLUOROBORATE see FDD125
HYDRONITRIC ACID see HHG500
2-HYDROPEROXY-2-METHYLPROPANE see BRM250
HYDROPHENOL see CPB750
HYDROQUINONE see HIH000
m-HYDROQUINONE see REA000
HYDROQUINONE, liquid or solid (DOT) see HIH000
HYDROQUINONE MONOMETHYL ETHER see MFC700
HYDROXYACETONITRILE see HIM500
1-HYDROXYBUTANE see BPW500
2-HYDROXYBUTANE see BPW750
1-HYDROXY-4-tert-BUTYLBENZENE see BSE500
HYDROXYCYCLOHEXANE see CPB750
3-HYDROXY-1,2-EPOXYPROPANE see GGW500
2-HYDROXYETHYLAMINE see EEC600
4-HYDROXY-2-KETO-4-METHYLPENTANE see DBF750
1-HYDROXY-2-METHYLBENZENE see CNX000
2-HYDROXYMETHYLFURAN see FPU000
1-HYDROXYMETHYLPROPANE see IIL000
2-HYDROXY-2-METHYLPROPIONITRILE see MLC750
3-HYDROXYPHENOL see REA000
o-HYDROXYPHENOL see CCP850
p-HYDROXYPHENOL see HIH000
1-HYDROXYPROPANE see PND000
3-HYDROXYPROPENE see AFV500
3-HYDROXYPROPIONIC ACID LACTONE see PMT100
2-HYDROXYPROPYL ACRYLATE see HNT600
o-HYDROXYTOLUENE see CNX000
2-HYDROXYTRIETHYLAMINE see DHO500
2-HYDROXY-1,3,5-TRINITROBENZENE see PID000
IGE (OSHA) see IPD000
2-IMIDAZOLIDINETHIONE see IAQ000
4,4'-(IMIDOCARBONYL)BIS(N,N-DIMETHYLAMINE) MONOHYDROCHLORIDE see IBA000
4,4'-(IMIDOCARBONYL)BIS(N,N-DIMETHYLANILINE) see IBB000
2,2'-IMINOBISETHANOL see DHF000
2,2'-IMINOBISETHYLAMINE see DJG600
2,2'-IMINODIETHANOL see DHF000
2,2'-IMINODI-N-NITROSOETHANOL see NKM000
INDENE see IBX000
INDIUM see ICF000
INDONAPHTHENE see IBX000
IODINE see IDM000
IODINE CRYSTALS see IDM000
IODINE SUBLIMED see IDM000
IODOFORM see IEP000
IODOMETHANE see MKW200
IPDI see IMG000
IPN see PHX550
IRON BIS(CYCLOPENTADIENE) see FBC000
IRON CARBONYL see IHG500
IRON CHLORIDE see FAU000
IRON COMPOUNDS see IGR499
IRON DICHLORIDE see FBI000
IRON GLUCONATE see FBK000
IRON MONOSULFATE see FBN100
IRON OXIDE see IHC450
IRON(III) OXIDE see IHC450
IRON PENTACARBONYL see IHG500
IRON PROTOCHLORIDE see FBI000
IRON(II) SULFATE (1:1) see FBN100
IRON TRICHLORIDE see FAU000
ISOACETOPHORONE see IMF400
ISOAMYL ACETATE see IHO850
ISOAMYL ALCOHOL see IHP000
ISOAMYL ALCOHOL see IHP010
ISOAMYL ETHANOATE see IHO850
ISOAMYLHYDRIDE see EIK000
ISOAMYL METHYL KETONE see MKW450
ISOBUTANOL (DOT) see IIL000
ISOBUTENYL METHYL KETONE see MDJ750
ISOBUTYL ACETATE see IIJ000
ISOBUTYL ALCOHOL see IIL000
ISOBUTYLAMINE see IIM000
ISOBUTYLCARBINOL see IHP000

ISOBUTYL KETONE see DNI800
ISOBUTYL METHYL CARBINOL see MKW600
ISOBUTYL METHYL KETONE see HFG500
ISOBUTYLTRIMETHYLETHANE see TLY500
ISOBUTYRONITRILE see IJX000
3-ISOCYANATOMETHYL-3,5,5-TRIMETHYLCYCLOHEXYLISOCYANATE see IMG000
ISOCYANIC ACID, METHYL ESTER see MKX250
ISOHEXANE see IKS600
ISONITROPROPANE see NIY000
ISOOCTANE (DOT) see TLY500
ISOOCTANOL see ILL000
ISOOCTYL ALCOHOL see ILL000
ISOPENTANE (DOT) see EIK000
ISOPENTYL ALCOHOL see IHP000
ISOPENTYL ALCOHOL ACETATE see IHO850
ISOPENTYL METHYL KETONE see MKW450
ISOPHORONE see IMF400
ISOPHORONE DIAMINE DIISOCYANATE see IMG000
ISOPHORONE DIISOCYANATE see IMG000
ISOPHTHALONITRILE see PHX550
ISOPROPANOL (DOT) see INJ000
ISOPROPENE CYANIDE see MGA750
ISOPROPENYLBENZENE see MPK250
ISOPROPENYLNITRILE see MGA750
2-ISOPROPOXYETHANOL see INA500
o-ISOPROPOXYPHENYL METHYLCARBAMATE see PMY300
2-ISOPROPOXYPROPANE see IOZ750
ISOPROPYL ACETATE see INE100
ISOPROPYLACETONE see HFG500
ISOPROPYL ALCOHOL see INJ000
ISOPROPYLAMINE see INK000
N-ISOPROPYLANILINE see INX000
ISOPROPYL BENZENE see COE750
ISOPROPYLBENZENE HYDROPEROXIDE see IOB000
ISOPROPYLCARBINOL see IIL000
ISOPROPYL CELLOSOLVE see INA500
ISOPROPYL CYANIDE see IJX000
ISOPROPYL ETHER see IOZ750
ISOPROPYL GLYCIDYL ETHER see IPD000
ISOPROPYL METHYL KETONE see MLA750
ISOPROPYL NITRILE see IJX000
3-ISOPROPYLOXYPROPYLENE OXIDE see IPD000
ISOVALERONE see DNI800
JEWELER'S ROUGE see IHC450
KEROSENE see KEK000
KETENE see KEU000
KETO-ETHYLENE see KEU000
KETOHEXAMETHYLENE see CPC000
KETONE PROPANE see ABC750
KIESELGUHR see DCJ800
LACTIC ACID, BUTYL ESTER see BRR600
LAUGHING GAS see NGU000
LAUREL CAMPHOR see CBA750
LAUROYL PEROXIDE see LBR000
LAURYL MERCAPTAN see LBX000
LEAD see LCF000
LEAD(II) AZIDE see LCM000
LEAD CHROMATE see LCR000
LEAD COMPOUNDS see LCT000
LIGROIN see PCT250
LIME see CAU500
LIME, BURNED see CAU500
LIME-NITROGEN (DOT) see CAQ250
LIMESTONE (FCC) see CAO000
LIME, UNSLAKED (DOT) see CAU500
LINDANE (ACGIH, DOT, USDA) see BBQ500
α-LINDANE see BBQ000
β-LINDANE see BBR000
LIQUEFIED PETROLEUM GAS see LGM000
LITHIUM FLUORIDE see LHF000
LITHIUM HYDRIDE see LHH000
LPG see LGM000
L.P.G. (OSHA, ACGIH) see LGM000
LYE see PLJ500
LYE (DOT) see SHS000
MAAC see HFJ000
MACE (lachrymator) see CEA750
MAGNESIA WHITE see CAX750
MAGNESIUM OXIDE see MAH500
MAGNESIUM OXIDE FUME (ACGIH) see MAH500
MALEIC ACID ANHYDRIDE (MAK) see MAM000
MALEIC ANHYDRIDE see MAM000
MALONIC DINITRILE see MAO250
MALONONITRILE see MAO250
MANGANESE see MAP750
MANGANESE COMPOUNDS see MAR500
MANGANESE CYCLOPENTADIENYL TRICARBONYL see CPV000
MANGANESE TRICARBONYL METHYLCYCLOPENTADIENYL see MAV750
MAOH see MKW600
MAPP (OSHA) see MFX600
MATTING ACID (DOT) see SOI500
MBK see HEV000

MBOCA see MJM200
MBOT see MJO250
MCB see CEJ125
MCT see CPV000
MDA see MJQ000
MDI see MJP400
MEK see MKA400
MEKP (OSHA) see MKA500
MEK PEROXIDE see MKA500
ME-MDA see MJO250
MEMPA see HEK000
MERCAPTOACETATE see TFJ100
2-MERCAPTOACETIC ACID see TFJ100
1-MERCAPTODODECANE see LBX000
2-MERCAPTOIMIDAZOLINE see IAQ000
1-MERCAPTOPROPANE see PML500
MERCURIC ACETATE see MCS750
MERCURIC BROMIDE see MCY000
MERCURIC BROMIDE, solid see MCY000
MERCURIC DIACETATE see MCS750
MERCURIC SULFOCYANATE see MCU250
MERCURIC THIOCYANATE see MCU250
MERCURY see MCW250
MERCURY BISULFATE see MDG500
MERCURY(II) BROMIDE (1:2) see MCY000
MERCURY COMPOUNDS, INORGANIC see MCZ000
MERCURY COMPOUNDS, ORGANIC see MDA000
MERCURY DITHIOCYANATE see MCU250
MERCURY, METALLIC (DOT) see MCW250
MERCURY(I) NITRATE (1:1) see MDE750
MERCURY(I) OXIDE see MDF750
MERCURY PERSULFATE see MDG500
MERCURY(II) SULFATE (1:1) see MDG500
MERCURY(II) THIOCYANATE see MCU250
MERCURYL ACETATE see MCS750
MESITYL OXIDE see MDJ750
METACETONE see DJN750
METALLIC ARSENIC see ARA750
METASILICIC ACID see SCL000
METHACETONE see DJN750
METHACRYLIC ACID see MDN250
METHACRYLIC ACID, METHYL ESTER (MAK) see MLH750
METHANAMIDE see FMY000
METHANECARBONITRILE see ABE500
METHANECARBOXAMIDE see AAI000
METHANE DICHLORIDE see MJP450
METHANE, TETRABROMIDE see CBX750
METHANE, TETRABROMO- see CBX750
METHANE TETRACHLORIDE see CBY000
METHANETHIOL see MLE650
METHANE TRICHLORIDE see CHJ500
METHANOIC ACID see FNA000
METHANOL see MGB150
METHENYL TRIBROMIDE see BNL000
METHOMYL see MDU600
o-METHOXYANILINE see AOV900
p-METHOXYANILINE see AOW000
2-METHOXYBENZENAMINE see AOV900
2-METHOXYETHANOL (ACGIH) see EJH500
METHOXY ETHER of PROPYLENE GLYCOL see PNL250
2-METHOXYETHYL ACETATE (ACGIH) see EJJ500
4-METHOXYPHENOL see MFC700
p-METHOXYPHENOL see MFC700
p-METHOXYPHENYLAMINE see AOW000
p-METHOXY-m-PHENYLENEDIAMINE see DBO000
1-METHOXY-2-PROPANOL see PNL250
METHYL ACETATE see MFW100
METHYL ACETIC ACID see PMU750
METHYL ACETONE (DOT) see MKA400
METHYL ACETYLENE see MFX590
METHYL ACETYLENE and PROPADIENE MIXTURES, stabilized (DOT) see MFX600
METHYL ACETYLENE-PROPADIENE MIXTURE see MFX600
β-METHYLACROLEIN see COB250
METHYL ACRYLATE see MGA500
METHYLACRYLONITRILE see MGA750
METHYLAL see MGA850
METHYL ALCOHOL see MGB150
METHYL ALDEHYDE see FMV000
METHYLAMINE see MGC250
METHYLAMINE (ACGIH,OSHA) see MGC250
2-METHYL-1-AMINOBENZENE see TGQ750
N-METHYLAMINOBENZENE see MGN750
METHYL AMYL ACETATE (DOT) see HFJ000
METHYL AMYL ALCOHOL see MKW600
METHYL n-AMYL KETONE see MGN500
METHYLANILINE see MGN750
4-METHYLANILINE see TGR000
m-METHYLANILINE see TGQ500
2-METHYLAZACYCLOPROPANE see PNL400
2-METHYLAZIRIDINE see PNL400
3-METHYLBENZENAMINE see TGQ500
N-METHYLBENZENAMINE see MGN750
o-METHYLBENZENAMINE see TGQ750
METHYLBENZENE see TGK750
METHYL BROMIDE see MHR200

2-METHYLBUTANE see EIK000
3-METHYL BUTANOL see IHP000
3-METHYL BUTAN-2-ONE (DOT) see MLA750
1-METHYLBUTYL ACETATE see AOD735
3-METHYL-1-BUTYL ACETATE see IHO850
1-METHYL-4-tert-BUTYLBENZENE see BSP500
p-METHYL-tert-BUTYLBENZENE see BSP500
METHYL n-BUTYL KETONE (ACGIH) see HEV000
METHYLCARBONYL FLUORIDE see ACM000
METHYL CELLOSOLVE (OSHA, DOT) see EJH500
METHYL CELLOSOLVE ACETATE (OSHA, DOT) see EJJ500
METHYL CHLORIDE see MIF765
1-METHYL-2-CHLOROBENZENE see CLK100
METHYL CHLOROFORM see MIH275
METHYL CHLOROMETHYL ETHER, anhydrous (DOT) see CIO250
METHYL CYANIDE see ABE500
METHYL 2-CYANOACRYLATE see MIQ075
METHYLCYCLOHEXANE see MIQ740
METHYLCYCLOHEXANOL see MIQ745
2-METHYLCYCLOHEXANONE see MIR500
1-METHYLCYCLOHEXAN-2-ONE see MIR500
METHYLCYCLOPENTADIENYL MANGANESE TRICARBONYL (OSHA) see MAV750
2-METHYLCYCLOPENTADIENYL MANGANESE TRICARBONYL (ACGIH) see MAV750
METHYLDINITROBENZENE see DVG600
2-METHYL-4,6-DINITROPHENOL see DUS700
METHYLENE ACETONE see BOY500
4,4′-METHYLENEBISANILINE see MJQ000
4,4′-METHYLENE BIS(2-CHLOROANILINE) see MJM200
METHYLENE-4,4′-BIS(o-CHLOROANILINE) see MJM200
p,p′-METHYLENEBIS(o-CHLOROANILINE) see MJM200
METHYLENE BIS(4-CYCLOHEXYLISOCYANATE) see MJM600
4,4′-METHYLENE BIS(N,N′-DIMETHYLANILINE) see MJN000
5,5′-METHYLENEBIS(2-ISOCYANATO)TOLUENE see MJN750
4,4′-METHYLENEBIS(2-METHYLANILINE) see MJO250
METHYLENE-BIS-ORTHOCHLOROANILINE see MJM200
METHYLENEBIS(p-PHENYLENE ISOCYANATE) see MJP400
METHYLENE BISPHENYL ISOCYANATE see MJP400
METHYLENE CHLORIDE see MJP450
METHYLENE CYANIDE see MAO250
4,4′-METHYLENEDIANILINE see MJQ000
METHYLENE DICHLORIDE see MJP450
METHYLENE DIMETHYL ETHER see MGA850
METHYLENE DI(PHENYLENE ISOCYANATE) (DOT) see MJP400
4,4′-METHYLENEDIPHENYL ISOCYANATE see MJP400
4,4′-METHYLENE DI-o-TOLUIDINE see MJO250
METHYL ETHANOATE see MFW100
1-METHYLETHYLAMINE see INK000
METHYLETHYLCARBINOL see BPW750
METHYL ETHYLENE OXIDE see PNL600
2-METHYLETHYLENIMINE see PNL400
METHYL ETHYL KETONE see MKA400
METHYL ETHYL KETONE HYDROPEROXIDE see MKA500
METHYL ETHYL KETONE PEROXIDE see MKA500
METHYLETHYLMETHANE see BOR500
N-(1-METHYLETHYL)-2-PROPANAMINE see DNM200
METHYLFLURETHER see EAT900
METHYL FORMATE see MKG750
METHYL GLYCOL ACETATE see EJJ500
METHYL GLYCOL MONOACETATE see EJJ500
3-METHYL-5-HEPTANONE see EGI750
5-METHYL-3-HEPTANONE see EGI750
5-METHYL-3-HEPTANONE (OSHA) see ODI000
2-METHYL-5-HEXANONE see MKW450
5-METHYLHEXAN-2-ONE (DOT) see MKW450
METHYL HYDRAZINE see MKN000
METHYLHYDRAZINE (DOT) see MKN000
METHYLHYDRAZINE HYDROCHLORIDE

see MKN250
METHYL IODIDE see MKW200
METHYL ISOAMYL KETONE see MKW450
METHYL ISOBUTENYL KETONE see MDJ750
METHYL ISOBUTYL CARBINOL see MKW600
METHYLISOBUTYLCARBINOL ACETATE see HFJ000
METHYL ISOBUTYL KETONE (ACGIH, DOT) see HFG500
METHYL ISOCYANATE see MKX250
N-METHYL-2-ISOPROPOXYPHENYLCARBAMATE see PMY300
METHYL ISOPROPYL KETONE see MLA750
METHYL KETONE see ABC750
2-METHYLLACTONITRILE see MLC750
METHYL MERCAPTAN see MLE650
METHYLMERCURY see MLF550
METHYL METHACRYLATE see MLH750
METHYL METHACRYLATE MONOMER, INHIBITED (DOT) see MLH750
N-METHYLMETHANAMINE see DOQ800
METHYL METHANOATE see MKG750
METHYL-N-((METHYLCARBAMOYL)OXY)THIOACETIMIDATE see MDU600
METHYL-2-METHYL-2-PROPENOATE see MLH750
2-METHYLNITROBENZENE see NMO525
3-METHYLNITROBENZENE see NMO500
p-METHYL NITROBENZENE see NMO550
N-METHYL-N-NITROSOANILINE see MMU250
N-METHYL-N-NITROSOBENZENAMINE see MMU250
N-METHYL-N-NITROSOMETHANAMINE see NKA600
METHYL ORTHOSILICATE (DOT) see MPI750
METHYL OXIRANE see PNL600
2-METHYLPENTANE see IKS600
2-METHYL-2,4-PENTANEDIOL see HFP875
4-METHYL-2-PENTANOL (MAK) see MKW600
4-METHYL-2-PENTANOL, ACETATE see HFJ000
2-METHYL-2-PENTANOL-4-ONE see DBF750
4-METHYL-2-PENTANONE (FCC) see HFG500
4-METHYL-3-PENTENE-2-ONE see MDJ750
METHYL PENTYL KETONE see MGN500
2-METHYLPHENOL see CNX000
N-METHYLPHENYLAMINE see MGN750
4-METHYL-m-PHENYLENEDIAMINE see TGL750
4-METHYL-PHENYLENE ISOCYANATE see TGM750
2-METHYL-m-PHENYLENE ISOCYANATE see TGM800
METHYL PHOSPHATE see TMD250
METHYL PHOSPHITE see TMD500
METHYL PHTHALATE see DTR200
2-METHYLPROPANENITRILE see IJX000
2-METHYL PROPANOL see IIL000
METHYL PROPENATE see MGA500
2-METHYLPROPENENITRILE see MGA750
METHYL PROPENOATE see MGA500
2-METHYLPROPENOIC ACID see MDN250
2-METHYLPROPYL ACETATE see IIJ000
1-METHYLPROPYLAMINE see BPY000
1-METHYL-5-PYRROLIDINONE see MPF200
METHYL PROPYL KETONE (ACGIH, DOT) see PBN250
N-METHYL-2-PYRROLIDINONE see MPF200
(S)-3-(1-METHYL-2-PYRROLIDINYL)PYRIDINE (9CI) see NDN000
N-METHYLPYRROLIDONE see MPF200
1-METHYL-2-PYRROLIDONE see MPF200
METHYL SILICATE see MPI750
METHYLSTYRENE see VQK650
α-METHYL STYRENE see MPK250
METHYL SULFATE (DOT) see DUD100
N-METHYL-N,2,4,6-TETRANITROANILINE see TEG250
METHYL TOLUENE see XGS000
2-METHYL-p-TOLUIDINE see XMS000
METHYLTRICHLOROMETHANE see MIH275
METHYL VINYL KETONE see BOY500
MIAK see MKW450
MIBC see MKW600
MIBK see HFG500
MIC see MKW600
MIC see MKX250
MICA see MQS250
MICHLER'S BASE see MJN000
MICHLER'S KETONE see MQS500
MIK see HFG500
MILK OF LIME see CAT225
MINERAL NAPHTHA see BBL250
MINERAL OIL see MQV750
MIPK see MLA750
MME see MFC700
MME see MLH750

MMH see MKN000
4-MMPD see DBO000
MMT see MAV750
MNA see MMU250
MNT see NMO500
MOCA see MJM200
MOLYBDATE see MRC250
MOLYBDENUM see MRC250
MOLYBDENUM COMPOUNDS see MRC750
MOLYBDENUM(VI) OXIDE see MRE000
MOLYBDENUM TRIOXIDE see MRE000
MOLYBDIC ACID DIAMMONIUM SALT see ANM750
MOLYBDIC ANHYDRIDE see MRE000
MOLYBDIC TRIOXIDE see MRE000
MONOAMMONIUM SULFAMATE see ANU650
MONOBROMOETHANE see EGV400
MONOBROMOMETHANE see MHR200
MONOBUTYLAMINE see BPX750
MONOCHLORETHANE see EHH000
MONOCHLOROACETALDEHYDE see CDY500
MONOCHLOROACETONE see CDN200
MONOCHLOROACETYL CHLORIDE see CEC250
MONOCHLOROBENZENE see CEJ125
MONOCHLORODIFLUOROMETHANE see CFX500
MONOCHLORODIMETHYL ETHER (MAK) see CIO250
MONOCHLORODIPHENYL OXIDE see MRG000
2-MONOCHLOROETHANOL see EIU800
MONOCHLOROETHYLENE (DOT) see VNP000
MONOCHLOROMETHANE see MIF765
MONO-CHLORO-MONO-BROMO-METHANE see CES650
MONOCHLOROMONOFLUOROMETHANE see CHI900
MONOCHLOROPENTAFLUOROETHANE (DOT) see CJI500
MONOCHLOROPHENYLETHER see MRG000
MONOETHANOLAMINE see EEC600
MONOETHYLAMINE (DOT) see EFU400
MONOETHYLENE GLYCOL see EJC500
MONOFLUOROETHYLENE see VPA000
MONOFLUOROTRICHLOROMETHANE see TIP500
MONOISOBUTYLAMINE see IIM000
MONOISOPROPYL ETHER of ETHYLENE GLYCOL see INA500
MONOMETHYLAMINE see MGC250
MONOMETHYL ANILINE (OSHA) see MGN750
MONOMETHYL ETHER of ETHYLENE GLYCOL see EJH500
MONOMETHYL ETHER HYDROQUINONE see MFC700
MONOMETHYL HYDRAZINE see MKN000
MONOPERACETIC ACID see PCL500
MONOSILANE see SDH575
MORPHOLINE see MRP750
MOTH BALLS (DOT) see NAJ500
MPK see PBN250
MTD see TGL750
MURIATIC ACID see HHL000
MURIATIC ETHER see EHH000
NAPHTHA see NAH600
NAPHTHA COAL TAR (OSHA) see NAH600
2-NAPHTHALAMINE see NBE500
NAPHTHALENE see NAJ500
1,5-NAPHTHALENE DIISOCYANATE see NAM500
NAPHTHALENE OIL see CMY825
NAPHTHALINE see NAJ500
NAPHTHA SAFETY SOLVENT see SLU500
1-NAPHTHYLAMINE see NBE700
β-NAPHTHYLAMINE see NBE500
2-NAPHTHYLPHENYLAMINE see PFT500
NDBA see BRY500
NDEA see NJW500
NDELA see NKM000
NDMA see NKA600
NDPA see NKB700
NEA see NKD000
NEM see ENL000
NEOHEXANE (DOT) see DQT200
NEOPENTANE see NCH000
NEOPRENE see NCI500
NICKEL see NCW500
NICKEL(II) CARBONATE (1:1) see NCY500
NICKEL CARBONYL see NCZ000
NICKEL COMPOUNDS see NDB000
NICKEL MONOXIDE see NDF500
NICKEL OXIDE (MAK) see NDF500
NICKEL SPONGE see NCW500
NICKEL TETRACARBONYL see NCZ000
NICOTINE see NDN000
NITRAMINE see TEG250
NITRAPYRIN (ACGIH) see CLP750
NITRIC ACID see NED500
NITRIC ACID, ALUMINUM SALT see AHD750

NITRIC ACID, BARIUM SALT see BAN250
NITRIC ACID, MERCURY(I) SALT see MDE750
NITRIC ACID, SILVER(1+) SALT see SDS000
NITRIC OXIDE see NEG100
NITRILOACETONITRILE see COO000
5-NITROACENAPHTHENE see NEJ500
p-NITROANILINE see NEO500
3-NITRO-3-AZAPENTANE-1,5-DIISOCYANATE see NHI500
4-NITROBENZENAMINE see NEO500
NITROBENZENE see NEX000
NITROBENZOL (DOT) see NEX000
p-NITROBIPHENYL see NFQ000
NITROCARBOL see NHM500
p-NITROCHLOROBENZENE see NFS525
NITROETHANE see NFY500
NITROGEN DIOXIDE see NGR500
NITROGEN FLUORIDE see NGW000
NITROGEN LIME see CAQ250
NITROGEN OXIDE see NGU000
NITROGEN PEROXIDE see NGR500
NITROGEN TRIFLUORIDE see NGW000
NITROGLYCERIN see NGY000
NITROGLYCERINE see NGY000
NITROIMINODIETHYLENEDIISOCYANIC ACID see NHI500
NITROMETHANE see NHM500
1-NITRONAPHTHALENE see NHQ000
2-NITRONAPHTHALENE see NHQ500
α-NITRONAPHTHALENE see NHQ000
β-NITRONAPHTHALENE see NHQ500
5-NITRONAPHTHALENE ETHYLENE see NEJ500
o-NITRO-p-PHENYLENEDIAMINE (MAK) see ALL750
1-NITROPROPANE see NIX500
2-NITROPROPANE see NIY000
1-NITROPYRENE see NJA000
3-NITROPYRENE see NJA000
N-NITROSODI-n-BUTYLAMINE (MAK) see BRY500
N-NITROSODIETHANOLAMINE (MAK) see NKM000
N-NITROSODIETHYLAMINE see NJW500
N-NITROSODIMETHYLAMINE see NKA600
N-NITROSQDI-N-PROPYLAMINE see NKB700
N-NITROSO-N-DIPROPYLAMINE see NKB700
N-NITROSO-N-ETHYL ANILINE see NKD000
N-NITROSOETHYLPHENYLAMINE (MAK) see NKD000
NITROSOIMINO DIETHANOL see NKM000
N-NITROSOMETHYLPHENYLAMINE (MAK) see MMU250
NITROSOMORPHOLINE see NKZ000
4-NITROSOMORPHOLINE see NKZ000
N-NITROSOMORPHOLINE (MAK) see NKZ000
1-NITROSOPIPERIDINE see NLJ500
N-NITROSOPIPERIDINE see NLJ500
N-NITROSO-N-PROPYL-1-PROPANAMINE see NKB700
1-NITROSOPYRROLIDINE see NLP500
N-NITROSOPYRROLIDINE see NLP500
2-NITROTOLUENE see NMO525
4-NITROTOLUENE see NMO550
m-NITROTOLUENE see NMO500
o-NITROTOLUENE see NMO525
p-NITROTOLUENE see NMO550
5-NITRO-o-TOLUIDINE see NMP500
3-NITROTOLUOL see NMO500
NITROTRICHLOROMETHANE see CKN500
NMA see MMU250
NMOR see NKZ000
NMP see MPF200
NONANE see NMX000
NORDHAUSEN ACID (DOT) see SOI500
1-NP see NIX500
2-NP see NIY000
NPIP see NLJ500
NPYR see NLP500
OCTACARBONYLDICOBALT see CNB500
OCTACHLOROCAMPHENE see CDV100
OCTACHLORONAPHTHALENE see OAP000
OCTAFLUORO-sec-BUTENE see OBM000
OCTAFLUOROISOBUTENE see OBM000
OCTANE see OCU000
3-OCTANONE see ODI000
OCTYL PHTHALATE see DVL700
ODB see DEP600
ODCB see DEP600
OIL of MIRBANE (DOT) see NEX000
OIL of TURPENTINE see TOD750
OIL MIST, MINERAL (OSHA, ACGIH) see MQV750
OIL of VITRIOL (DOT) see SOI500
ONT see NMO525
ORTHODICHLOROBENZENE see DEP600
ORTHODICHLOROBENZOL see DEP600
ORTHOPHOSPHORIC ACID see PHB250
ORTHOTELLURIC ACID see TAI750

OSMIC ACID see OKK000
OSMIUM(VIII) OXIDE see OKK000
OSMIUM TETROXIDE see OKK000
OXACYCLOPENTANE see TCR750
OXACYCLOPROPANE see EJN500
OXALIC ACID see OLA000
OXALIC ACID DINITRILE see COO000
OXALONITRILE see COO000
2-OXOBORNANE see CBA750
2-OXOHEXAMETHYLENIMINE see CBF700
OXYBENZENE see PDN750
4,4'-OXYBISBENZENAMINE see OPM000
4,4'-OXYDIANILINE see OPM000
4,4'-OXYDIPHENYLAMINE see OPM000
OXYFUME see EJN500
OXYGEN DIFLUORIDE see ORA000
OXYGEN FLUORIDE see ORA000
o-OXYTOLUENE see CNX000
OZONE see ORW000
PARA CRYSTALS see DEP800
PARADICHLOROBENZENE see DEP800
PARAFFIN see PAH750
PARAFFIN OIL see MQV750
PARAFFIN WAX see PAH750
PARAFFIN WAX FUME (ACGIH) see PAH750
PARANAPHTHALENE see APG500
PBNA see PFT500
PCB see PJL750
PCBs see PJL750
PCL see HCE500
PCM see PCF300
PCP see PAX250
PDB see DEP800
PDCB see DEP800
m-PDN see PHX550
PEAR OIL see IHO850
PENTA see PAX250
PENTABORANE(9) see PAT750
PENTABORANE (ACGIH,DOT,OSHA) see PAT750
PENTACHLOROANTIMONY see AQD000
PENTACHLORO DIPHENYL OXIDE see PAW250
PENTACHLOROETHANE see PAW500
PENTACHLORONAPHTHALENE see PAW750
PENTACHLOROPHENOL see PAX250
2,3,4,5,6-PENTACHLOROPHENOL see PAX250
PENTACHLOROPHENYL CHLORIDE see HCC500
PENTAERYTHRITE see PBB750
PENTAERYTHRITOL see PBB750
PENTAFLUOROANTIMONY see AQF250
PENTALIN see PAW500
PENTAMETHYLENE see CPV750
n-PENTANAL see VAG000
n-PENTANE see PBK250
tert-PENTANE see NCH000
1,5-PENTANEDIONE see GFQ000
1-PENTANETHIOL see PBM000
3-PENTANOL see IHP010
2-PENTANONE see PBN250
3-PENTANONE see DJN750
PENTOLE see CPU500
2-PENTYL ACETATE see AOD735
n-PENTYL ACETATE see AOD725
PENTYL MERCAPTAN see PBM000
PERCHLORETHYLENE see PCF275
PERCHLOROBENZENE see HCC500
PERCHLOROBUTADIENE see HCD250
PERCHLOROCYCLOPENTADIENE see HCE500
PERCHLOROETHANE see HCI000
PERCHLOROETHYLENE see PCF275
PERCHLOROMETHANE see CBY000
PERCHLOROMETHYL MERCAPTAN see PCF300
PERCHLORYL FLUORIDE see PCF750
PERFLUOROAMMONIUM OCTANOATE see ANP625
PERFLUOROISOBUTYLENE (ACGIH) see OBM000
2-PERHYDROAZEPINONE see CBF700
PERK see PCF275
PERMANGANATE of POTASH (DOT) see PLP000
PEROXYACETIC ACID see PCL500
PEROXYDISULFURIC ACID DIPOTASSIUM SALT see DWQ000
PETROLATUM, liquid see MQV750
PETROLEUM ASPHALT see ARO500
PETROLEUM BENZIN see NAH600
PETROLEUM DISTILLATE see PCS250
PETROLEUM DISTILLATES (NAPHTHA) see NAH600
PETROLEUM GAS, LIQUEFIED see LGM000
PETROLEUM SPIRIT (DOT) see PCT250
PETROLEUM SPIRITS see PCT250
PFIB see OBM000
PGDN see PNL000
PGE see PFF360
PHENACYL CHLORIDE see CEA750
PHENANTHRENE see PCW250
PHENANTRIN see PCW250
PHENOL see PDN750

PHENOL GLYCIDYL ETHER (MAK) see PFF360
PHENOTHIAZINE see PDP250
3-PHENOXY-1,2-EPOXYPROPANE see PFF360
PHENTHIAZINE see PDP250
3-(1'-PHENYL-2'-ACETYLETHYL)-4-HYDROXYCOUMARIN see WAT200
PHENYLAMINE see AOQ000
N-PHENYLANILINE see DVX800
N-PHENYLBENEZENAMINE see DVX800
PHENYLBENZENE see BGE000
PHENYL CHLORIDE see CEJ125
1,2-PHENYLENEDIAMINE (DOT) see PEY250
o-PHENYLENEDIAMINE see PEY250
p-PHENYLENEDIAMINE see PEY500
PHENYLENEDIAMINE, PARA, solid (DOT) see PEY500
m-PHENYLENEDIMETHYLENE ISOCYANATE see XIJ000
o-PHENYLENEDIOL see CCP850
m-PHENYLENE ISOCYANATE see BBP000
PHENYL-2,3-EPOXYPROPYL ETHER see PFF360
PHENYLETHANE see EGP500
PHENYL ETHER see PFA850
PHENYL ETHER-BIPHENYL MIXTURE see PFA860
PHENYL ETHER, HEXACHLORO derivative (8CI) see CDV175
PHENYL ETHER PENTACHLORO see PAW250
PHENYL ETHER TETRACHLORO see TBP250
PHENYL GLYCIDYL ETHER see PFF360
PHENYLHYDRAZINE see PFI000
PHENYLHYDRAZINE HYDROCHLORIDE see PFI250
PHENYLHYDRAZINE MONOHYDRO-CHLORIDE see PFI250
PHENYL HYDRIDE see BBL250
PHENYL MERCAPTAN see PFL850
PHENYLMETHANE see TGK750
N-PHENYLMETHYLAMINE see MGN750
PHENYLMETHYLNITROSAMINE see MMU250
PHENYL-2-NAPHTHYLAMINE see PFT500
N-PHENYL-β-NAPHTHYLAMINE see PFT500
4-PHENYL-NITROBENZENE see NFQ000
PHENYL PERCHLORYL see HCC500
PHENYLPHOSPHINE see PFV250
2-PHENYLPROPANE see COE750
2-PHENYLPROPENE see MPK250
2-PHENYLPROPYLENE see MPK250
PHENYLTRICHLOROMETHANE see BFL250
PHOSGENE see PGX000
PHOSPHINE see PGY000
PHOSPHORIC ACID see PHB250
PHOSPHORIC ACID, TRIMETHYL ESTER see TMD250
PHOSPHORIC ACID, TRIPHENYL ESTER see TMT750
PHOSPHORIC CHLORIDE see PHR500
PHOSPHORIC SULFIDE see PHS000
PHOSPHORIC TRIS(DIMETHYLAMIDE) see HEK000
PHOSPHORODITHIOIC ACID-S,S'-1,4-DIOXANE-2,3-DIYL-O,O,O',O'-TETRAETHYL ESTER see DVQ709
PHOSPHOROUS (WHITE) see PHP010
PHOSPHORUS (red) see PHO500
PHOSPHORUS (yellow) see PHP010
PHOSPHORUS, amorphous (DOT) see PHO500
PHOSPHORUS ACID, TRIMETHYL ESTER see TMD500
PHOSPHORUS CHLORIDE see PHT275
PHOSPHORUS OXYCHLORIDE see PHQ800
PHOSPHORUS OXYTRICHLORIDE see PHQ800
PHOSPHORUS PENTACHLORIDE see PHR500
PHOSPHORUS PENTAOXIDE see PHS250
PHOSPHORUS PENTASULFIDE see PHS000
PHOSPHORUS PENTOXIDE see PHS250
PHOSPHORUS PERCHLORIDE see PHR500
PHOSPHORUS PERSULFIDE see PHS000
PHOSPHORUS TRICHLORIDE see PHT275
PHOSPHORUS TRIHYDRIDE see PGY000
PHOSPHORYL CHLORIDE see PHQ800
PHTHALIC ACID ANHYDRIDE see PHW750
PHTHALIC ACID, DIETHYL ESTER see DJX000
PHTHALIC ACID DIOCTYL ESTER see DVL700
PHTHALIC ACID METHYL ESTER see DTR200
PHTHALIC ANHYDRIDE see PHW750
m-PHTHALODINITRILE see PHX550
PICRIC ACID see PID000
PICRYLNITROMETHYLAMINE see TEG250
PIMELIC KETONE see CPC000
PIPERAZINE DIHYDROCHLORIDE see PIK000

PIPERAZINE HYDROCHLORIDE see PIK000
PLASTER of PARIS see CAX500
PLATINIC AMMONIUM CHLORIDE see ANF250
PLATINUM see PJD500
PLATINUM BLACK see PJD500
PLATINUM(IV) CHLORIDE see PJE250
PLATINUM SPONGE see PJD500
PLATINUM TETRACHLORIDE see PJE250
PNA see NEO500
PNB see NFQ000
PNCB see NFS525
PNT see NMO550
POLYCHLORCAMPHENE see CDV100
POLYCHLORINATED BIPHENYLS see PJL750
POLYCHLOROBIPHENYL see PJL750
POLY(CHLOROETHYLENE) see PKQ059
POLYVINYL CHLORIDE see PKQ059
PORTLAND CEMENT see PKS750
POTASSIUM ACID FLUORIDE see PKU250
POTASSIUM BIFLUORIDE, solid or solution (DOT) see PKU250
POTASSIUM CYANIDE see PLC500
POTASSIUM FLUORIDE see PKU250
POTASSIUM FLUORIDE see PLF500
POTASSIUM FLUOSILICATE see PLH750
POTASSIUM FLUOTANTALATE see PLH000
POTASSIUM HEPTAFLUOROTANTALATE see PLH000
POTASSIUM HEXAFLUOROSILICATE see PLH750
POTASSIUM HYDRATE (DOT) see PLJ500
POTASSIUM HYDROGEN FLUORIDE see PKU250
POTASSIUM HYDROXIDE see PLJ500
POTASSIUM PERMANGANATE see PLP000
POTASSIUM PEROXYDISULFATE see DWQ000
POTASSIUM PERSULFATE (DOT) see DWQ000
POTASSIUM SILICOFLUORIDE (DOT) see PLH750
POTASSIUM SILVER CYANIDE see PLS250
PPD see PEY500
PRECIPITATED SILICA see SCL000
2-PROPANAMINE see INK000
1-PROPANAMINE, 2-METHYL- see IIM000
2-PROPANAMINE, N-(1-METHYLETHYL)- see DNM200
PROPANE see PMJ750
PROPANEDINITRILE see MAO250
PROPANEDINTRILE((2-CHLOROPHENYL) METHYLENE) see CEQ600
1,2-PROPANEDIOL-1-ACRYLATE see HNT600
PROPANENITRILE see PMV750
1-PROPANETHIOL see PML500
1,2,3-PROPANETRIOL see GGA000
2-PROPANOL see INJ000
n-PROPANOL see PND000
2-PROPANONE see ABC750
PROPARGYL ALCOHOL see PMN450
2-PROPENAL see ADR000
PROPENAMIDE see ADS250
2-PROPENAMIDE see ADS250
2-PROPENENITRILE see ADX500
PROPENOIC ACID see ADS750
2-PROPENOIC ACID, ETHYL ESTER (MAK) see EFT000
PROPENOIC ACID METHYL ESTER see MGA500
PROPENOL see AFV500
PROPENYL ALCOHOL see AFV500
2-PROPENYL CHLORIDE see AGB250
((2-PROPENYLOXY)METHYL)OXIRANE see AGH150
PROPINE see MFX590
1,3-PROPIOLACTONE see PMT100
β-PROPIOLACTONE see PMT100
PROPIONE see DJN750
PROPIONIC ACID see PMU750
PROPIONIC ACID (ACGIH,DOT,OSHA) see PMU750
PROPIONIC NITRILE see PMV750
PROPIONONITRILE see PMV750
PROPOXUR see PMY300
2-PROPYL ACETATE see INE100
n-PROPYL ACETATE see PNC250
n-PROPYL ALCOHOL see PND000
PROPYL CYANIDE see BSX250
PROPYLENE CHLORIDE see PNJ400
PROPYLENE DICHLORIDE see PNJ400
PROPYLENE GLYCOL DINITRATE see PNL000
1,2-PROPYLENE GLYCOL DINITRATE see PNL000
PROPYLENE GLYCOL METHYL ETHER see PNL250
PROPYLENE GLYCOL MONOACRYLATE see HNT600
PROPYLENE GLYCOL MONOMETHYL ETHER see PNL250
PROPYLENE IMINE see PNL400
PROPYLENE OXIDE see PNL600
PROPYLIC ALCOHOL see PND000

PROPYL KETONE see DWT600
PROPYL MERCAPTAN see PML500
PROPYLMETHANOL see BPW500
n-PROPYL NITRATE see PNQ500
N-PROPYLTHIOL see PML500
PROPYNE (OSHA) see MFX590
1-PROPYNE-3-OL see PMN450
3-PROPYNOL see PMN450
2-PROPYNYL ALCOHOL see PMN450
PRUSSIC ACID see HHS000
PRUSSITE see COO000
PSEUDOCUMIDINE see TLG250
PURE QUARTZ see SCI500
PURE QUARTZ see SCJ500
PVC (MAK) see PKQ059
PYRANTON see DBF750
PYRENE see PON250
PYRETHRINS see POO250
PYRETHRUM (ACGIH) see POO250
PYRIDINE see POP250
α-PYRIDYLAMINE see AMI000
PYROCATECHOL see CCP850
PYROPENTYLENE see CPU500
QUARTZ see SCJ500
QUICKLIME (DOT) see CAU500
QUICK SILVER see MCW250
QUINONE see QQS200
R12B2 (DOT) see DKG850
R13B1 (DOT) see TJY100
R 20 (refrigerant) see CHJ500
R22 (DOT) see CFX500
R114 (DOT) see DGL600
R1132a (DOT) see VPP000
RDX see CPR800
REFRIGERANT 12 see DFA600
REFRIGERANT 113 see FOO000
REFRIGERANT 112a see TBP000
RESORCINE see REA000
RESORCINOL see REA000
RHODIUM see RHF000
ROAD ASPHALT (DOT) see ARO500
ROAD TAR (DOT) see ARO500
RONNEL see RMA500
ROTENONE see RNZ000
SALICYLOYLOXYTRIBUTYLSTANNANE see SAM000
SAND see SCI500, SCJ500
SDMH see DSF600
SELENINYL CHLORIDE see SBT500
SELENIOUS ACID, DISODIUM SALT see SJT500
SELENIUM see SBO500
SELENIUM CHLORIDE OXIDE see SBT500
SELENIUM FLUORIDE see SBS000
SELENIUM HEXAFLUORIDE see SBS000
SELENIUM HYDRIDE see HIC000
SELENIUM METAL POWDER, NON-PYROPHORIC (DOT) see SBO500
SELENIUM OXYCHLORIDE see SBT500
SILANE see SDH575
SILICA, AMORPHOUS-DIATOMACEOUS EARTH (UNCALCINED) (ACGIH) see DCJ800
SILICA, AMORPHOUS FUMED see SCH002
SILICA, AMORPHOUS-FUSED (ACGIH) see SCK600
SILICA, AMORPHOUS HYDRATED see SCI000
SILICA, CRYSTALLINE see SCI500
SILICA, CRYSTALLINE–CRISTOBALITE see SCJ000
SILICA, CRYSTALLINE–QUARTZ see SCJ500
SILICA, CRYSTALLINE–TRIDYMITE see SCK000
SILICA FLOUR (powdered crystalline silica) see SCJ500
SILICA, FUSED see SCK600
SILICA GEL see SCI000
SILICA GEL see SCL000
SILICA, GEL and AMORPHOUS–PRECIPITATED see SCL000
SILICIC ACID see SCI000
SILICIC ACID see SCL000
SILICON see SCP000
SILICON CARBIDE see SCQ000
SILICON DIOXIDE see SCI500
SILICON DIOXIDE see SCK600
SILICON DIOXIDE (FCC) see SCH002
SILICON FLUORIDE see SDF650
SILICON MONOCARBIDE see SCQ000
SILICON POWDER, amorphous (DOT) see SCP000
SILICON TETRAFLUORIDE (DOT) see SDF650
SILICON TETRAHYDRIDE see SDH575
SILVER see SDI500
SILVER AMMONIUM NITRATE see SDL500
SILVER CYANIDE see SDP000
SILVER NITRATE (DOT) see SDS000
SILVER(I) NITRATE (1:1) see SDS000
SILVER POTASSIUM CYANIDE see PLS250
SLAKED LIME see CAT225
SODA LYE see SHS000
SODIUM ACID SULFITE see SFE000
SODIUM AZIDE see SFA000
SODIUM BISULFITE see SFE000
SODIUM BORATE see SFE500

SODIUM BORATE anhydrous see SFE500
SODIUM BORATE DECAHYDRATE see SFF000
SODIUM CYANIDE see SGA500
SODIUM DISULFITE see SII000
SODIUM FLUORIDE see SHF500
SODIUM FLUOROACETATE see SHG500
SODIUM HYDRATE (DOT) see SHS000
SODIUM HYDROFLUORIDE see SHF500
SODIUM HYDROGEN SULFITE see SFE000
SODIUM HYDROXIDE see SHS000
SODIUM METABISULFITE see SII000
SODIUM METAVANADATE see SKP000
SODIUM MONOFLUORIDE see SHF500
SODIUM MONOFLUOROACETATE see SHG500
SODIUM ORTHOVANADATE see SIY250
SODIUM PYROPHOSPHATE (FCC) see TEE500
SODIUM PYROSULFITE see SII000
SODIUM SELENITE see SJT500
SODIUM TCA SOLUTION see TII250
SODIUM TELLURATE see SKC000
SODIUM TETRABORATE see SFF000
SODIUM TETRABORATE DECAHYDRATE see SFF000
SODIUM TUNGSTATE see SKN500
SODIUM VANADATE see SIY250
SODIUM VANADATE see SKP000
SODIUM VANADIUM OXIDE see SIY250
SODIUM WOLFRAMATE see SKN500
SPENT SULFURIC ACID (DOT) see SOI500
SPIRIT of TURPENTINE see TOD750
STANNOUS CHLORIDE (FCC) see TGC000
STANNOUS FLUORIDE see TGD100
STEARIC ACID, ZINC SALT see ZMS000
STIBINE see SLQ000
STIBIUM see AQB750
STODDARD SOLVENT see SLU500
STRAIGHT-RUN KEROSENE see KEK000
STRONTIUM CHROMATE (1:1) see SMH000
STRYCHNINE see SMN500
STYRENE see SMQ000
STYRENE MONOMER, inhibited (DOT) see SMQ000
SUBTILISINS (ACGIH) see BAB750
SUCCINIC ACID DINITRILE see SNE000
SUCCINONITRILE see SNE000
SULFINYL CHLORIDE see TFL000
SULFOTEP see SOD100
SULFOTEPP see SOD100
SULFUR CHLORIDE OXIDE see TFL000
SULFUR DECAFLUORIDE see SOQ450
SULFUR DIOXIDE see SOH500
SULFUR FLUORIDE see SOI000
SULFUR HEXAFLUORIDE see SOI000
SULFURIC ACID see SOI500
SULFURIC ACID, BARIUM SALT (1:1) see BAP000
SULFURIC ACID, CALCIUM(2+) SALT, DIHYDRATE see CAX750
SULFURIC ACID, DIMETHYL ESTER see DUD100
SULFURIC OXYFLUORIDE see SOU500
SULFUR MONOCHLORIDE see SON510
SULFUROUS ACID ANHYDRIDE see SOH500
SULFUROUS ACID, MONOSODIUM SALT see SFE000
SULFUROUS ANHYDRIDE see SOH500
SULFUROUS OXIDE see SOH500
SULFUROUS OXYCHLORIDE see TFL000
SULFUR OXIDE see SOH500
SULFUR PENTAFLUORIDE see SOQ450
SULFUR SUBCHLORIDE see SON510
SULFUR TETRAFLUORIDE see SOR000
SULFURYL CHLORIDE FLUORIDE see SOT500
SULFURYL FLUORIDE see SOU500
SULFURYL FLUOROCHLORIDE see SOT500
SULPHOCARBONIC ANHYDRIDE see CBV500
SUPER GLUE see MIQ075
SUPER VMP see NAH600
T4 see CPR800
TALC see TAB750
TALCUM see TAB750
TANTALUM see TAE750
TANTALUM POTASSIUM FLUORIDE see PLH000
TAR OIL see CMY825
TBE see ACK250
TBP see TIA250
TBT see BSP500
TCA see TII250
TCE see TBQ100
1,1,1-TCE see MIH275
TDI (OSHA) see TGM750
2,6-TDI see TGM800
TEDP see SOD100
TEDP (OSHA) see SOD100
TEDTP see SOD100
TEL see TCF000
TELLURATE see TAI750
TELLURIC ACID see TAI750
TELLURIUM see TAJ000
TELLURIUM HEXAFLUORIDE see TAK250

TELLURIUM HYDROXIDE see TAI750
TEN see TJO000
TEPP (ACGIH) see TCF250
o-TERPHENYL see TBC640
TERPHENYLS see TBD000
TETRABROMIDE METHANE see CBX750
TETRABROMOACETYLENE see ACK250
1,1,2,2-TETRABROMOETHANE see ACK250
TETRABROMOMETHANE see CBX750
TETRACARBONYL NICKEL see NCZ000
1,1,1,2-TETRACHLORO-2,2-DIFLUOROETHANE see TBP000
1,1,2,2-TETRACHLORO-1,2-DIFLUOROETHANE see TBP050
TETRACHLORODIPHENYL OXIDE see TBP250
TETRACHLOROETHANE see TBP750
1,1,2,2-TETRACHLOROETHANE see TBQ100
TETRACHLOROETHYLENE (DOT) see PCF275
TETRACHLOROMETHANE see CBY000
TETRACHLORONAPHTHALENE see TBR000
TETRACHLOROPHENYL ETHER see TBP250
TETRAETHYL DITHIONOPYROPHOSPHATE see SOD100
TETRAETHYL DITHIOPYROPHOSPHATE see SOD100
TETRAETHYL LEAD see TCF000
TETRAETHYL ORTHOSILICATE (DOT) see EPF550
TETRAETHYLPLUMBANE see TCF000
TETRAETHYL PYROPHOSPHATE see TCF250
TETRAETHYL SILICATE (DOT) see EPF550
TETRAETHYLSTANNANE see TCF750
TETRAETHYL TIN see TCF750
TETRAFLUOROBORIC ACID see FDD125
TETRAFLUORODICHLOROETHANE see DGL600
1,1,2,2-TETRAFLUORO-1,2-DICHLOROETHANE see FOO509
TETRAFLUOROSILANE see SDF650
TETRAFLUOROSULFURANE see SOR000
1,2,3,4-TETRAHYDROBENZENE see CPC579
TETRAHYDRO-p-DIOXIN see DVQ000
TETRAHYDROFURAN see TCR750
TETRAHYDRO-p-ISOXAZINE see MRP750
TETRAHYDRO-N-NITROSOPYRROLE see NLP500
1,2,3,4-TETRAHYDROSTYRENE see CPD750
TETRAKIS(HYDROXYMETHYL)METHANE see PBB750
TETRAMETHOXYSILANE see MPI750
TETRAMETHYLDIAMINOBENZOPHENONE see MQS500
TETRAMETHYLDIAMINODIPHENYLACETIMINE see IBB000
4,4'-TETRAMETHYLDIAMINODIPHENYL METHANE see MJN000
TETRAMETHYLENE CYANIDE see AER250
TETRAMETHYLENE OXIDE see TCR750
TETRAMETHYL LEAD see TDR500
TETRAMETHYL SILICATE see MPI750
TETRAMETHYLSUCCINONITRILE see TDW250
TETRANITROMETHANE see TDY250
N,2,4,5-TETRANITRO-N-METHYLANILINE see TEG250
TETRASODIUM DIPHOSPHATE see TEE500
TETRASODIUM PYROPHOSPHATE see TEE500
TETRIL see TEG250
TETRYL see TEG250
THALLIC OXIDE see TEL050
THALLIUM see TEI000
THALLIUM ACETATE see TEI250
THALLIUM BROMIDE see TEI750
THALLIUM(I) CARBONATE (2:1) see TEJ000
THALLIUM CHLORIDE see TEJ250
THALLIUM COMPOUNDS see TEJ500
THALLIUM MONOACETATE see TEI250
THALLIUM MONOCHLORIDE see TEJ250
THALLIUM MONONITRATE see TEK750
THALLIUM NITRATE see TEK750
THALLIUM(III) OXIDE see TEL050
THALLIUM PEROXIDE see TEL050
THALLIUM SESQUIOXIDE see TEL050
THALLOUS ACETATE see TEI250
THALLOUS CARBONATE see TEJ000
THALLOUS CHLORIDE see TEJ250
THALLOUS NITRATE see TEK750
THERMAL ACETYLENE BLACK see CBT750
THF see TCR750
4,4'-THIOBISBENZENAMINE see TFI000
4,4'-THIOBIS(6-tert-BUTYL-m-CRESOL) see TFC600
4,4'-THIOBIS(2-tert-BUTYL-5-METHYLPHENOL) see TFC600

THIOCARBAMIZINE see TFD750
4,4'-THIODIANILINE see TFI000
THIODIPHENYLAMINE see PDP250
THIODI-p-PHENYLENEDIAMINE see TFI000
THIOGLYCOLIC ACID see TFJ100
THIOGLYCOLLIC ACID see TFJ100
2-THIOL-DIHYDROGLYOXALINE see IAQ000
THIOMETHANOL see MLE650
THIONYL CHLORIDE see TFL000
THIOPHENOL (DOT) see PFL850
THIOPHOSPHORIC ANHYDRIDE see PHS000
THIOSULFUROUS DICHLORIDE see SON510
TIN see TGB250
TIN(II) CHLORIDE (1:2) see TGC000
TIN COMPOUNDS see TGC500
TIN DICHLORIDE see TGC000
TIN DIFLUORIDE see TGD100
TIN FLUORIDE see TGD100
TITANIUM DIOXIDE see TGG760
TITANIUM OXIDE see TGG760
TL 212 see SOT500
TLA see BPV100
TMA see TKV000
TMA see TLD500
TML see TDR500
TMP see TMD250
TMSN see TDW250
TNM see TDY250
TNT (OSHA) see TMN490
TOCP see TMO600
TOFK see TMO600
o-TOLIDIN see TGJ750
o-TOLIDINE see TGJ750
TOLUENE see TGK750
TOLUENE-2,4-DIAMINE see TGL750
2,4-TOLUENEDIAMINE (DOT) see TGL750
TOLUENE-2,4-DIISOCYANATE see TGM750
TOLUENE-2,6-DIISOCYANATE see TGM800
2,6-TOLUENE DIISOCYANATE see TGM800
TOLUENE HEXAHYDRIDE see MIQ740
m-TOLUIDINE see TGQ500
o-TOLUIDINE see TGQ750
p-TOLUIDINE see TGR000
TOLUOL (DOT) see TGK750
o-TOLUOL see CNX000
2,4-TOLUYLENEDIAMINE (DOT) see TGL750
m-TOLYLAMINE see TGQ500
TOLYL CHLORIDE see BEE375
o-TOLYL CHLORIDE see CLK100
2,4-TOLYLENEDIISOCYANATE see TGM750
o-TOLYL PHOSPHATE see TMO600
TOTP see TMO600
TOXAPHENE see CDV100
TPP see TMT750
TREMOLITE see ARM250
TRIATOMIC OXYGEN see ORW000
1,3,5-TRIAZINE, HEXAHYDRO-1,3,5-TRINITRO-(9CI) see CPR800
TRIAZOIC ACID see HHG500
TRIBROMOALUMINUM see AGX750
TRIBROMOMETHANE see BNL000
TRIBUTYL(METHACRYLOXY) STANNANE see THZ000
(TRIBUTYL)PEROXIDE see BSC750
TRIBUTYL PHOSPHATE see TIA250
TRI-n-BUTYL PHOSPHATE see TIA250
TRIBUTYLSTANNYL METHACRYLATE see THZ000
TRIBUTYLTIN METHACRYLATE see THZ000
TRIBUTYLTIN SALICYLATE see SAM000
TRI-N-BUTYLTIN SALICYLATE see SAM000
TRICARBONYL(METHYLCYCLOPENTADIENYL)MANGANESE see MAV750
TRICHLOROACETIC ACID see TII250
TRICHLOROALUMINUM see AGY750
1,2,4-TRICHLOROBENZENE see TIK250
2,3,4-TRICHLOROBUTENE-1 see TIL360
TRICHLORO DIPHENYL OXIDE see CDV175
1,1,1-TRICHLOROETHANE see MIH275
1,1,2-TRICHLOROETHANE see TIN000
TRICHLOROETHANOIC ACID see TII250
TRICHLOROETHYLENE see TIO750
1,2,2-TRICHLOROETHYLENE see TIO750
TRICHLOROFLUOROMETHANE see TIP500
TRICHLOROHYDRIN see TJB600
TRICHLOROMETHANE see CHJ500
TRICHLOROMETHANE SULFENYL CHLORIDE see PCF300
TRICHLOROMETHYLBENZENE see BFL250
TRICHLORONAPHTHALENE see TIT500
TRICHLORONITROMETHANE see CKN500
TRICHLOROOXOVANADIUM see VDP000

2,4,5-TRICHLOROPHENOL, O-ESTER with O,O-DIMETHYL PHOSPHOROTHIOATE see RMA500
1,2,3-TRICHLOROPROPANE see TJB600
TRICHLOROSTIBINE see AQC500
TRICHLOROTRIFLUOROETHANE see FOO000
1,1,2-TRICHLORO-1,2,2-TRIFLUOROETHANE (OSHA, ACGIH, MAK) see FOO000
TRICRESOL see CNW500
TRI-o-CRESYL PHOSPHATE see TMO600
TRI(DIMETHYLAMINO)PHOSPHINE OXIDE see HEK000
TRIDYMITE see SCK000
TRIETHYLAMINE see TJO000
TRIFLUOROBROMOMETHANE see TJY100
(2,2,2-TRIFLUOROETHOXY)ETHENE see TKB250
2,2,2-TRIFLUOROETHYL VINYL ETHER see TKB250
TRIFLUOROMONOBROMOMETHANE see TJY100
1,2,3-TRIHYDROXYPROPANE see GGA000
TRIIODOMETHANE see IEP000
TRIISOCYANATOISOCYANURATE, solution, 70%, by weight (DOT) see IMG000
TRIMELLIC ACID ANHYDRIDE see TKV000
TRIMELLITIC ANHYDRIDE see TKV000
TRIMETHYLAMINE see TLD500
1,2,4-TRIMETHYL-5-AMINOBENZENE see TLG250
TRIMETHYLAMINOMETHANE see BPY250
2,4,5-TRIMETHYLANILINE see TLG250
2,4,5-TRIMETHYLBENZENAMINE see TLG250
TRIMETHYL BENZENE see TLL250
1,7,7-TRIMETHYLBICYCLO(2.2.1)-2-HEPTANONE see CBA750
TRIMETHYLCARBINOL see BPX000
1,1,3-TRIMETHYL-3-CYCLOHEXENE-5-ONE see IMF400
TRIMETHYLENETRINITRAMINE see CPR800
sym-TRIMETHYLENETRINITRAMINE see CPR800
2,2,4-TRIMETHYLPENTANE see TLY500
TRIMETHYL PHOSPHATE see TMD250
TRIMETHYL PHOSPHITE see TMD500
TRINITROCYCLOTRIMETHYLENE TRIAMINE see CPR800
2,4,7-TRINITROFLUORENONE (MAK) see TMM250
2,4,7-TRINITROFLUOREN-9-ONE see TMM250
TRINITROGLYCEROL see NGY000
TRINITROPHENOL (UN 0154) (DOT) see PID000
2,4,6-TRINITROPHENYL (OSHA) see PID000
2,4,6-TRINITROPHENYLMETHYLNITRAMINE see TEG250
1,3,6-TRINITROPYRENE see TMN000
TRINITROTOLUENE see TMN490
2,4,6-TRINITROTOLUENE see TMN490
1,3,5-TRINITRO-1,3,5-TRIAZACYCLOHEXANE see CPR800
TRIORTHOCRESYL PHOSPHATE see TMO600
TRIPHENYL see TBD000
TRIPHENYLAMINE see TMQ500
TRIPHENYL PHOSPHATE see TMT750
TRIPOLI see SCI500
TRIPOLI see TMX500
TRISODIUM ORTHOVANADATE see SIY250
TRI-o-TOLYL PHOSPHATE see TMO600
TRONA see BMG400
TSPP see TEE500
TUNGSTEN see TOA750
TUNGSTEN COMPOUNDS see TOC500
TURPENTINE see TOD750
TURPENTINE STEAM DISTILLED see TOD750
TYRANTON see DBF750
UDMH (DOT) see DSF400
URANIUM see UNS000
(p-UREIDOBENZENEARSYLENEDITHIO)DI-o-BENZOIC ACID see TFD750
URETHANE see UVA000
VAC see VLU250
n-VALERALDEHYDE see VAG000
VALERIC ACID ALDEHYDE see VAG000
VALERIC ALDEHYDE see VAG000
VALERONE see DNI800
Δ-VALEROSULTONE see BOU250
VANADIC ACID, AMMONIUM SALT see ANY250
VANADIC ACID, MONOSODIUM SALT see SKP000
VANADIC ANHYDRIDE see VDU000
VANADIC OXIDE see VEA000
VANADIUM see VCP000
VANADIUM CHLORIDE see VEF000
VANADIUM COMPOUNDS see VCZ000
VANADIUM OXIDE see VEA000

VANADIUM OXYTRICHLORIDE see VDP000
VANADIUM PENTOXIDE (dust) see VDU000
VANADIUM SESQUIOXIDE see VEA000
VANADIUM TETRACHLORIDE see VEF000
VANADIUM TRICHLORIDE see VEP000
VANADIUM TRICHLORIDE OXIDE see VDP000
VANADIUM TRIOXIDE see VEA000
VANADYL SULFATE see VEZ000
VCM see VNP000
VDF see VPP000
VINEGAR ACID see AAT250
VINEGAR NAPHTHA see EFR000
VINYL ACETATE see VLU250
VINYL AMIDE see ADS250
VINYLBENZENE see SMQ000
VINYLBENZOL see SMQ000
VINYL BROMIDE see VMP000
VINYLCARBINOL see AFV500
VINYL CHLORIDE see VNP000
VINYL CHLORIDE HOMOPOLYMER see PKQ059
VINYL CHLORIDE MONOMER see VNP000
VINYL CHLORIDE POLYMER see PKQ059
VINYL C MONOMER see VNP000
VINYL CYANIDE see ADX500
4-VINYLCYCLOHEXENE see CPD750
4-VINYL-1-CYCLOHEXENE see CPD750
VINYL CYCLOHEXENE DIEPOXIDE see VOA000
VINYL CYCLOHEXENE DIOXIDE see VOA000
4-VINYL-1-CYCLOHEXENE DIOXIDE (MAK) see VOA000
VINYLETHYLENE see BOP500
VINYL FLUORIDE see VPA000
VINYLFORMIC ACID see ADS750
VINYLIDENE CHLORIDE see VPK000
VINYLIDENE DICHLORIDE see VPK000
VINYLIDENE DIFLUORIDE see VPP000
VINYLIDENE FLUORIDE see VPP000
VINYLIDINE CHLORIDE see VPK000
VINYL METHYL KETONE see BOY500
m-VINYLSTYRENE see DXQ745
VINYL TOLUENE see VQK650
VINYL TRICHLORIDE see TIN000
VITRIOL, OIL OF (DOT) see SOI500
VM & P NAPHTHA see PCT250
VM&P NAPHTHA see PCT250
VM and P NAPHTHA see PCT250
VM & P NAPHTHA (ACGIH,OSHA) see PCT250
WARFARIN see WAT200
WELDING FUMES see WBJ000
WHITE MINERAL OIL see MQV750
WHITE PHOSPHORUS see PHP010
WHITE SPIRITS see SLU500
WOLFRAM see TOA750
WOOD ALCOHOL (DOT) see MGB150
WOOD NAPHTHA see MGB150
WOOD SPIRIT see MGB150
m-XDI see XIJ000
m-XYLENE DIISOCYANATE see XIJ000
XYLENES see XGS000
XYLIDINE see XMA000
2,4-XYLIDINE see XMS000
m-XYLIDINE see XMS000
XYLOL (DOT) see XGS000
YELLOW PHOSPHORUS see PHP010
YTTRIUM see YEJ000
ZINC BUTTER see ZFA000
ZINC CHLORIDE see ZFA000
ZINC CHROMATE see ZFJ100
ZINC COMPOUNDS see ZFS000
ZINC DICHLORIDE see ZFA000
ZINC DISTEARATE see ZMS000
ZINC FLUOROSILICATE (DOT) see ZIA000
ZINC FLUOSILICATE see ZIA000
ZINC HEXAFLUOROSILICATE see ZIA000
ZINC OCTADECANOATE see ZMS000
ZINC OXIDE see ZKA000
ZINC OXIDE FUME (MAK) see ZKA000
ZINC STEARATE see ZMS000
ZIRCONIUM see ZOA000
ZIRCONIUM COMPOUNDS see ZQA000

Appendix II
CAS Number Cross-Reference

50-00-0 see FMV000
50-32-8 see BCS750
50-78-2 see ADA725
51-79-6 see UVA000
53-96-3 see FDR000
54-11-5 see NDN000
55-18-5 see NJW500
55-63-0 see NGY000
56-23-5 see CBY000
56-81-5 see GGA000
57-12-5 see COI500
57-14-7 see DSF400
57-24-9 see SMN500
57-57-8 see PMT100
58-89-9 see BBQ500
59-88-1 see PFI250
59-89-2 see NKZ000
60-11-7 see DOT300
60-29-7 see EJU000
60-34-4 see MKN000
60-35-5 see AAI000
62-53-3 see AOQ000
62-74-8 see SHG500
62-75-9 see NKA600
64-17-5 see EFU000
64-18-6 see FNA000
64-19-7 see AAT250
64-67-5 see DKB110
67-56-1 see MGB150
67-63-0 see INJ000
67-64-1 see ABC750
67-66-3 see CHJ500
67-72-1 see HCI000
68-11-1 see TFJ100
68-12-2 see DSB000
71-23-8 see PND000
71-36-3 see BPW500
71-43-2 see BBL250
71-55-6 see MIH275
74-83-9 see MHR200
74-86-2 see ACI750
74-89-5 see MGC250
74-90-8 see HHS000
74-93-1 see MLE650
74-96-4 see EGV400
74-97-5 see CES650
74-98-6 see PMJ750
74-99-7 see MFX590
75-00-3 see EHH000
75-01-4 see VNP000
75-02-5 see VPA000
75-04-7 see EFU400
75-05-8 see ABE500
75-07-0 see AAG250
75-09-2 see MJP450
75-12-7 see FMY000
75-15-0 see CBV500
75-21-8 see EJN500
75-25-2 see BNL000
75-31-0 see INK000
75-34-3 see DFF809
75-35-4 see VPK000
75-38-7 see VPP000
75-43-4 see DFL000
75-44-5 see PGX000
75-45-6 see CFX500
75-47-8 see IEP000
75-50-3 see TLD500
75-52-5 see NHM500
75-55-8 see PNL400
75-56-9 see PNL600
75-61-6 see DKG850
75-63-8 see TJY100
75-64-9 see BPY250
75-65-0 see BPX000
75-69-4 see TIP500
75-71-8 see DFA600
75-74-1 see TDR500
75-83-2 see DQT200
75-86-5 see MLC750
75-91-2 see BRM250
75-99-0 see DGI400
76-01-7 see PAW500
76-03-9 see TII250
76-06-2 see CKN500
76-11-9 see TBP000
76-12-0 see TBP050
76-13-1 see FOO000
76-14-2 see FOO509
76-15-3 see CJI500
76-22-2 see CBA750
76-44-8 see HAR000
77-47-4 see HCE500
77-73-6 see DGW000
77-78-1 see DUD100
78-00-2 see TCF000
78-10-4 see EPF550
78-30-8 see TMO600
78-34-2 see DVQ709
78-59-1 see IMF400

78-78-4 see EIK000
78-81-9 see IIM000
78-82-0 see IJX000
78-83-1 see IIL000
78-87-5 see PNJ400
78-88-4 see MKW200
78-92-2 see BPW750
78-93-3 see MKA400
78-94-4 see BOY500
78-95-5 see CDN200
79-00-5 see TIN000
79-01-6 see TIO750
79-04-9 see CEC250
79-06-1 see ADS250
79-09-4 see PMU750
79-10-7 see ADS750
79-20-9 see MFW100
79-21-0 see PCL500
79-24-3 see NFY500
79-27-6 see ACK250
79-29-8 see DQT400
79-34-5 see TBQ100
79-41-4 see MDN250
79-44-7 see DQY950
79-46-9 see NIY000
80-15-9 see IOB000
80-62-6 see MLH750
81-81-2 see WAT200
83-79-4 see RNZ000
84-15-1 see TBC640
84-66-2 see DJX000
84-74-2 see DEH200
85-01-8 see PCW250
85-44-9 see PHW750
86-57-7 see NHQ000
87-68-3 see HCD250
87-86-5 see PAX250
88-10-8 see DIW400
88-72-2 see NMO525
88-89-1 see PID000
89-72-5 see BSE000
90-04-0 see AOV900
90-94-8 see MQS500
91-08-7 see TGM800
91-20-3 see NAJ500
91-59-8 see NBE500
91-71-4 see TFD750
91-93-0 see DCJ400
91-94-1 see DEQ600
92-52-4 see BGE000
92-84-2 see PDP250
92-87-5 see BBX000
92-93-3 see NFQ000
94-36-0 see BDS000
95-13-6 see IBX000
95-48-7 see CNX000
95-49-8 see CLK100
95-50-1 see DEP600
95-53-4 see TGQ750
95-54-5 see PEY250
95-68-1 see XMS000
95-69-2 see CLK220
95-79-4 see CLK225
95-80-7 see TGL750
96-12-8 see DDL800
96-18-4 see TJB600
96-22-0 see DJN750
96-33-3 see MGA500
96-45-7 see IAQ000
96-69-5 see TFC600
98-00-0 see FPU000
98-01-1 see FPQ875
98-07-7 see BFL250
98-51-1 see BSP500
98-54-4 see BSE500
98-82-8 see COE750
98-83-9 see MPK250
98-87-3 see BAY300
98-95-3 see NEX000
99-08-1 see NMO500
99-55-8 see NMP500
99-65-0 see DUQ200
99-99-0 see NMO550
100-00-5 see NFS525
100-01-6 see NEO500
100-25-4 see DUQ600
100-37-8 see DHO500
100-40-3 see CPD750
100-41-4 see EGP500
100-42-5 see SMQ000
100-44-7 see BEE375
100-61-8 see MGN750
100-63-0 see PFI000
100-74-3 see ENL000
100-75-4 see NLJ500
101-14-4 see MJM200
101-61-1 see MJN000
101-68-8 see MJP400
101-77-9 see MJQ000
101-80-4 see OPM000
101-84-8 see PFA850
102-54-5 see FBC000
102-81-8 see DDU600
104-94-9 see AOW000
105-46-4 see BPV000
105-60-2 see CBF700
105-74-8 see LBR000
106-35-4 see EHA600
106-46-7 see DEP800
106-49-0 see TGR000

106-50-3 see PEY500
106-51-4 see QQS200
106-68-3 see ODI000
106-87-6 see VOA000
106-89-8 see EAZ500
106-92-3 see AGH150
106-93-4 see EIY500
106-97-8 see BOR500
106-99-0 see BOP500
107-02-8 see ADR000
107-03-9 see PML500
107-05-1 see AGB250
107-06-2 see EIY600
107-07-3 see EIU800
107-12-0 see PMV750
107-13-1 see ADX500
107-15-3 see EEA500
107-16-4 see HIM500
107-18-6 see AFV500
107-19-7 see PMN450
107-20-0 see CDY500
107-21-1 see EJC500
107-30-2 see CIO250
107-31-3 see MKG750
107-41-5 see HFP875
107-49-3 see TCF250
107-66-4 see DEG700
107-71-1 see BSC250
107-83-5 see IKS600
107-87-9 see PBN250
107-98-2 see PNL250
108-03-2 see NIX500
108-05-4 see VLU250
108-10-1 see HFG500
108-11-2 see MKW600
108-18-9 see DNM200
108-20-3 see IOZ750
108-21-4 see INE100
108-24-7 see AAX500
108-31-6 see MAM000
108-44-1 see TGQ500
108-46-3 see REA000
108-57-6 see DXQ745
108-83-8 see DNI800
108-84-9 see HFJ000
108-87-2 see MIQ740
108-88-3 see TGK750
108-90-7 see CEJ125
108-91-8 see CPF500
108-93-0 see CPB750
108-94-1 see CPC000
108-95-2 see PDN750
108-98-5 see PFL850
109-59-1 see INA500
109-60-4 see PNC250
109-66-0 see PBK250
109-73-9 see BPX750
109-74-0 see BSX250
109-77-3 see MAO250
109-79-5 see BRR900
109-86-4 see EJH500
109-87-5 see MGA850
109-89-7 see DHJ200
109-94-4 see EKL000
109-99-9 see TCR750
110-05-4 see BSC750
110-12-3 see MKW450
110-19-0 see IIJ000
110-43-0 see MGN500
110-49-6 see EJJ500
110-54-3 see HEN000
110-61-2 see SNE000
110-62-3 see VAG000
110-66-7 see PBM000
110-80-5 see EES350
110-82-7 see CPB000
110-83-8 see CPC579
110-86-1 see POP250
110-91-8 see MRP750
111-15-9 see EES400
111-30-8 see GFQ000
111-31-9 see HES000
111-40-0 see DJG600
111-42-2 see DHF000
111-44-4 see DFJ050
111-65-9 see OCU000
111-69-3 see AER250
111-76-2 see BPJ850
111-84-2 see NMX000
112-07-2 see BPM000
112-55-0 see LBX000
114-26-1 see PMY300
115-77-5 see PBB750
115-86-6 see TMT750
117-81-7 see DVL700
118-52-5 see DFE200
118-74-1 see HCC500
118-96-7 see TMN490
119-90-4 see DCJ200
119-93-7 see TGJ750
120-12-7 see APG500
120-80-9 see CCP850
120-82-1 see TIK250
121-44-8 see TJO000
121-45-9 see TMD500
121-69-7 see DQF800
121-82-4 see CPR800
122-39-4 see DVX800
122-60-1 see PFF360
123-19-3 see DWT600

123-31-9 see HIH000
123-42-2 see DBF750
123-51-3 see IHP000
123-61-5 see BBP000
123-86-4 see BPU750
123-91-1 see DVQ000
123-92-2 see IHO850
124-09-4 see HEO000
124-38-9 see CBU250
124-40-3 see DOQ800
126-73-8 see TIA250
126-98-7 see MGA750
126-99-8 see NCI500
127-18-4 see PCF275
127-19-5 see DOO800
128-37-0 see BFW750
129-00-0 see PON250
129-79-3 see TMM250
131-11-3 see DTR200
134-32-7 see NBE700
135-88-6 see PFT500
137-05-3 see MIQ075
137-17-7 see TLG250
138-22-7 see BRR600
139-25-3 see MJN750
139-65-1 see TFI000
140-11-4 see BDX000
140-88-5 see EFT000
141-32-2 see BPW100
141-43-5 see EEC600
141-78-6 see EFR000
141-79-7 see MDJ750
142-64-3 see PIK000
142-82-5 see HBC500
143-33-9 see SGA500
144-62-7 see OLA000
150-76-5 see MFC700
151-50-8 see PLC500
151-56-4 see EJM900
156-59-2 see DFI200
156-62-7 see CAQ250
287-92-3 see CPV750
299-29-6 see FBK000
299-84-3 see RMA500
302-01-2 see HGS000
319-84-6 see BBQ000
319-85-7 see BBR000
334-88-3 see DCP800
353-50-4 see CCA500
382-21-8 see OBM000
406-90-6 see TKB250
409-21-2 see SCQ000
420-04-2 see COH500
460-19-5 see COO000
463-51-4 see KEU000
463-82-1 see NCH000
479-45-8 see TEG250
492-80-8 see IBB000
504-29-0 see AMI000
506-61-6 see PLS250
506-64-9 see SDP000
506-77-4 see COO750
509-14-8 see TDY250
512-56-1 see TMD250
528-29-0 see DUQ400
532-27-4 see CEA750
534-52-1 see DUS700
540-59-0 see DFI210
540-73-8 see DSF600
540-84-1 see TLY500
540-88-5 see BPV100
541-85-5 see EGI750
542-75-6 see DGG950
542-88-1 see BIK000
542-92-7 see CPU500
544-92-3 see CNL000
552-30-7 see TKV000
556-52-5 see GGW500
557-05-1 see ZMS000
557-99-3 see ACM000
558-13-4 see CBX750
563-68-8 see TEI250
563-80-4 see MLA750
581-89-5 see NHQ500
583-60-8 see MIR500
584-02-1 see IHP010
584-84-9 see TGM750
591-78-6 see HEV000
592-01-8 see CAQ500
592-85-8 see MCU250
593-60-2 see VMP000
593-70-4 see CHI900
594-42-3 see PCF300
594-72-9 see DFU000
597-64-8 see TCF750
598-78-7 see CKS750
600-25-9 see CJE000
602-87-9 see NEJ500
603-34-9 see TMQ500
612-64-6 see NKD000
612-83-9 see DEQ800
614-00-6 see MMU250
615-05-4 see DBO000
621-64-7 see NKB700
624-83-9 see MKX250
626-17-5 see PHX550
626-38-0 see AOD735
627-13-4 see PNQ500
628-63-7 see AOD725
628-96-6 see EJG000

630-08-0 see CBW750
638-21-1 see PFV250
680-31-9 see HEK000
681-84-5 see MPI750
684-16-2 see HCZ000
764-41-0 see DEV000
768-52-5 see INX000
818-08-6 see DEF400
822-06-0 see DNJ800
838-88-0 see MJO250
872-50-4 see MPF200
924-16-3 see BRY500
930-55-2 see NLP500
999-61-1 see HNT600
1116-54-7 see NKM000
1300-73-8 see XMA000
1303-86-2 see BMG000
1303-96-4 see SFE500
1303-96-4 see SFF000
1304-82-1 see BKY000
1305-62-0 see CAT225
1305-78-8 see CAU500
1309-37-1 see IHC450
1309-48-4 see MAH500
1309-64-4 see AQF000
1310-58-3 see PLJ500
1310-73-2 see SHS000
1313-27-5 see MRE000
1313-99-1 see NDF500
1314-13-2 see ZKA000
1314-32-5 see TEL050
1314-34-7 see VEA000
1314-56-3 see PHS250
1314-62-1 see VDU000
1314-80-3 see PHS000
1317-65-3 see CAO000
1317-95-9 see TMX500
1319-77-3 see CNW500
1320-37-2 see DGL600
1321-64-8 see PAW750
1321-65-9 see TIT500
1327-53-3 see ARI750
1330-20-7 see XGS000
1332-21-4 see ARM250
1333-86-4 see CBT750
1335-87-1 see HCK500
1335-88-2 see TBR000
1336-21-6 see ANK250
1336-36-3 see PJL750
1338-23-4 see MKA500
1341-49-7 see ANJ000
1344-28-1 see AHE250
1344-95-2 see CAW850
1395-21-7 see BAB750
1569-69-3 see CPB625
1600-27-7 see MCS750
1633-83-6 see BOU250
1639-09-4 see HBD500
1929-82-4 see CLP750
2039-87-4 see CLE750
2155-70-6 see THZ000
2179-59-1 see AGR500
2234-13-1 see OAP000
2238-07-5 see DKM200
2426-08-6 see BRK750
2431-50-7 see TIL360
2465-27-2 see IBA000
2528-36-1 see DEG600
2551-62-4 see SOI000
2644-70-4 see HGV000
2698-41-1 see CEQ600
2699-79-8 see SOU500
2781-10-4 see BJQ250
3033-62-3 see BJH750
3173-72-6 see NAM500
3333-52-6 see TDW250
3333-67-3 see NCY500
3634-83-1 see XIJ000
3689-24-5 see SOD100
3825-26-1 see ANP625
4016-14-2 see IPD000
4098-71-9 see IMG000
4170-30-3 see COB250
4342-30-7 see SAM000
5124-30-1 see MJM600
5307-14-2 see ALL750
5522-43-0 see NJA000
5714-22-7 see SOQ450
6423-43-4 see PNL000
6533-73-9 see TEJ000
7046-61-9 see NHI500
7339-53-9 see MKN250
7429-90-5 see AGX000
7439-92-1 see LCF000
7439-96-5 see MAP750
7439-97-6 see MCW250
7439-98-7 see MRC250
7440-02-0 see NCW500
7440-06-4 see PJD500
7440-16-6 see RHF000
7440-21-3 see SCP000
7440-22-4 see SDI500
7440-25-7 see TAE750
7440-28-0 see TEI000
7440-31-5 see TGB250
7440-33-7 see TOA750
7440-36-0 see AQB750
7440-38-2 see ARA750
7440-39-3 see BAH250
7440-41-7 see BFO750

7440-43-9 see CAD000
7440-44-0 see CBT500
7440-47-3 see CMI750
7440-48-4 see CNA250
7440-50-8 see CNI000
7440-61-1 see UNS000
7440-62-2 see VCP000
7440-65-5 see YEJ000
7440-67-7 see ZOA000
7440-74-6 see ICF000
7446-09-5 see SOH500
7446-70-0 see AGY750
7553-56-2 see IDM000
7572-29-4 see DEN600
7580-67-8 see LHH000
7616-94-6 see PCF750
7631-86-9 see SCI000
7631-90-5 see SFE000
7632-51-1 see VEF000
7637-07-2 see BMG700
7646-85-7 see ZFA000
7647-01-0 see HHL000
7647-01-0 see HHX000
7647-18-9 see AQD000
7664-38-2 see PHB250
7664-39-3 see HHU500
7664-41-7 see AMY500
7664-93-9 see SOI500
7681-49-4 see SHF500
7681-57-4 see SII000
7697-37-2 see NED500
7699-41-4 see SCL000
7705-08-0 see FAU000
7718-98-1 see VEP000
7719-09-7 see TFL000
7719-12-2 see PHT275
7720-78-7 see FBN100
7722-64-7 see PLP000
7722-84-1 see HIB050
7722-88-5 see TEE500
7723-14-0 see PHO500
7723-14-0 see PHP010
7726-95-6 see BMP000
7727-15-3 see AGX750
7727-18-6 see VDP000
7727-21-1 see DWQ000
7727-43-7 see BAP000
7727-54-0 see ANR000
7738-94-5 see CMH250
7758-94-3 see FBI000
7758-97-6 see LCR000
7761-88-8 see SDS000
7772-99-8 see TGC000
7773-06-0 see ANU650
7778-18-9 see CAX500
7782-41-4 see FEZ000
7782-49-2 see SBO500
7782-50-5 see CDV750
7782-65-2 see GEI100
7782-79-8 see HHG500
7783-06-4 see HIC500
7783-07-5 see HIC000
7783-35-9 see MDG500
7783-41-7 see ORA000
7783-47-3 see TGD100
7783-54-2 see NGW000
7783-60-0 see SOR000
7783-61-1 see SDF650
7783-70-2 see AQF250
7783-79-1 see SBS000
7783-80-4 see TAK250
7784-18-1 see AHB000
7784-42-1 see ARK250
7787-32-8 see BAM000
7787-71-5 see BMQ325
7789-06-2 see SMH000
7789-23-3 see PLF500
7789-24-4 see LHF000
7789-29-9 see PKU250
7789-40-4 see TEI750
7789-47-1 see MCY000
7790-91-2 see CDX750
7791-12-0 see TEJ250
7791-23-3 see SBT500
7803-51-2 see PGY000
7803-52-3 see SLQ000
7803-55-6 see ANY250
7803-62-5 see SDH575
7803-68-1 see TAI750
8001-35-2 see CDV100
8001-58-9 see CMY825
8002-05-9 see PCS250
8002-74-2 see PAH750
8003-34-7 see POO250
8004-13-5 see PFA860
8006-61-9 see GBY000
8006-64-2 see TOD750
8007-45-2 see CMY800
8008-20-6 see KEK000
8012-95-1 see MQV750
8030-30-6 see NAH600
8032-32-4 see PCT250
8052-41-3 see SLU500
8052-42-4 see ARO500
9002-86-2 see PKQ059
10022-31-8 see BAN250
10024-97-2 see NGU000
10025-67-9 see SON510
10025-87-3 see PHQ800
10025-91-9 see AQC500

10026-13-8 see PHR500
10028-15-6 see ORW000
10034-93-2 see HGW500
10035-10-6 see HHJ000
10049-04-4 see CDW450
10101-41-4 see CAX750
10101-83-4 see SKC000
10102-18-8 see SJT500
10102-43-9 see NEG100
10102-44-0 see NGR500
10102-45-1 see TEK750
10108-64-2 see CAE250
10210-68-1 see CNB500
10217-52-4 see HGU500
10294-33-4 see BMG400
10294-40-3 see BAK250
10415-75-5 see MDE750
12001-26-2 see MQS250
12079-65-1 see CPV000
12108-13-3 see MAV750
12125-02-9 see ANE500
12604-58-9 see FBP000
13007-92-6 see HCB000
13106-76-8 see ANM750
13424-46-9 see LCM000
13454-96-1 see PJE250
13463-39-3 see NCZ000
13463-40-6 see IHG500
13463-67-7 see TGG760
13472-45-2 see SKN500
13473-90-0 see AHD750
13494-80-9 see TAJ000
13530-65-9 see ZFJ100
13637-84-8 see SOT500
13718-26-8 see SKP000
13721-39-6 see SIY250
13826-83-0 see ANH000
13838-16-9 see EAT900
13952-84-6 see BPY000
14464-46-1 see SCJ000
14807-96-6 see TAB750
14808-60-7 see SCJ500
15468-32-3 see SCK000
15829-53-5 see MDF750
16219-75-3 see ELO500
16752-77-5 see MDU600
16842-03-8 see CNC230
16871-71-9 see ZIA000
16871-90-2 see PLH750
16872-11-0 see FDD125
16919-58-7 see ANF250
16924-00-8 see PLH000
16940-81-1 see HDE000
17702-41-9 see DAE400
18810-58-7 see BAI000
19287-45-7 see DDI450
19624-22-7 see PAT750
20325-40-0 see DOA800
20816-12-0 see OKK000
21351-79-1 see CDD750
22967-92-6 see MLF550
23606-32-8 see SDL500
25013-15-4 see VQK650
25321-14-6 see DVG600
25322-20-7 see TBP750
25551-13-7 see TLL250
25639-42-3 see MIQ745
26140-60-3 see TBD000
26628-22-8 see SFA000
26952-21-6 see ILL000
27774-13-6 see VEZ000
28675-08-3 see DFE800
31242-93-0 see CDV175
31242-94-1 see TBP250
34590-94-8 see DWT200
42279-29-8 see PAW250
55398-86-2 see MRG000
59355-75-8 see MFX600
60676-86-0 see SCK600
61788-32-7 see HHW800
61790-53-2 see DCJ800
65996-93-2 see CMZ100
65997-15-1 see PKS750
68476-85-7 see LGM000
75321-19-6 see TMN000
112945-52-5 see SCH002

Appendix III
DOT Number Cross-Reference

NA 1247 see MLH750
NA 1649 see TCF000
NA 1807 see PHS250
NA 1811 see PKU250
NA 1911 see DDI450
NA 1999 see ARO500
NA 2212 see ARM250
NA 2672 see ANK250
NA 2783 see TCF250
NA 2809 see MCW250
NA 2821 see PDN750
NA 3018 see TCF250
NA 9202 see CBW750
NA 9260 see AGX000
UN 0072 see CPR800
UN 0118 see CPR800
UN 0129 see LCM000
UN 0143 see NGY000
UN 0144 see NGY000
UN 0154 see PID000
UN 0208 see TEG250
UN 0209 see TMN490
UN 0224 see BAI000
UN 0391 see CPR800
UN 0483 see CPR800
UN 1001 see ACI750
UN 1005 see AMY500
UN 1008 see BMG700
UN 1009 see TJY100
UN 1011 see BOR500
UN 1013 see CBU250
UN 1016 see CBW750
UN 1017 see CDV750
UN 1018 see CFX500
UN 1020 see CJI500
UN 1026 see COO000
UN 1028 see DFA600
UN 1029 see DFL000
UN 1032 see DOQ800
UN 1036 see EFU400
UN 1037 see EHH000
UN 1040 see EJN500
UN 1045 see FEZ000
UN 1048 see HHJ000
UN 1050 see HHL000
UN 1052 see HHU500
UN 1053 see HIC500
UN 1060 see MFX600
UN 1061 see MGC250
UN 1063 see MIF765
UN 1064 see MLE650
UN 1070 see NGU000
UN 1075 see LGM000
UN 1076 see PGX000
UN 1079 see SOH500
UN 1080 see SOI000
UN 1083 see TLD500
UN 1085 see VMP000
UN 1086 see VNP000
UN 1089 see AAG250
UN 1090 see ABC750
UN 1091 see ABC750
UN 1092 see ADR000
UN 1093 see ADX500
UN 1098 see AFV500
UN 1100 see AGB250
UN 1104 see AOD725, AOD735
UN 1110 see MGN500
UN 1114 see BBL250
UN 1123 see BPU750, BPV000, BPV100
UN 1125 see BPX750
UN 1131 see CBV500
UN 1134 see CEJ125
UN 1135 see EIU800
UN 1143 see COB250
UN 1145 see CPB000
UN 1146 see CPV750
UN 1148 see DBF750
UN 1154 see DHJ200
UN 1155 see EJU000
UN 1156 see DJN750
UN 1157 see DNI800
UN 1158 see DNM200
UN 1159 see IOZ750
UN 1160 see DOQ800
UN 1163 see DSF400
UN 1165 see DVQ000
UN 1170 see EFU000
UN 1171 see EES350
UN 1172 see EES400
UN 1173 see EFR000
UN 1175 see EGP500
UN 1184 see EIY600
UN 1185 see EJM900
UN 1188 see EJH500
UN 1189 see EJJ500
UN 1190 see EKL000
UN 1193 see MKA400
UN 1198 see FMV000

UN 1199 see FPQ875
UN 1203 see GBY000
UN 1204 see NGY000
UN 1206 see HBC500
UN 1208 see HEN000
UN 1212 see IIL000
UN 1213 see IIJ000
UN 1214 see IIM000
UN 1219 see INJ000
UN 1220 see INE100
UN 1221 see INK000
UN 1223 see KEK000
UN 1228 see HBD500, HES000, LBX000, PBM000, PML500
UN 1229 see MDJ750
UN 1230 see MGB150
UN 1231 see MFW100
UN 1233 see HFJ000
UN 1234 see MGA850
UN 1235 see MGC250
UN 1239 see CIO250
UN 1243 see MKG750
UN 1244 see MKN000
UN 1245 see HFG500
UN 1249 see PBN250
UN 1251 see BOY500
UN 1255 see NAH600
UN 1256 see NAH600
UN 1257 see GBY000
UN 1259 see NCZ000
UN 1261 see NHM500
UN 1262 see OCU000
UN 1265 see EIK000, PBK250
UN 1268 see PCS250
UN 1270 see NAH600
UN 1271 see PCT250
UN 1274 see PND000
UN 1276 see PNC250
UN 1279 see PNJ400
UN 1280 see PNL600
UN 1282 see POP250
UN 1292 see EPF550
UN 1294 see TGK750
UN 1296 see TJO000
UN 1297 see TLD500
UN 1299 see TOD750
UN 1300 see TOD750
UN 1301 see VLU250
UN 1303 see VPK000
UN 1307 see XGS000
UN 1309 see AGX000
UN 1334 see NAJ500
UN 1338 see PHO500
UN 1340 see PHS000
UN 1344 see PID000
UN 1346 see SCP000
UN 1356 see TMN490
UN 1358 see ZOA000
UN 1361 see CBT500
UN 1362 see CBT500
UN 1380 see PAT750
UN 1381 see PHP010
UN 1396 see AGX000
UN 1400 see BAH250
UN 1403 see CAQ250
UN 1414 see LHH000
UN 1438 see AHD750
UN 1444 see ANR000
UN 1446 see BAN250
UN 1490 see PLP000
UN 1492 see DWQ000
UN 1493 see SDS000
UN 1510 see TDY250
UN 1541 see MLC750
UN 1547 see AOQ000
UN 1558 see ARA750
UN 1561 see ARI750
UN 1567 see BFO750
UN 1571 see BAI000
UN 1575 see CAQ500
UN 1578 see NFS525
UN 1580 see CKN500
UN 1583 see CKN500
UN 1587 see CNL000
UN 1589 see COO750
UN 1591 see DEP600
UN 1592 see DEP800
UN 1593 see MJP450
UN 1594 see DKB110
UN 1595 see DUD100
UN 1597 see DUQ200, DUQ400, DUQ600
UN 1604 see EEA500
UN 1605 see EIY500
UN 1613 see HHS000
UN 1614 see HHS000
UN 1627 see MDE750
UN 1629 see MCS750
UN 1646 see MCU250
UN 1648 see ABE500
UN 1650 see NBE500
UN 1654 see NDN000
UN 1660 see NEG100
UN 1661 see NEO500
UN 1662 see NEX000
UN 1664 see NMO500, NMO525, NMO550
UN 1669 see PAW500
UN 1670 see PCF300
UN 1671 see PDN750

UN 1673 see PEY250, PEY500
UN 1680 see PLC500
UN 1684 see SDP000
UN 1687 see SFA000
UN 1689 see SGA500
UN 1690 see SHF500
UN 1692 see SMN500
UN 1695 see CDN200
UN 1697 see CEA750
UN 1702 see TBP750
UN 1704 see SOD100
UN 1708 see TGQ500, TGQ750, TGR000
UN 1709 see TGL750
UN 1710 see TIO750
UN 1715 see AAX500
UN 1725 see AGX750
UN 1726 see AGY750
UN 1727 see ANJ000
UN 1730 see AQD000
UN 1731 see AQD000
UN 1732 see AQF250
UN 1733 see AQC500
UN 1738 see BEE375
UN 1744 see BMP000
UN 1746 see BMQ325
UN 1749 see CDX750
UN 1752 see CEC250
UN 1759 see FBI000
UN 1760 see FBI000
UN 1773 see FAU000
UN 1775 see FDD125
UN 1779 see FNA000
UN 1782 see HDE000
UN 1783 see HEO000
UN 1788 see HHJ000
UN 1789 see HHL000
UN 1790 see HHU500
UN 1805 see PHB250
UN 1806 see PHR500
UN 1809 see PHT275
UN 1810 see PHQ800
UN 1812 see PLF500
UN 1813 see PLJ500
UN 1814 see PLJ500
UN 1823 see SHS000
UN 1824 see SHS000
UN 1830 see SOI500
UN 1832 see SOI500
UN 1836 see TFL000
UN 1839 see TII250
UN 1840 see ZFA000
UN 1845 see CBU250
UN 1846 see CBY000
UN 1848 see PMU750
UN 1859 see SDF650
UN 1860 see VPA000
UN 1865 see PNQ500
UN 1868 see DAE400
UN 1885 see BBX000
UN 1886 see BAY300
UN 1887 see CES650
UN 1888 see CHJ500
UN 1891 see EGV400
UN 1897 see PCF275
UN 1910 see CAU500
UN 1911 see DDI450
UN 1915 see CPC000
UN 1916 see DFJ050
UN 1917 see EFT000
UN 1918 see COE750
UN 1919 see MGA500
UN 1921 see PNL400
UN 1932 see ZOA000
UN 1935 see COI500
UN 1940 see TFJ100
UN 1941 see DKG850
UN 1958 see DGL600
UN 1959 see VPP000
UN 1966 see BFO750
UN 1978 see PMJ750
UN 1986 see EFU000
UN 1987 see EFU000
UN 1991 see NCI500
UN 1994 see IHG500
UN 2008 see ZOA000
UN 2009 see ZOA000
UN 2014 see HIB050
UN 2022 see CNW500
UN 2023 see EAZ500
UN 2029 see HGS000
UN 2030 see HGU500
UN 2031 see NED500
UN 2038 see DVG600
UN 2044 see NCH000
UN 2047 see DGG950
UN 2048 see DGW000
UN 2053 see MKW600
UN 2054 see MRP750
UN 2055 see SMQ000
UN 2056 see TCR750
UN 2058 see VAG000
UN 2074 see ADS250
UN 2076 see CNX000
UN 2077 see NBE700
UN 2079 see DJG600
UN 2186 see HHL000
UN 2187 see CBU250
UN 2188 see ARK250
UN 2190 see ORA000

UN 2191 see SOU500
UN 2192 see GEI100
UN 2194 see SBS000
UN 2195 see TAK250
UN 2199 see PGY000
UN 2201 see NGU000
UN 2202 see HIC000
UN 2203 see SDH575
UN 2205 see AER250
UN 2206 see TGM750
UN 2207 see TGM750, TGM800, XIJ000
UN 2209 see FMV000
UN 2214 see PHW750
UN 2215 see MAM000
UN 2218 see ADS750
UN 2219 see AGH150
UN 2226 see BFL250
UN 2232 see CDY500
UN 2238 see CLK100
UN 2249 see BIK000
UN 2253 see DQF800
UN 2256 see CPC579
UN 2262 see DQY950
UN 2265 see DSB000
UN 2270 see EFU400
UN 2271 see ODI000
UN 2279 see HCD250
UN 2280 see HEO000
UN 2281 see DNJ800
UN 2284 see IJX000
UN 2290 see IMG000
UN 2294 see MGN750
UN 2296 see MIQ740
UN 2302 see MKW450
UN 2303 see MPK250
UN 2304 see NAJ500
UN 2312 see PDN750
UN 2315 see PJL750
UN 2329 see TMD500
UN 2331 see ZFA000
UN 2337 see PFL850
UN 2347 see BRR900
UN 2348 see BPW100
UN 2357 see CPF500
UN 2362 see DFF809
UN 2369 see BPJ850
UN 2382 see DSF600
UN 2397 see MLA750
UN 2404 see PMV750
UN 2411 see BSX250
UN 2417 see CCA500
UN 2418 see SOR000
UN 2420 see HCZ000
UN 2431 see AOV900
UN 2443 see VDP000
UN 2444 see VEF000
UN 2447 see PHP010
UN 2451 see NGW000
UN 2457 see DQT400
UN 2471 see OKK000
UN 2475 see VEP000
UN 2478 see TGM750
UN 2480 see MKX250
UN 2489 see MJP400
UN 2491 see EEC600
UN 2511 see CKS750
UN 2515 see BNL000
UN 2516 see CBX750
UN 2531 see MDN250
UN 2553 see NAH600
UN 2564 see TII250
UN 2572 see PFI000
UN 2580 see AGX750
UN 2581 see AGY750
UN 2582 see FAU000
UN 2587 see QQS200
UN 2606 see MPI750
UN 2608 see NIY000
UN 2617 see MIQ745
UN 2618 see VQK650
UN 2629 see SHG500
UN 2630 see SJT500
UN 2644 see MKW200
UN 2646 see HCE500
UN 2647 see MAO250
UN 2650 see DFU000
UN 2651 see MJQ000
UN 2655 see PLH750
UN 2658 see SBO500
UN 2662 see HIH000
UN 2671 see AMI000
UN 2676 see SLQ000
UN 2681 see CDD750
UN 2682 see CDD750
UN 2686 see DHO500
UN 2692 see BMG400
UN 2710 see DWT600
UN 2717 see CBA750
UN 2727 see TEK750
UN 2729 see HCC500
UN 2733 see BPY000, BPY250
UN 2734 see BPY000, BPY250
UN 2789 see AAT250
UN 2790 see AAT250
UN 2805 see LHH000
UN 2817 see ANJ000
UN 2831 see MIH275
UN 2842 see NFY500
UN 2855 see ZIA000

UN 2858 see ZOA000
UN 2859 see ANY250
UN 2862 see VDU000
UN 2871 see AQB750
UN 2872 see DDL800
UN 2873 see DDU600
UN 2874 see FPU000
UN 2876 see REA000
UN 2879 see SBT500
UN 2906 see IMG000
UN 2931 see VEZ000
UN 2979 see UNS000
UN 3054 see CPB625
UN 3064 see NGY000
UN 3071 see HBD500, HES000, LBX000, PBM000, PML500
UN 3080 see TGM750, XIJ000
UN 3083 see PCF750
UN 3149 see PCL500

Printed in the United States
54973LVS00003B/76

9 780471 355427